MATHEMATICS RESEARCH DEVELOPMENTS

NEW DEVELOPMENTS IN APPLIED STATISTICS

MATHEMATICS RESEARCH DEVELOPMENTS

Additional books in this series can be found on Nova's website under the Series tab.

Additional E-books in this series can be found on Nova's website under the E-books tab.

MATHEMATICS RESEARCH DEVELOPMENTS

NEW DEVELOPMENTS IN APPLIED STATISTICS

MOHAMMAD AHSANULLAH
EDITOR

Nova Science Publishers, Inc.
New York

For permission to use material from this book please contact us:
Telephone 631-231-7269; Fax 631-231-8175
Web Site: http://www.novapublishers.com

LIBRARY OF CONGRESS CATALOGING-IN-PUBLICATION DATA

New developments in applied statistics / editor, Mohammad Ahsanullah.
p. cm.
Includes index.
ISBN 978-1-61324-648-1 (hardcover)
1. Statistics. I. Ahsanullah, M. (Mohammad)
QA276.A12N49 2011
519.5--dc23

2011014547

Published by Nova Science Publishers, Inc. † New York

CONTENTS

PREFACE

This book presents current research in the field of applied statistics. Topics discussed include the estimation of finite population mean using random non-response in survey sampling; balanced groups ranked set samples for estimating the population median; Marshall-Olkin exponential distribution; inference and prediction for a generalized logistics distribution; minimum perpendicular distance squared method estimation; shrinkage estimators for calculating reliability and analyzing incomplete categorical data. (Imprint: Nova)

Chapter 1 proposes some general estimators for finite population mean in presence of random non-response using an auxiliary variable. All possible non-response cases are considered here. The properties of these estimators are studied. Finally, a simulation study has been done to compare their performance.

Chapter 2 gives some recurrence relations satisfied by marginal and joint moment generating functions of generalized order statistics from the generalized exponential distribution. These recurrence relations would hence enable one to compute all the higher order moments of the generalized order statistics from those of the lower order in a simple recursive way.

In Chapter 3, the problem of estimating the number of signals under a colored noise model is solved when the data are received from many different places that are widely apart by using modified information criterion proposed by the authors. This criterion involves the eigenvalues of the sample canonical correlation matrix obtained from data consisting of both noise and signals as well as from the data consisting of noise only. Some simulation results are also presented.

In Chapter 4, balanced groups ranked set samples method (BGRSS) is suggested for estimating the population median with samples of size $m = 3k$ where $(k = 1, 2, \ldots)$. The BGRSS sample median is considered as estimator of the population median. It is found that, the BGRSS produces unbiased estimators with smaller variance than the commonly used simple random sampling (SRS) when the underlying distribution is symmetric. For asymmetric distributions, the BGRSS estimators have a small bias. A real data is used to illustrate the BGRSS method.

Many models on loss and delay queueing system with no passing have been studied in the past. Due to restriction of no passing the customers are allowed to depart from the system in the chronological order of their arrival and the customers are classified into two classes (elective or emergency) depending upon whether they can wait or are lost when all servers are busy. The arrival and service processes are taken to be independent in most of the models. In Chapter 5, we consider a finite population , loss and delay queueing system with no passing where arrival and service process are interdependent. In addition to this interdependence, an operating strategy , called controllable arrival rates , is incorporated in the model. Such a model is much useful in controlling congestion in buffers in the computer communication system. The steady state solutions and the analytical expressions for the expected waiting time of a customer in the system are derived explicitly. The numerical results for the expected waiting times of the two classes of customers for varying arrival rates(faster and slower) , the effect of dependence parameter between inter arrival and service times and the effect of switching over to varying arrival rates have been provided in the tables.

A new Bayesian method for estimating moving average (MA) models is proposed in Chapter 6. The proposed methodology is based on replacing lagged errors of the original MA model with appropriately lagged residuals from a long autoregression. Unlike Broemeling and Shaarawy (1988), the exact structure of the approximation error when replacing true errors with corresponding residuals is derived and used in deriving the posterior distribution of the model parameters. In addition, a modified version of Broemeling and Shaarawy's (1988) method is suggested. The proposed method and original Broemeling and Shaarawy's method and its modified version are compared using several simulation studies and a real data.

Likelihood-based and bootstrap confidence intervals for the quantiles of the generalized exponential distribution based on record values are developed in Chapter 7. Intervals for the difference of quantiles based on two independent samples are also considered. Simulation studies are carried out to investigate and compare the performance of the intervals.

Order statistics from Marshall-Olkin exponential distribution is studied in Chapter 8. We derive exact forms for means, variances and covariances of order statistics from Marshall-Olkin exponential distribution. We also establish several recurrence relations satisfied by the single moments and the product moments for order statistics from the Marshall-Olkin exponential distribution. These relations are used in a simple recursive manner in order to evaluate the single and the product moments of all order statistics for all sample sizes.

Chapter 9 deals with the problem of the shrinkage test - estimation for the Measure of Dispersion of the inverse Gaussian distribution when a prior point information of the parameter is available. The performances of the shrinkage test - estimators based on a single stage and double stage samples are studied.

In Chapter 10, the minimum variance unbiased estimator, the maximum likelihood estimator and the Bayesian estimator for the parameter of the generalized logistic distribution are obtained based on *k*th lower record values. The Bayes estimators are obtain using both the symmetric squared error, squared log error loss function and the asymmetric LINEX and General Entropy loss functions and two new loss functions which we call Modified General Entropy (MGE) loss function and Kullback-Leibler divergence type loss function (KLD). Interval prediction for future *k*th upper record values is also presented from a Bayesian point of view. Numerical computations are given to illustrate these procedures.

The number of parameters in a periodic autoregressive model is a multiple of the period. Large sample tests on the parameters of interest may be misleading when the sample size is small and the period is large. The modified profile likelihood proposed in Chapter 11 aims at reducing the adverse effect of the nuisance parameters on the inference of seasonality. The likelihood ratio test, the Wald test and the score test are compared by means of a simulation study. The technique is illustrated by the analysis of two real data sets.

Using the first k components of a random sample of n components from Rayleigh and Inverse Rayleigh lifetime models, this paper presents Bayesian prediction bounds for the remaining (n-k) components for each of the above models. The prior belief of the models is represented by its respective natural conjugate prior. It is indicated in Chapter 12 that the inverse Rayleigh distribution could be used as an alternative to the lognormal distribution in modelling the survival time of certain diseases. Numerical examples are given to illustrate the results derived.

In degradation analysis, the degradation measurements are often taken over time, and these measurements can be used to get information about the product's reliability. The time-to-failure is defined as the time when the degradation of a unit reaches a critical level. In Chapter 13, we used a Ranked Set Sampling (RSS) technique to estimate the parameters of a time-to-failure distribution of the linear degradation model. We obtained the maximum likelihood estimates (MLE's) of the parameters based on a RSS and based on a simple random sample (SRS). The performance of these estimators are compared in terms of their efficiencies. Simulation results showed that the MLE's based on a RSS are more efficient than the MLE's based on a SRS. Application to a real data set is given.

In Chapter 14, the book "Reliability and Risk: A Bayesian Perspective", by Nozer D. Singpurwalla, Wiley, New York (2006) is reviewed.

In Chapter 15, a new approach 'minimum perpendicular distance squared method' has been suggested to estimate the regression coefficient and constant, instead of traditionally least squared method to fit a regression line, with an aim to get a better fitted regression line, so that the estimated line will be most closest one to the observed points. Mathematical form of the estimator for the parameters have been derived and a logical stand has been placed behind the relationship between the slope parameters of the lines $y = \alpha x + \beta$ and $x = \alpha' y + \beta'$. Finally, a simulation study has been conducted to justify the validity of the suggested new method against the traditionally least squared method.

In Chapter 16 four bivariate exponential (BVE) distributions with subject to right censoring samples are presented. We have opted gamma distribution as a prior for the parameters of the model on which the hyper parameters have been estimated based on the method of moments. A simulation was used to compute the mean and variance of the moment estimators. Furthermore, it is observed from the simulation study that the posterior mode of parameters obtained by moment method are very close to the true values. We estimate the parameters of posterior pdf using EM algorithm in BVE models. The estimations and standard errors of parameters have also been obtained. We compare the Bayesian estimates with MLEs.

Chapter 17 considers the problem of estimating the population mean $\overline{Y}$ of the study variate y in the situation when information on the study variate y is not available for the initial sample while information on the auxiliary variable x is available for initial sample (i.e. complete sample) but the population mean $\overline{X}$ of the auxiliary character is not known. A family of estimators for $\overline{Y}$ has been suggested and its properties are studied. Usual unbiased estimator, ratio, product and difference type estimators are identified as members of the proposed family. Realistic conditions are obtained in which the proposed family is better than conventional estimators. Optimum sampling and sub sampling aspects are given. For illustration an empirical study is carried out to demonstrate the performance of various estimators.

The estimation of the reliability $R = P(Y < X)$, assuming X and Y to be independent and identically distributed random variables, has been extensively studied in the literature. Distribution functions such as Normal, Burr Type X, Exponential, Gamma, and Weibull have been considered. In Chapter 18, we consider the Weibull type of distributions. Utilizing the Maximum Likelihood Estimation (MLE), Uniformly Minimum Variance Unbiased Estimation (UMVUE), Shrinkage Estimation Procedures, Method of Moments, and other methods of estimation, using simulation, we will estimate the unknown parameters of the Weibull distribution. In particular, we will observe how well the shrinkage estimators will do compared with the MLE based on the mean squared error (MSE). Also, a comparison among the estimators found by the different methods will be done to choose the best results in calculating R. Those results will be made based on the ratio of the parameters and the sample sizes.

The curve of concentration approach for examining the extent of variation and/or inequality in a system has been extended for reliability analysis by replacing the incomplete first order moment by the standardized cumulative reliability value. In this process, its connection with the life distribution classes has been explored in general and also with specific reference to the Weibull distribution. Characterization results for the exponential distribution have been reported based on zero concentration and zero reliability index, and those for the Lomax and related distributions have been studied based on area equivalence with the Gini's measure. Further in Chapter 19, weighted distribution has been considered for an extended development of the curve of concentration.

In Chapter 20, the two-parameter Pareto distribution is assumed as a stress-strength model. Estimation of the function $R=Pr(Y<X)$ is obtained based on generalized order statistics. The Bayes method of estimation is used under squared error and linex loss functions. Monte carlo simulation technique is used to compare the Bayes estimates with their corresponding maximum likelihood.

During the process of collecting data, we might not able to get the fully observed data. This will cause us to have the incomplete data. An inappropriate conclusion and confusion occur when the researchers ignore, truncate, censored or collapse those data as it might contain important information. Maximum likelihood estimate and EM algorithm are the most common ways for researchers to solve for the missing data. Here is an alternative method is proposed. In Chapter 21, we consider a logistic regression with the conditional probabilities to obtain the cell probabilities instead of MLE and EM algorithm. We consider a set of data from a clinical investigation of 1502 infants. In this trial, subjects were given either Palivizumab or a placebo and were followed for five months to see if they were hospitalized for respiratory syncytial virus (RSV). The estimation procedure is illustrated in this paper on the basis of the proposed technique of estimating missing values under various assumptions. We consider logistic regression model to estimate the incomplete categorical data for three way contingency table where we let one of the variable to be the outcome variable and the other two as independent variable. In our work, we demonstrate the use of the logistic regression model to estimate the missing independent variable 1, missing independent variable 2 and missing independent variable 1 and independent variable 2. We compare the expected value of the missing value for the logistic regression model with EM algorithm. An alternative approach is proposed in this paper based on the conditional probabilities as logit link functions for estimating missing cell frequencies. This provides an alternative to the existing techniques based on the MLE and the EM algorithm. The proposed method appears to be more flexible and the prediction because it does not require additional assumption on the underlying probability distributions. In addition, the proposed technique provides estimates similar to that of the estimates based on existing techniques.

The most of the material of Chapter 22 is an overview of principle results in and related to the computing of the product moments (of any order) of bivariate (and trivariate) order statistics arisen from any arbitrary bivariate (and trivariate) distribution function. The exact explicit expression, as well as formulae to compute easily these moments are presented. Some recent analytic bootstrap-type methods for estimating these moments are exhibited. In addition, universal distribution -free bounds for the differences of any two different product moments and general recurrence relations satisfied by the product moments (of any order) are discussed. Finally, a recent general bivariate and trivariate versions of the basic triangle rule for order statistics arisen from univariate distribution function is presented.

Chapter 23 deals with, the types of possible nondegenerate limit laws of distributions that could arise as limiting distributions of suitably power normalized record values. Domain of attraction criteria for each of the six types of limit laws can be derived by appealing to the limiting record value distributions to the extreme value distributions under power normalizations.

Chapter 24 studies the preliminary test and shrinkage estimators based on the Kalman filtering procedure applied to a dynamic linear state space regression model. The performance of these estimators, with respect to mean square error, was investigated. It was revealed that under certain conditions both the preliminary test and shrinkage estimators proposed outperform the Kalman filter. This out-performance was not uniform. Further, the shrinkage estimator was found to be superior to the preliminary test estimator over large regions. The results presented in this paper invalidates the global minimum mean square error property of the Kalman filter that is widely used by the engineers for estimation of the parameters of linear state space models.

Chapter 25 presents Bayesian estimation of Lorenz curve and Gini-index in the presence of outliers and using record values for the classical Pareto distribution. Here we considered two important models for accommodating outliers viz the identified outliers model introduced by Veale (1975) and the exchangeable model by Kale (1976). Bayes estimators have been developed under squared error loss function as well as under LINEX loss functions. We demonstrate the use of the proposed outlier estimation procedure with the tournament earnings data given in Arnold (1983). Based on a Monte Carlo simulation study, comparison are made between the proposed estimators.

In: New Developments in Applied Statistics
Editor: Mohammad Ahsanullah
ISBN: 978-1-61324-648-1

Chapter 1

ESTIMATION OF FINITE POPULATION MEAN USING RANDOM NON-RESPONSE IN SURVEY SAMPLING

***M.S. Ahmed*[a], *Omar Al-Titi*[b], *Ziad Al-Rawi*[b] *and Walid Abu-Dayyeh*[a]**

[a] Department of Mathematics and Statistics, Sultan Qaboos University, Oman
[b] Department of Statistics, Yarmouk University, Jordan

Abstract

This paper proposes some general estimators for finite population mean in presence of random non-response using an auxiliary variable. All possible non-response cases are considered here. The properties of these estimators are studied. Finally, a simulation study has been done to compare their performance.

Keywords: Randomized Response, Ratio and Regression Estimators, Simple Random, Sample without Replacement.

1. Introduction

Ratio and product methods are two popular comprehensive techniques for the estimation of the population mean in survey sampling when an auxiliary characteristic correlated with the study characteristic is available (see, e.g., Sukhatme et al., 1984). The ratio (product) method, generally, provide biased but more efficient estimator in comparison with the traditional unbiased estimator (the sample mean) if the correlation between the auxiliary characteristic and the study characteristic is positive (negative). Both methods of estimation assume that the sample data contain no missing observation and the population mean of auxiliary characteristic is known. One or both of these specifications may not be tenable in many practical application; see, e.g., Little and Rubin (1987) for an excellent exposition.

When no observation is missing while the population mean of auxiliary characteristic is not available, it is customary to make use of a large preliminary sample for finding an estimate of it. If the circumstances do not permit to have the preliminary sample due to some practical difficulties or otherwise, an alternative estimator for the population of auxiliary characteristic based on the given sample data may be utilized (Srivastava & Bhatnagar, 1981).

On the other hand, when some observations are missing but the population mean of auxiliary characteristic is available; Tracy and Osahan (1994) studied the effect of random non-response on the ratio estimator of the population mean in two situations: (i) nonresponse in the study as well as the auxiliary variables and (ii) nonresponse in the study variable only.

Singh et al. (2000) studied the three regression type estimators in the presence of random non-response in different situations under the assumption that the number of sampling units on which information can not be obtained due to random non-response follows some distribution.

In this paper, we study the effect of random non-response on the study and auxiliary variables of several estimators of the men under the three strategies considered by Singh et al. (2000) for estimating the population variance:

Strategy I: Random non-response exists on both the study variable y and the auxiliary variable x and the population mean $\overline{X}$ of the auxiliary character is known.

Strategy II: Information on variable y could not be obtained for r units while information on variable x is available and the population mean of the auxiliary variable is known.

Strategy III: Information on variable y could not be obtained for r units while information on variable x is available, but the mean of the auxiliary variable is not known.

For each strategy, we propose a number of estimators of the population mean and for each estimator we derive the approximate bias, mean square error up to first order approximation and give an estimator of the mean square error.

Let $\omega:(v_1,v_2,....,v_N)$ denote the population of N units from which a simple random sample of size n is drawn without replacement. If $r\,(r=0,1,2,....(n-2))$ denotes the number of sampling units on which information could not be obtained due to random non-response, then the remaining $(n-r)$ units, in the sample can be treated as a SRSWOR from ω . Since we are considering the problem of unbiased estimator of finite population mean, therefore we are assuming that r should be less than $(n-1)$ i.e. $0 \le r \le n-2$. We will use the discrete distribution of Singh et al. (2000) for the number of non-responses among the $(n-2)$ possible values which is given by:

$$P(r)=\frac{(n-r)}{nq+2p}\,{}^{n-2}C_r\,p^r q^{n-2-r}, \qquad (1.1)$$

where $q=1-p$, $r=0,1,2,....,(n-2)$ and ${}^{n-2}C_r$, denote the total number of ways of r non-responses out of totals possible $(n-2)$ responses.

Let us define

$$\varepsilon = \frac{\bar{y}_{n-r}}{\bar{Y}} - 1, \quad \delta = \frac{\bar{x}_{n-r}}{\bar{X}} - 1 \quad \text{and} \quad \eta = \frac{\bar{x}}{\bar{X}} - 1$$

where

$$\bar{y}_{n-r} = (n-r)^{-1}\sum_{i=1}^{n-r} y_i \;, \; \bar{x}_{n-r} = (n-r)^{-1}\sum_{i=1}^{n-r} x_i \;\text{ and }\; \bar{x} = n^{-1}\sum_{i=1}^{n} x_i$$

have their usual meaning (Tracy and Osahan, 1994). Under the probability model given by (1.1), we have the following results:

$$E(\varepsilon) = E(\delta) = E(\eta) = 0,\; E(\varepsilon^2) = \left(\frac{1}{nq+2p} - \frac{1}{N}\right)C_y^2,$$

$$E(\delta^2) = \left(\frac{1}{nq+2p} - \frac{1}{N}\right)C_x^2,$$

$$E(\eta^2) = \left(\frac{1}{n} - \frac{1}{N}\right)C_x^2 \quad E(\varepsilon\delta) = \left(\frac{1}{nq+2p} - \frac{1}{N}\right)\rho C_y C_x,\; E(\delta\eta) = \left(\frac{1}{n} - \frac{1}{N}\right)C_x^2,$$

$$E(\varepsilon\eta) = \left(\frac{1}{n} - \frac{1}{N}\right)\rho C_y C_x \text{ and } E(\varepsilon^k \eta^s) \approx 0 \text{ for all } k,s>1$$

where $C_y^2 = S_y^2/\bar{Y}^2$, $C_x^2 = S_x^2/\bar{X}^2$, $\rho = S_{xy}/(S_x S_y)$ and S_y^2, S_x^2 and S_{xy} have their usual meanings. Also, define

$$s_{xy}^* = (n-r-1)^{-1}\sum_{i=1}^{n-r}(y_i - \bar{y}_{n-r})(x_i - \bar{x}_{n-r}) \;,\; s_x^{*2} = (n-r-1)^{-1}\sum_{i=1}^{n-r}(x_i - \bar{x}_{n-r})^2$$

$$s_y^{*2} = (n-r-1)^{-1}\sum_{i=1}^{n-r}(y_i - \bar{y}_{n-r})^2 \;,\; \hat{\rho} = s_{xy}^*/(s_x^* s_y^*),\; c_y^2 = s_y^{*2}/\bar{y}_{n-r}^2 \;,\; c_x^2 = s_x^{*2}/\bar{x}_{n-r}^2 \;.$$

Then $\hat{p} = \dfrac{(n-1+r) - \sqrt{(n-1+r)^2 - \dfrac{4rn(n-3)}{(n-2)}}}{2(n-3)}$ is a maximum likelihood estimator of p (Singh and Joarder (1998)) and $\hat{\rho} = s_{xy}^*/(s_x^* s_y^*)$. Note that under this distribution of random non-response, the exact bias and mean square error expressions, up to first order of approximation, exist for the proposed strategies and hence comparisons with Singh et al. (2000) estimators are possible.

2. Proposed Estimators

Under Strategy I: Singh et al. (2000) proposed the following regression estimator of the finite population mean of the study variable (μ_y):

$$\bar{y}_1 = \bar{y}_{n-r} + \alpha_1(\bar{X} - \bar{x}_{n-r}) \tag{2.1}$$

For this case, we propose the following estimators of μ_y :

$$\bar{y}_2 = \bar{y}_{n-r}\left(\frac{\bar{X}}{\bar{x}_{n-r}}\right)^{\beta_1} \tag{2.2}$$

$$\bar{y}_3 = \frac{\bar{y}_{n-r}\bar{X}}{\theta_1\,\bar{x}_{n-r} + (1-\theta_1)\bar{X}} \tag{2.3}$$

Next, we will give the biases and mean square errors of the proposed estimators. The proofs are similar and therefore we will only prove one of them.

Theorem 2.1:

(a). *The bias of* $\bar{y}_2$ *which is given by (2.2) is given by:*

$$B(\bar{y}_2) \approx \left(\frac{1}{nq+2p} - \frac{1}{N}\right)\bar{Y}\left[\frac{\beta_1(\beta_1+1)}{2}C_x^2 - \beta_1\rho C_y C_x\right] \tag{2.4}$$

(b). *The mean squared error of* $\bar{y}_2$ *is given by*

$$M(\bar{y}_2) \approx \left(\frac{1}{nq+2p} - \frac{1}{N}\right)\bar{Y}^2(C_y^2 + \beta_1^2 C_x^2 - 2\beta_1\rho C_y C_x) \tag{2.5}$$

(c). *The minimum mean squared error of* $\bar{y}_2$ *is given by:*

$$M(\bar{y}_2)_{\min} \approx \left(\frac{1}{nq+2p} - \frac{1}{N}\right)S_y^2(1-\rho^2) \tag{2.6}$$

which is attained at $\beta_1 = \rho\dfrac{C_y}{C_x}$

Proof:

(a). The estimator $\bar{y}_2$ in terms of ε, δ and η can be written as

$$\bar{y}_2 = \bar{Y}(1+\varepsilon)(1+\delta)^{-\beta_1}$$
$$= \bar{Y}(1+\varepsilon)\left[1-\beta_1\delta+\frac{\beta_1(\beta_1+1)}{2}\delta^2+\ldots\right] \approx \bar{Y}\left\{1+\varepsilon-\beta_1\delta-\beta_1\varepsilon\delta+\frac{\beta_1(\beta_1+1)}{2}\delta^2\right\}.$$

Then the bias of $\bar{y}_2$ is given by

$$B(\bar{y}_2) = E(\bar{y}_2-\bar{Y})$$
$$\approx \bar{Y}\,E\left[\varepsilon-\beta_1\delta-\beta_1\varepsilon\delta+\frac{\beta_1(\beta_1+1)}{2}\delta^2\right] = \bar{Y}\left[-\beta_1 E(\varepsilon\delta)+\frac{\beta_1(\beta_1+1)}{2}E(\delta^2)\right]$$
$$= \bar{Y}\left[-\beta_1\left(\frac{1}{nq+2p}-\frac{1}{N}\right)\rho C_y C_x+\frac{\beta_1(\beta_1+1)}{2}\left(\frac{1}{nq+2p}-\frac{1}{N}\right)C_x^2\right]$$
$$= \left(\frac{1}{nq+2p}-\frac{1}{N}\right)\bar{Y}\left[\frac{\beta_1(\beta_1+1)}{2}C_X^2-\beta_1\rho C_y C_x\right]$$

(b). The mean square error of $\bar{y}_2$ is given by

$$M(\bar{y}_2) = E(\bar{y}_2-\bar{Y})^2$$
$$\approx \bar{Y}^2 E(\varepsilon-\beta_1\delta)^2$$
$$= \bar{Y}^2\left[E(\varepsilon^2)-2\beta_1 E(\varepsilon\delta)+\beta_1^2 E(\delta^2)\right]$$
$$= \bar{Y}^2\left[\left(\frac{1}{nq+2p}-\frac{1}{N}\right)C_y^2-2\beta_1\left(\frac{1}{nq+2p}-\frac{1}{N}\right)\rho C_y C_x+\beta_1^2\left(\frac{1}{nq+2p}-\frac{1}{N}\right)C_x^2\right]$$
$$= \left(\frac{1}{nq+2p}-\frac{1}{N}\right)\bar{Y}^2(C_y^2+\beta_1^2 C_x^2-2\beta_1\rho C_y C_x)$$

To find the minimum mean square error of $\bar{y}_2$, we differentiate (2.5) with respect to β_1 and equate to zero. Then we get:

$$\beta_{11} = \rho\frac{C_y}{C_x}$$

On substituting the optimum value of β_{11} in (2.5)), we get

$$M(\bar{y}_2)_{\min} \approx \left(\frac{1}{nq+2p} - \frac{1}{N}\right) S_y^2 (1-\rho^2)$$

***Remark* 2.1**

(a). *For the optimum value of* β_{11} *the bias of* $\bar{y}_2$ *is:*

$$B(\bar{y}_2) \approx \left(\frac{1}{nq+2p} - \frac{1}{N}\right) \bar{Y} \left[\frac{\rho C_y (C_x - \rho C_y)}{2}\right] \quad (2.7)$$

(b). *We can find the estimate of* β_1 *from the sample information, that is*

$$\hat{\beta}_1 = \hat{\rho} \frac{c_y}{c_x} \quad (2.8)$$

Theorem 2.2

(a). *The bias of the proposed estimator* $\bar{y}_3$ *in (2.3) is given by:*

$$B(\bar{y}_3) \approx \left(\frac{1}{nq+2p} - \frac{1}{N}\right) \bar{Y} (\theta_1^2 C_x^2 - \theta_1 \rho C_y C_x) \quad (2.9)$$

(b). *The mean squared error of* $\bar{y}_3$ *is given by:*

$$M(\bar{y}_3) \approx \left(\frac{1}{nq+2p} - \frac{1}{N}\right) \bar{Y}^2 (C_y^2 + \theta_1^2 C_x^2 - 2\theta_1 \rho C_y C_x) \quad (2.10)$$

(c). *The minimum mean squared error of* $\bar{y}_3$ *is given by:*

$$M(\bar{y}_3)_{\min} \approx \left(\frac{1}{nq+2p} - \frac{1}{N}\right) S_y^2 (1-\rho^2) \quad (2.11)$$

which is attained at: $\theta_1 = \rho \dfrac{C_y}{C_x}$

Remark 2.2:

(a). *At the optimum value* θ_1 *the bias of* $\bar{y}_3$ *is :*

$$B(\bar{y}_3) \approx 0 \quad (2.12)$$

(b). The estimator of θ_{11} is:

$$\hat{\theta}_{11} = \hat{\rho}\frac{c_y}{c_x} \tag{2.13}$$

Remark 2.3:

The estimators $\bar{y}_1$, $\bar{y}_2$ *and* $\bar{y}_3$ have the same mean square error as up to first order approximation.

Under Strategy II:

Singh et al. (2000) proposed the following estimator of finite population mean

$$\bar{y}_4 = \bar{y}_{n-r} + \alpha_2(\bar{X} - \bar{x}\ ) \tag{2.14}$$

For this case, we propose the following estimators of finite population mean

$$\bar{y}_5 = \bar{y}_{n-r}\left(\frac{\bar{X}}{\bar{x}}\right)^{\beta_2} \tag{2.15}$$

$$\bar{y}_6 = \frac{\bar{y}_{n-r}\bar{X}}{\theta_2\,\bar{x} + (1-\theta_2)\bar{X}} \tag{2.16}$$

Next, we will give the biases and mean square errors of the proposed estimators. The proofs are similar and therefore we will only prove one of them.

Theorem 2.3:

(a). *The bias of the* $\bar{y}_5$ *in (2.17) is given by:*

$$B(\bar{y}_5) \approx \left(\frac{1}{n} - \frac{1}{N}\right)\bar{Y}\left[\frac{\beta_2(\beta_2+1)}{2}C_x^2 - \beta_2\rho C_y C_x\right] \tag{2.17}$$

(b). *The mean squared error of* $\bar{y}_5$ *is given by:*

$$M(\bar{y}_5) \approx \bar{Y}^2\left[\left(\frac{1}{nq+2p} - \frac{1}{N}\right)C_y^2 + \beta_2^2\left(\frac{1}{n} - \frac{1}{N}\right)C_x^2 - 2\beta_2\left(\frac{1}{n} - \frac{1}{N}\right)\rho C_y C_x\right] \tag{2.18}$$

(c). *The minimum mean squared error of* $\bar{y}_5$ *is given by:*

$$M(\bar{y}_5)_{\min} \approx \left(\frac{1}{nq+2p} - \frac{1}{N}\right)S_y^2 - \left(\frac{1}{n} - \frac{1}{N}\right)\frac{S_{xy}^2}{S_x^2} \tag{2.19}$$

which is attained at $\beta_{22} = \rho\dfrac{C_y}{C_x}$

***Remark* 2.4:**

(a). *At the optimum value* β_{22}, *the bias of* $\bar{y}_5$ is:

$$B(\bar{y}_5) \approx \left(\frac{1}{n} - \frac{1}{N}\right)\bar{Y}\left[\frac{\rho C_y (C_x - \rho C_y)}{2}\right] \tag{2.20}$$

(b). *The estimate of* β_2 *is given by:*

$$\hat{\beta}_2 = \hat{\rho}\frac{c_y}{c_x} \tag{2.21}$$

Theorem 2.4:

(a). *The bias of the estimator* $\bar{y}_6$ *in (2.18) is given by:*

$$B(\bar{y}_6) \approx \left(\frac{1}{n} - \frac{1}{N}\right)\bar{Y}\,(\theta_2^2 C_x^2 - \theta_2 \rho C_y C_x) \tag{2.22}$$

(b). *The mean squared error of* $\bar{y}_6$ *is given by:*

$$M(\bar{y}_6) \approx \bar{Y}^2\left[\left(\frac{1}{nq+2p} - \frac{1}{N}\right)C_y^2 + \theta_2^2\left(\frac{1}{n} - \frac{1}{N}\right)C_x^2 - 2\theta_2\left(\frac{1}{n} - \frac{1}{N}\right)\rho C_y C_x\right] \tag{2.23}$$

(c). *The minimum mean squared error of* $\bar{y}_6$ *is given by*

$$M(\bar{y}_6)_{\min} \approx \left(\frac{1}{nq+2p} - \frac{1}{N}\right)S_y^2 - \left(\frac{1}{n} - \frac{1}{N}\right)\frac{S_{xy}^2}{S_x^2} \tag{2.24}$$

which is attained at :

$$\theta_{22} = \rho \frac{C_y}{C_x} \tag{2.25}$$

Remark 2.5:

(a). *At the optimum value of* θ_{22}, the bias of $\bar{y}_6$ *is:*

$$B(\bar{y}_6) \approx 0 \tag{2.26}$$

(b). The estimate of θ_{22} is:,

$$\hat{\theta}_2 = \hat{\rho} \frac{c_y}{c_x} \tag{2.27}$$

Remark 2.6:

The estimators $\bar{y}_4$, $\bar{y}_5$ *and* $\bar{y}_6$ have the same mean as up square error to first order approximation.

Under Strategy III:

Singh et al. (2000) proposed the following estimator of finite population mean

$$\bar{y}_7 = \bar{y}_{n-r} + \alpha_3 (\bar{x}_{n-r} - \bar{x}) \tag{2.28}$$

For this case, we propose the following estimators of finite population mean

$$\bar{y}_8 = \bar{y}_{n-r} \left(\frac{\bar{x}}{\bar{x}_{n-r}} \right)^{\beta_3} \tag{2.29}$$

$$\bar{y}_9 = \frac{\bar{y}_{n-r} \bar{x}}{\theta_3 \, \bar{x}_{n-r} + (1 - \theta_3) \bar{x}} \tag{2.30}$$

Theorem 2.5:

(a). *The bias of* $\bar{y}_8$ *in (2.32) is given* by:

$$B(\bar{y}_8) \approx \left(\frac{1}{nq + 2p} - \frac{1}{n} \right) \bar{Y} \left[\frac{\beta_3 (\beta_3 + 1)}{2} C_x^2 - \beta_3 \rho C_y C_x \right] \tag{2.31}$$

(b). The mean squared error of $\bar{y}_8$ is given by:

$$M(\bar{y}_8) \approx \bar{Y}^2\left[\left(\frac{1}{nq+2p}-\frac{1}{N}\right)C_y^2 + \beta_3^2\left(\frac{1}{nq+2p}-\frac{1}{n}\right)C_x^2 - 2\beta_3\left(\frac{1}{nq+2p}-\frac{1}{n}\right)\rho C_y C_x\right] \quad (2.32)$$

(c). The minimum mean squared error of $\bar{y}_8$ is given by

$$M(\bar{y}_8)_{\min} \approx \left[\left(\frac{1}{nq+2p}-\frac{1}{N}\right)S_y^2 - \left(\frac{1}{nq+2p}-\frac{1}{n}\right)\frac{S_{xy}^2}{S_x^2}\right] \quad (2.33)$$

which is attained at $\beta_{33} = \rho\dfrac{C_y}{C_x}$

Remark 2.7:

(a). For optimum value of β_{33}, the bias of $\bar{y}_8$ is:

$$B(\bar{y}_8) \approx \left(\frac{1}{nq+2p}-\frac{1}{n}\right)\bar{Y}\left[\frac{\rho C_y(C_x - \rho C_y)}{2}\right] \quad (2.34)$$

(b). The estimate of β_{33} *is*

$$\hat{\beta}_3 = \hat{\rho}\frac{c_y}{c_x} \quad (2.35)$$

Theorem 2.6:

(a). *The bias of the proposed estimator* $\bar{y}_9$ *in (2.33) is given by*

$$B(\bar{y}_9) \approx \left(\frac{1}{nq+2p}-\frac{1}{n}\right)\bar{Y}\,(\theta_3^2 C_x^2 - \theta_3 \rho C_y C_x) \quad (2.36)$$

(b). The mean squared error of $\bar{y}_9$ is given by:

$$M(\bar{y}_9) \approx \bar{Y}^2\left[\left(\frac{1}{nq+2p}-\frac{1}{N}\right)C_y^2 + \theta_3^2\left(\frac{1}{nq+2p}-\frac{1}{n}\right)C_x^2 - 2\theta_3\left(\frac{1}{nq+2p}-\frac{1}{n}\right)\rho C_y C_x\right] \quad (2.41)$$

(c). The minimum mean squared error of $\bar{y}_9$ *is given by*

$$M(\bar{y}_9)_{\min} \approx \left(\frac{1}{nq+2p}-\frac{1}{N}\right)S_y^2 - \left(\frac{1}{nq+2p}-\frac{1}{n}\right)\frac{S_{xy}^2}{S_x^2} \quad (2.37)$$

which is attained at:

$$\theta_{33} = \rho \frac{C_y}{C_x} \tag{2.38}$$

Remark 2.8:

(a). *For* optimum value of θ_{33}, the bias of $\bar{y}_9$ is:

$$B(\bar{y}_9) \approx 0 \tag{2.39}$$

(b). *The estimate of* θ_{33} *is:*

$$\hat{\theta}_{33} = \hat{\rho} \frac{c_y}{c_x} \tag{2.40}$$

Remark 2.9:

The estimators of $\bar{y}_7$, $\bar{y}_8$ and $\bar{y}_9$ have the same mean squared error to first order approximation.

3. Numerical Illustration with Simulation, Conclusions and Recommendations

The relative comparison among these estimators is given by using a real data set. The data for this illustration has been taken from the Department of Statistics (Jordan), Healthcare Utilization and Expenditure Survey, 2000. The population that we like to study contains 8306 households. We consider the variables y and x where y is the expenditure of the household and x is the income of the household. Throughout this study, calculations are used based on the simulation study of 50,000 repeated samples without replacement. We compute the bias and the MSE for all estimators. We take the following value of the probability of no-response: $p = 0.10$. The following values were obtained using the whole data set:

$$S_y^2 = 338006\,,\ S_x^2 = 862017\ \ S_{xy} = 281892\,,\ \bar{Y} = 253.75,\ \bar{X} = 343.316$$

$$C_y = 2.29116,$$

$$C_x = 2.70436,\ \rho = 0.522231,\ \alpha_1 = \alpha_2 = \alpha_3 = 0.327014$$

$$\theta_1 = \theta_2 = \theta_3 = \beta_1 = \beta_2 = \beta_3 = 0.442441$$

Simulation Work

The following steps summarize the simulation procedures to find the bias and MSE of any the estimators that introduced that are under consideration:

- **Step 1.** We select a SRSWOR of size 200 from the 8306 real data set.
- **Step 2.** We select value of r, say $r_\circ$.
- **Step 3.** We drop $r_\circ$ units randomly from each sample in step 1.
- **Step 4.** We find the value of the estimator based on the $(n-r_\circ)$ observation say $\hat{y}$.
- **Step 5.** We repeat step (1), (2), (3) and (4) 50000 times. Thus, we obtain 50000 values for $\hat{y}$ say $\hat{y}_1,\ldots,\hat{y}_{50000}$.
- **Step 6.** The bias of $\hat{y}$ is obtained by $B(\hat{y})=\frac{1}{50000}\sum_{i=1}^{50000}\hat{y}_i-\bar{Y}$
- **Step 7.** The MSE of $\hat{y}$ is obtained by $MSE(\hat{y})=\frac{1}{50000}\sum_{i=1}^{50000}(\hat{y}_i-\bar{Y})^2$

All calculations have been done using Mathematica (4). The results are summarized in Tables (8), (9(and (10) for strategies I, II and III respectively. The tables contain the bias , MSE , and efficiency of the estimators where the efficiency of the estimator $\hat{y}$ with respect to $\bar{y}_1$ is defined by: $e(\hat{y})=\frac{MSE(\bar{y}_1)}{MSE(\hat{y})}$.

Based on the simulation results we may conclude:

(i) $\bar{y}_1$ and $\bar{y}_3$ are approximately unbiased while $\bar{y}_2$ is biased

(ii) $\bar{y}_2$ and $\bar{y}_3$ have the same efficiency and more efficient than $\bar{y}_1$. Therefore we recommend the use of $\bar{y}_3$ under strategy I.

Table 3.1. Bias and MSE for estimators of Strategy I under study based on simulation

Estimator	MSE	Efficiency	Bias
$\bar{y}_1$	1328.14	1	0.0213273
$\bar{y}_2$	1254.94	1.058329	1.2236300
$\bar{y}_3$	1242.07	1.069296	0.0304842

Table 3.2. Bias and MSE for estimators of Strategy II under study based on simulation

Estimator	MSE	Efficiency	Bias
$\bar{y}_4$	1380.41	1	0.0083099
$\bar{y}_5$	1309.85	1.053869	1.0989000
$\bar{y}_6$	1297.84	1.063621	0.0209209

Table 3.3. Bias and MSE for estimators of Strategy III under study based on simulation

Estimator	MSE	Efficiency	Bias
$\bar{y}_7$	1769.97	1	0.00759638
$\bar{y}_8$	1765.09	1.002765	0.12113300
$\bar{y}_9$	1761.67	1.004711	-0.00280604

References

[1] Little, R. J. A. and Rubin, D. B. (1987). *Statistical Analysis with Missing Data*. New York: Wiley.

[2] Singh, S. and Joarder, AH. (1998). Estimation of finite population variance in the presence of random non-response. *Metrika,* **47**:241-249.

[3] Singh, S., Joarder AH., and Tracy D. S. (2000). Regression Type Estimators for Random Non-Response in Survey Sampling. *Statistica,* **1**:39-44.

[4] Srivastava, V. K. and Bhatnagar, S. (1981). Ratio and product methods of estimation when $\bar{X}$ is not known. *Journal of Statistical Research* **15**: 29-39.

[5] Sukhatme, P. V., Sukhatme, B. V., Sukhatme, S. and Asok, C. (1984). *Sampling Theory of Surveys with Applications*, Iowa State University Press, Iowa.

[6] Tracy, D. S. and Osahan, S. S. (1994). Random non-response on study variable versus on study as well as auxiliary variables. *Statistica,* **54**: 163-168.

In: New Developments in Applied Statistics
Editor: Mohammad Ahsanullah

ISBN: 978-1-61324-648-1

Chapter 2

RECURRENCE RELATIONS FOR MARGINAL AND JOINT MOMENT GENERATING FUNCTIONS OF GENERALIZED ORDER STATISTICS FROM GENERALIZED EXPONENTIAL DISTRIBUTION

Jagdish Saran and Aparna Pandey
Department of Statistics, University of Delhi, Delhi-110007 (India)

Abstract

In this paper we give some recurrence relations satisfied by marginal and joint moment generating functions of generalized order statistics from the generalized exponential distribution. These recurrence relations would hence enable one to compute all the higher order moments of the generalized order statistics from those of the lower order in a simple recursive way.

Keywords and Phrases: Order statistics; record values; generalized order statistics; recurrence relations; moment generating function; single moments; product moments; generalized exponential distribution; exponential distribution .

1. Introduction

Let $\{X_n , n \geq 1\}$ be a sequence of independent and identically distributed random variables with cdf F(x) and pdf f(x). Let $X_{j:n}$ denote the j-th order statistic of a sample (X_1, $X_2, \ldots, X_n$). Assume that $k > 0$, $n \in N$, $m \in Z$ and $\gamma_r = k + (n-r)(m+1) > 0$. If the random variables X(r,n,m,k), r = 1,2,…,n, possess a joint pdf of the form

$$f^{X(1,n,m,k),\ldots,X(n,n,m,k)}(x_1, x_2, \ldots, x_n) = k\left(\prod_{r=1}^{n-1}\gamma_r\right)\left(\prod_{j=1}^{n-1}(1-F(x_j))^m f(x_j)\right)(1-F(x_n))^{k-1} f(x_n),$$
$$F^{-1}(0+) < x_1 \le \ldots \le x_n < F^{-1}(1), \tag{1.1}$$

then they are called generalized order statistics of a sample from a distribution with cdf $F(x)$.

For convenience, let us define $X(0,n,m,k) = 0$. It can be seen that for the case when $m = -1$, we get the joint pdf of the k-th upper record values, and for $m = 0$, $k = 1$ we obtain the joint pdf of the ordinary order statistics. The pdf of the r-th generalized order statistic obtained from (1.1) is given by

$$f^{X(r,n,m,k)}(x) = \frac{c_{r-1}}{(r-1)!}(1-F(x))^{\gamma_r - 1} f(x)\ g_m^{r-1}(F(x)) \quad , x \in R, \tag{1.2}$$

where

$$c_{r-1} = \prod_{j=1}^{r}\gamma_j \quad , \quad r = 1,2,\ldots, n,$$
$$g_m(x) = h_m(x) - h_m(0) \quad , \quad x \in [0,1),$$
$$h_m(x) = \begin{cases} -\dfrac{1}{m+1}(1-x)^{m+1} & , m \ne -1 \\ -\log(1-x) & , m = -1 \end{cases}.$$

The joint pdf of $X(r,n,m,k)$ and $X(s,n,m,k)$, $r < s$, is given by

$$f^{X(r,n,m,k),X(s,n,m,k)}(x, y) = \frac{c_{s-1}}{(r-1)!(s-r-1)!}(1-F(x))^m f(x)\ g_m^{r-1}(F(x))\,[h_m(F(y)) - h_m(F(x))]^{s-r-1} \tag{1.3}$$
$$.(1-F(y))^{\gamma_s - 1} f(y) \quad , \quad x < y,$$

(cf. Kamps (1995)).

A random variable X is said to have a generalized exponential distribution if its pdf is of the form

$$f(x) = (1-\alpha x)^{\frac{1}{\alpha}-1} \quad , \quad 0 \le x \le \tfrac{1}{\alpha} \quad , \quad 0 \le \alpha < 1, \tag{1.4}$$

and the cdf is given by

$$F(x) = 1 - (1-\alpha x)^{\frac{1}{\alpha}} \quad , \quad 0 \le x \le \tfrac{1}{\alpha} \quad , \quad 0 \le \alpha < 1. \tag{1.5}$$

If we let the shape parameter $\alpha \to 0$, the pdf in (1.4) and the cdf in (1.5) become e^{-x} and $1-e^{-x}$, respectively, for $x \geq 0$. This is the standard exponential distribution discussed in detail by Joshi (1978,1982).

The family of generalized exponential distribution is an IFR (Increasing Failure Rate) family. Therefore, the generalized exponential distribution discussed in this paper is quite useful as a life-span model (see Cohen and Whitten, 1988). Its applications include use in the analysis of extreme events, as a failure-time distribution in reliability studies, in the modeling of large insurance claims, and in any situation in which the exponential distribution might be used but in which some robustness is required against heavier tailed or lighter tailed alternatives.

From (1.4) and (1.5), we can easily see that the characterizing differential equation for the generalized exponential population is

$$\left(1-\alpha x\right) f(x) = 1 - F(x)\,. \tag{1.6}$$

In this paper, the relation in (1.6) will be utilized to derive the recurrence relations for marginal and joint moment generating functions of generalized order statistics from the generalized exponential distribution. We shall restrict our attention to the case $0 \leq \alpha < 1/2$, for both practical and theoretical reasons.When the generalized exponential distribution is used as an alternative to the exponential distribution, it is likely that values of α close to zero will be of much interest, since the exponential distribution is a special case of generalized exponential distribution with $\alpha = 0$. Generalized exponential distribution with $\alpha > 1/2$ have finite end points with $f(x) > 0$ at each end point, and such shapes rarely occur in statistical applications. Also generalized exponential distribution with $\alpha \geq 1/2$ have infinte variance, and this too is unusual in statistical applications.

If we let $\alpha \to 0$, the derived results reduce to the corresponding results for the standard exponential distribution.

The recurrence relations obtained in this paper can be utilized for obtaining characterization results for generalized exponential distribution.

2. Relations for Marginal Moment Generating Functions

Let us denote the marginal moment generating function of $X(r,n,m,k)$ by $M_{X(r,n,m,k)}(t)$ and its j-th derivative by

$$M^{(j)}_{X(r,n,m,k)}(t)\,.$$

We shall first establish the existence of $M_{X(r,n,m,k)}(t)$. Using (1.2), we obtain

$$M_{X(r,n,m,k)}(t) = \frac{c_{r-1}}{(r-1)\,!}\int_{-\infty}^{\infty} e^{tx}\,(1-F(x))^{\gamma_r - 1}\; f(x)\;\; g_m^{r-1}(F(x))\,dx\,. \tag{2.1}$$

Further, on using (1.4) and (1.5) in (2.1) , we get

$$M_{X(r,n,m,k)}(t)=\frac{c_{r-1}}{(r-1)!}\left(\frac{-1}{m+1}\right)^{r-1}\sum_{j=0}^{r-1}(-1)^j\binom{r-1}{j}\int_0^{\frac{1}{\alpha}}e^{tx}(1-\alpha x)^{\frac{1}{\alpha}\{(m+1)(r-1-j)+\gamma_r\}-1}\,dx$$

$$=\frac{c_{r-1}}{(r-1)!}\left(\frac{-1}{m+1}\right)^{r-1}\frac{1}{\alpha}\sum_{j=0}^{r-1}(-1)^j\binom{r-1}{j}e^{\frac{t}{\alpha}}\int_0^1 e^{-\frac{tz}{\alpha}}z^{\frac{1}{\alpha}\{(m+1)(r-1-j)+\gamma_r\}-1}\,dz$$

$$=\frac{c_{r-1}}{(r-1)!}\left(\frac{-1}{m+1}\right)^{r-1}\frac{1}{\alpha}e^{\frac{t}{\alpha}}\sum_{p=0}^{\infty}\sum_{j=0}^{r-1}(-1)^j\binom{r-1}{j}\frac{(-t/\alpha)^p}{p!\left[p+\frac{1}{\alpha}\{(m+1)(r-1-j)+\gamma_r\}\right]}.$$

Applying D' Alembert's ratio test for absolute convergence, $M_{X(r,n,m,k)}(t)$ exists $\forall$ t, $-\infty<t<\infty$ and is analytic in t .

It may be mentioned that, for $\alpha = 0$, the expression for $M_{X(r,n,m,k)}(t)$ given above will reduce to the result (2.4) of Ahsanullah (2000) .

Theorem 2.1: Fix a positive integer k . For $n\in N$, $m\in Z$, $1<r\le n$ and $j = 1,2,\ldots$,

$$(\gamma_r+\alpha(i+1)-t)\,M^{(i+1)}_{X(r,n,m,k)}(t)=\gamma_r M^{(i+1)}_{X(r-1,n,m,k)}(t)+(i+1)M^{(i)}_{X(r,n,m,k)}(t)-\alpha t\,M^{(i+2)}_{X(r,n,m,k)}(t) \quad .\tag{2.2}$$

Proof : For $n\in N$, $m\in Z$, $1<r\le n$, it follows from (1.2) that

$$M_{X(r,n,m,k)}(t)=\frac{c_{r-1}}{(r-1)!}\int_{-\infty}^{\infty}e^{tx}(1-F(x))^{\gamma_r-1}f(x)\;g_m^{r-1}(F(x))\,dx \quad .\tag{2.3}$$

Integrating (2.3) by parts taking $\{(1-F(x))^{\gamma_r-1}f(x)\}$ as the part to be integrated, we get

$$M_{X(r,n,m,k)}(t)=M_{X(r-1,n,m,k)}(t)+\frac{c_{r-1}}{(r-1)!}\frac{t}{\gamma_r}\int_{-\infty}^{\infty}e^{tx}(1-F(x))^{\gamma_r}\;g_m^{r-1}(F(x))\,dx \quad ,$$

the constant of integration vanishes since the integral considered in (2.3) is a definite integral. On using (1.6), we obtain

$$M_{X(r,n,m,k)}(t)=M_{X(r-1,n,m,k)}(t)+\frac{c_{r-1}}{(r-1)!}\frac{t}{\gamma_r}\int_0^{\frac{1}{\alpha}}e^{tx}(1-F(x))^{\gamma_r-1}\;(1-\alpha x)\,f(x)\;g_m^{r-1}(F(x))\,dx$$

$$=M_{X(r-1,n,m,k)}(t)+\frac{t}{\gamma_r}M_{X(r,n,m,k)}(t)-\frac{\alpha t}{\gamma_r}M^{(1)}_{X(r,n,m,k)}(t) \quad .\tag{2.4}$$

Differentiating both sides of (2.4) (i+1) times with respect to t, we obtain for $1 < r \leq n$, $n \in N$, $m \in Z$ that

$$M^{(i+1)}_{X(r,n,m,k)}(t) = M^{(i+1)}_{X(r-1,n,m,k)}(t) + \frac{t}{\gamma_r} M^{(i+1)}_{X(r,n,m,k)}(t) + \frac{(i+1)}{\gamma_r} M^{(i)}_{X(r,n,m,k)}(t)$$

$$-\frac{\alpha t}{\gamma_r} M^{(i+2)}_{X(r,n,m,k)}(t) - \frac{\alpha(i+1)}{\gamma_r} M^{(i+1)}_{X(r,n,m,k)}(t) \quad ,$$

which , when rewritten , gives the recurrence relation in equation (2.2) .

Remark 2.1: Putting $\alpha = 0$ in (2.2), we obtain the recurrence relation for marginal moment generating functions of generalized order statistics from the standard exponential distribution as follows :

$$(\gamma_r - t) M^{(i+1)}_{X(r,n,m,k)}(t) = \gamma_r M^{(i+1)}_{X(r-1,n,m,k)}(t) + (i+1) M^{(i)}_{X(r,n,m,k)}(t) \quad .$$

Remark 2.2: Putting t = 0 in (2.2) we can deduce the following recurrence relation for the single moments of generalized order statistics from the generalized exponential distribution:

$$(\gamma_r + (i+1)\alpha)\, E[X^{i+1}(r,n,m,k)] = (i+1)\, E[X^{i}(r,n,m,k)] + \gamma_r\, E[X^{i+1}(r-1,n,m,k)]. \tag{2.5}$$

A similar result can also be obtained by using result (3.5) of Al-Hussaini (2004) for generalized exponential distribution .

Remark 2.3: Putting k = 1, m = 0 and t = 0 in (2.2), we get the recurrence relation for single moments of order statistics from a generalized exponential distribution, thus verifying the results of Saran and Pushkarna (2000) .

Remark 2.4: Under the assumptions of Theorem 2.1, and setting t = 0 with $\alpha \to 0$, we shall obtain the following recurrence relation for single moments of generalized order statistics from the standard exponential distribution, which agrees with the result (3.3) of Al-Hussaini (2004):

$$E\left[X^{i+1}(r,n,m,k)\right] = \frac{(i+1)}{\gamma_r} E\left[X^{i}(r,n,m,k)\right] + E\left[X^{i+1}(r-1,n,m,k)\right]. \tag{2.6}$$

Further, on setting k = 1, m = 0 in (2.6), we shall deduce the recurrence relation for single moments of order statistics from the standard exponential distribution established by Joshi (1978,1982). Also see Balakrishnan, Malik and Ahmed (1988, pp.2672).

Remark 2.5: On letting k = 1, m = −1 and t = 0 in (2.2), we get the recurrence relation for single moments of 1- record values from the standard exponential distribution (Balakrishnan and Ahsanullah, 1995).

3. Recursive Computational Algorithm for Evaluation of Single Moments

We shall first obtain an exact expression for E[X(r,n,m,k)] . On using (1.2), we obtain

$$E\left[X(r,n,m,k)\right]=\frac{c_{r-1}}{(r-1)!}\int_0^{\frac{1}{\alpha}} x\left(1-F(x)\right)^{r-1} f(x)\; g_m^{r-1}\left(F(x)\right)\, dx, \tag{3.1}$$

which on using (1.5) and (1.6) and simplifying gives

$$E[X(r,n,m,k)]=\frac{c_{r-1}}{(r-1)!}\int_0^{\frac{1}{\alpha}} x(1-\alpha x)^{\frac{\gamma_r}{\alpha}-1}\left[\frac{-(1-F(x))^{m+1}}{m+1}+\frac{1}{m+1}\right]^{r-1} dx.$$

Now, expanding $\left[\frac{-(1-F(x))^{m+1}}{m+1}+\frac{1}{m+1}\right]^{r-1}$ binomially and using (1.5), we get

$$E[X(r,n,m,k)]=\frac{c_{r-1}}{(r-1)!}\left(\frac{1}{m+1}\right)^{r-1}\sum_{i=0}^{r-1}\binom{r-1}{i}(-1)^i\int_0^{\frac{1}{\alpha}} x\;(1-\alpha x)^{\frac{1}{\alpha}\{(m+1)i+\gamma_r\}-1}\,dx,$$

which on further simplification reduces to

$$E[X(r,n,m,k)]=\frac{c_{r-1}}{(r-1)!}\left(\frac{1}{m+1}\right)^{r-1}\sum_{i=0}^{r-1}\binom{r-1}{i}(-1)^i\frac{1}{((m+1)i+\gamma_r)\,((m+1)i+\gamma_r+\alpha)}. \tag{3.2}$$

Starting with the means of generalized order statistics obtained from their exact and explicit expression given in (3.2), the recurrence relation given in (2.5) could be used in a simple recursive way in order to compute the higher order single moments of generalized order statistics from the generalized exponential distribution.

Remark 3.1: Setting m = 0, k = 1, and r = 1 in (3.2), we shall deduce the explicit expression for the mean of the first order statistic in a sample of size n from a generalized exponential distribution, thus verifying the result given in Corollary 3.3 of Saran and Pushkarna (2000, pp.592) .

Remark 3.2: Taking α = 0 in (3.2), we get the expected value of X(r,n,m,k) from the standard exponential distribution given by

$$E\left[X(r,n,m,k)\right]=\frac{c_{r-1}}{(r-1)!}\left(\frac{1}{m+1}\right)^{r-1}\sum_{i=0}^{r-1}\binom{r-1}{i}(-1)^i\frac{1}{\left((m+1)i+\gamma_r\right)^2}. \tag{3.3}$$

Ahsanullah (2000), by using some distributional properties of X(r,n,m,k) for standard exponential distribution, obtained in equation (2.5), an alternative expression for E[X(r,n,m,k)] as given below:

$$E\left[X(r,n,m,k)\right]=\sum_{j=1}^{r}\frac{1}{\gamma_j}. \tag{3.4}$$

Thus, on equating (3.3) and (3.4), we get the following interesting combinatorial identity:

$$\frac{c_{r-1}}{(r-1)!}\left(\frac{1}{m+1}\right)^{r-1}\sum_{i=0}^{r-1}\binom{r-1}{i}(-1)^i\frac{1}{\left((m+1)i+\gamma_r\right)^2}=\sum_{j=1}^{r}\frac{1}{\gamma_j}. \tag{3.5}$$

4. Relations for Joint Moment Generating Functions

Let $M_{X(r,n,m,k),X(s,n,m,k)}(t_1,t_2)$ denote the joint moment generating function of $X(r,n,m,k)$ and $X(s,n,m,k)$ and its i,j-th partial derivative by $M^{(i,j)}_{X(r,n,m,k),X(s,n,m,k)}(t_1,t_2)$.

Theorem 4.1: For $1\le r\le s-2\le n$, $n\in N$, $m\in Z$, $i, j = 0, 1, 2, \ldots\ldots$, and a fixed positive integer $k\ge 1$,

$$\begin{aligned}\{\gamma_s-t_2+(j+1)\alpha\}\,&M^{(i,j+1)}_{X(r,n,m,k),X(s,n,m,k)}(t_1,t_2)\\&=\gamma_s M^{(i,j+1)}_{X(r,n,m,k),X(s\text{-}1,n,m,k)}(t_1,t_2)-\alpha t_2 M^{(i,j+2)}_{X(r,n,m,k),X(s,n,m,k)}(t_1,t_2)\\&\quad+(j+1)\,M^{(i,j)}_{X(r,n,m,k),X(s,n,m,k)}(t_1,t_2),\end{aligned} \tag{4.1}$$

and, for $1\le r\le n$,

$$\begin{aligned}\{\gamma_{r+1}-t_2+(j+1)\alpha\}\,&M^{(i,j+1)}_{X(r,n,m,k),X(r+1,n,m,k)}(t_1,t_2)\\&=\gamma_{r+1}M^{(i+j+1)}_{X(r,n,m,k)}(t_1+t_2)-\alpha t_2 M^{(i,j+2)}_{X(r,n,m,k),X(r+1,n,m,k)}(t_1,t_2)\\&\quad+(j+1)M^{(i,j)}_{X(r,n,m,k),X(r+1,n,m,k)}(t_1,t_2).\end{aligned} \tag{4.2}$$

Proof : Using (1.3) , the joint moment generating function of X(r,n,m,k) and X(s,n,m,k) is given by

$$M_{X(r,n,m,k),X(s,n,m,k)}(t_1,t_2)=\frac{c_{s-1}}{(r-1)!(s-r-1)!}\iint_{x<y} e^{t_1x+t_2y}[1-F(x)]^m f(x)\ g_m^{r-1}(F(x))[h_m(F(y))-h_m(F(x))]^{s-r-1}$$
$$.[1-F(y)]^{\gamma_s-1} f(y)\,dy\,dx$$

$$=\frac{c_{s-1}}{(r-1)!(s-r-1)!}\int_{-\infty}^{\infty}[1-F(x)]^m\ f(x)\quad g_m^{r-1}(F(x))\ I(x)\ dx\quad , \tag{4.3}$$

where

$$I(x)=\int_x^{\infty} e^{t_1x+t_2y}[h_m(F(y))-h_m(F(x))]^{s-r-1}[1-F(y)]^{\gamma_s-1} f(y)\,dy\quad .$$

Solving the integral in I(x) by parts and substituting the resulting expression in (4.3) , we get

$$M_{X(r,n,m,k),X(s,n,m,k)}(t_1,t_2)=M_{X(r,n,m,k),X(s-1,n,m,k)}(t_1,t_2)$$
$$+\frac{t_2\,c_{s-1}}{(r-1)!(s-r-1)!\gamma_s}\iint_{x<y} e^{t_1x+t_2y}[1-F(x)]^m f(x)\ g_m^{r-1}(F(x))$$
$$.[h_m(F(y))-h_m(F(x))]^{s-r-1}[1-F(y)]^{\gamma_s}\ dy\ dx\ ,$$

the constant of integration vanishes since the integral in I(x) is a definite integral. On using the relation (1.6), we obtain

$$M_{X(r,n,m,k),X(s,n,m,k)}(t_1,t_2)=M_{X(r,n,m,k),X(s-1,n,m,k)}(t_1,t_2)+\frac{t_2}{\gamma_s}M_{X(r,n,m,k),X(s,n,m,k)}(t_1,t_2)$$
$$-\frac{\alpha t_2}{\gamma_s}M^{(0,1)}_{X(r,n,m,k),X(s,n,m,k)}(t_1,t_2)\ . \tag{4.4}$$

Differentiating both sides of (4.4) i times with respect to t_1 and then (j+1) times with respect to

t_2, we get

$$M^{(i,j+1)}_{X(r,n,m,k),X(s,n,m,k)}(t_1,t_2)=M^{(i,j+1)}_{X(r,n,m,k),X(s-1,n,m,k)}(t_1,t_2)+\frac{t_2}{\gamma_s}M^{(i,j+1)}_{X(r,n,m,k),X(s,n,m,k)}(t_1,t_2)$$
$$+\frac{(j+1)}{\gamma_s}M^{(i,j)}_{X(r,n,m,k),X(s,n,m,k)}(t_1,t_2)-\frac{\alpha t_2}{\gamma_s}M^{(i,j+2)}_{X(r,n,m,k),X(s,n,m,k)}(t_1,t_2)$$
$$-\frac{\alpha(j+1)}{\gamma_s}M^{(i,j+1)}_{X(r,n,m,k),X(s,n,m,k)}(t_1,t_2),$$

which, when rewritten, gives the recurrence relation in (4.1). Proceeding in a similar manner for the case s = r+1, the recurrence relation given in (4.2) can easily be established. One can also note that Theorem 2.1 can be deduced from Theorem 4.1 by letting t_1 tend to zero .

Remark 4.1: Letting t_1 and t_2 tend to 0 in (4.1) and (4.2), we obtain the following recurrence relations for product moments of generalized order statistics from the generalized exponential distribution for the cases $1 \le r \le s-2 \le n$ and $1 \le r \le n$, respectively.

For $n \in N$, $m \in Z$, $1 \le r \le s-2 \le n$ and $i,j = 0,1,2,\ldots$,

$$(\gamma_s + (j+1)\alpha)\, E\left[X^i(r,n,m,k)X^{j+1}(s,n,m,k)\right] = (j+1)\, E\left[X^i(r,n,m,k)X^j(s,n,m,k)\right] + \gamma_s E\left[X^i(r,n,m,k)X^{j+1}(s-1,n,m,k)\right], \tag{4.5}$$

and, for $1 \le r \le n$, $i, j = 0,1,2,\ldots$,

$$(\gamma_{r+1} + (j+1)\alpha)\, E\left[X^i(r,n,m,k)X^{j+1}(r+1,n,m,k)\right] = (j+1)\, E\left[X^i(r,n,m,k)X^j(r+1,n,m,k)\right] + \gamma_{r+1} E\left[X^{i+j+1}(r,n,m,k)\right]. \tag{4.6}$$

Remark 4.2: Putting $k = 1$, $m = 0$ and $t_1 = t_2 = 0$ in Theorem 4.1, we shall obtain the recurrence relations for product moments of order statistics from a generalized exponential distribution (Saran and Pushkarna, 2000).

Remark 4.3: Under the assumptions of Theorem 4.1 and setting $t_1 = t_2 = 0$, with $\alpha \to 0$, we shall deduce the following recurrence relations for product moments of generalized order statistics from the standard exponential distribution:

For $1 \le r \le s-2 \le n$,

$$E\left[X^i(r,n,m,k)X^{j+1}(s,n,m,k)\right] = \frac{(j+1)}{\gamma_s} E\left[X^i(r,n,m,k)X^j(s,n,m,k)\right] + E\left[X^i(r,n,m,k)X^{j+1}(s-1,n,m,k)\right], \tag{4.7}$$

and, for $1 \le r \le n$,

$$E\left[X^i(r,n,m,k)X^{j+1}(r+1,n,m,k)\right] = \frac{(j+1)}{\gamma_{r+1}} E\left[X^i(r,n,m,k)X^j(r+1,n,m,k)\right] + E\left[X^{i+j+1}(r,n,m,k)\right]. \tag{4.8}$$

Further, on setting $k = 1$, $m = 0$ in (4.7) and (4.8), we shall obtain the recurrence relations for product moments of order statistics from the standard exponential distribution established by Joshi (1978,1982). Also see Balakrishnan, Malik and Ahmed (1988, pp.2672) .

Remark 4.4: Setting $k = 1$, $m = -1$ and letting $t_1 = t_2 = 0$ in Theorem 4.1, we shall obtain the recurrence relation for product moments of 1- record values from the standard exponential distribution, thus verifying the results of Balakrishnan and Ahsanullah (1995).

5. Recursive Computational Algorithm for Evaluation of Product Moments

On using (1.3), we obtain

$$E[X(r,n,m,k)X(s,n,m,k)]=\frac{c_{s-1}}{(s-r-1)!\,(r-1)!}\int_0^{\frac{1}{\alpha}}\int_x^{\frac{1}{\alpha}} xy\,(1-F(y))^{\gamma_s-1}g_m^{r-1}(F(x))\,[h_m(F(y))-h_m(F(x))]^{s-r-1}$$

$$.(1-F(x))^m f(y)\,f(x)\,dydx.$$

Further,

$$E[X(r,n,m,k)X(s,n,m,k)]=\frac{c_{s-1}}{(s-r-1)!\,(r-1)!}\int_0^{\frac{1}{\alpha}}\int_x^{\frac{1}{\alpha}} xy\frac{(1-F(y))^{\gamma_s}}{(1-\alpha y)}\left[\frac{-(1-F(y))^{m+1}}{m+1}+\frac{(1-F(x))^{m+1}}{m+1}\right]^{s-r-1}$$

$$.\left[\frac{-(1-F(x))^{m+1}}{m+1}+\frac{1}{m+1}\right]^{r-1}\frac{(1-F(x))^{m+1}}{(1-\alpha x)}\,dydx,$$

on using (1.6).

On expanding $\left[\frac{-(1-F(y))^{m+1}}{m+1}+\frac{(1-F(x))^{m+1}}{m+1}\right]^{s-r-1}$ and $\left[\frac{-(1-F(x))^{m+1}}{m+1}+\frac{1}{m+1}\right]^{r-1}$

binomially and using (1.5), we obtain

$$E[X(r,n,m,k)X(s,n,m,k)]=\frac{c_{s-1}}{(s-r-1)!\,(r-1)!}\left(\frac{1}{m+1}\right)^{s-2}\sum_{i=0}^{r-1}\sum_{j=0}^{s-r-1}\binom{r-1}{i}\binom{s-r-1}{j}(-1)^{i+j}\int_0^{\frac{1}{\alpha}}\int_x^{\frac{1}{\alpha}}\frac{xy}{(1-\alpha x)}$$

$$.\frac{(1-F(y))^{(\gamma_s+(m+1)j)}}{(1-\alpha y)}(1-F(x))^{(m+1)(s-r+i-j)}\,dydx$$

$$=\frac{c_{s-1}}{(s-r-1)!\,(r-1)!}\left(\frac{1}{m+1}\right)^{s-2}\sum_{i=0}^{r-1}\sum_{j=0}^{s-r-1}\binom{r-1}{i}\binom{s-r-1}{j}(-1)^{i+j}\int_0^{\frac{1}{\alpha}}x\,G(x)$$
$$.(1-\alpha x)^{\frac{1}{\alpha}(m+1)(s-r+i-j)-1}\,dx, \tag{5.1}$$

where

$$G(x)=\int_x^{\frac{1}{\alpha}}y\,(1-\alpha y)^{\frac{1}{\alpha}\{\gamma_s+(m+1)j\}-1}\,dy.$$

On solving the integral in G(x), we have

$$G(x)=\frac{x\,(1-\alpha x)^{\frac{1}{\alpha}\{\gamma_s+(m+1)j\}}}{\{\gamma_s+(m+1)j\}}+\frac{(1-\alpha x)^{\frac{1}{\alpha}\{\gamma_s+(m+1)j\}+1}}{\{\gamma_s+(m+1)j\}\{\gamma_s+(m+1)j+\alpha\}}.$$

Substituting the above expression for G(x) in (5.1), we obtain

$$E[X(r,n,m,k)X(s,n,m,k)] = \frac{c_{s-1}}{(s-r-1)!\,(r-1)!}\left(\frac{1}{m+1}\right)^{s-2}\sum_{i=0}^{r-1}\sum_{j=0}^{s-r-1}\binom{r-1}{i}\binom{s-r-1}{j}(-1)^{i+j}H(i,j), \tag{5.2}$$

where

$$H(i,j) = \frac{1}{\{\gamma_s+(m+1)\,j\}}\int_0^{\frac{1}{\alpha}} x^2(1-\alpha x)^{\frac{1}{\alpha}\{\gamma_s+(m+1)(s-r+i)\}-1}\,dx$$

$$+\ \frac{1}{\{\gamma_s+(m+1)\,j\}\{\gamma_s+(m+1)\,j+\alpha\}}\int_0^{\frac{1}{\alpha}} x\,(1-\alpha x)^{\frac{1}{\alpha}\{\gamma_s+(m+1)(s-r+i)\}}\,dx.$$

On solving and simplifying the integrals in the above expression for H(i,j), we get

$$H(i,j) = \frac{2}{\{\gamma_s+(m+1)j\}\{\gamma_s+(m+1)(s-r+i)\}\{\gamma_s+(m+1)(s-r+i)+\alpha\}\{\gamma_s+(m+1)(s-r+i)+2\alpha\}}$$

$$+\frac{1}{\{\gamma_s+(m+1)\,j\}\{\gamma_s+(m+1)\,j+\alpha\}\{\gamma_s+(m+1)(s-r+i)+\alpha\}\{\gamma_s+(m+1)(s-r+i)+2\alpha\}}.$$

Further, on substituting for H(i,j) in (5.2), an exact and explicit expression for the product moments can be obtained.

Next, by using the exact and explicit expression for the product moments obtained in (5.2) and also the recurrence relations given in equations (4.5) and (4.6) in a simple recursive way, one can easily obtain the variances, covariances and higher order product moments of all generalized order statistics from a generalized exponential distribution.

Acknowledgments

The authors are grateful to the referee for giving valuable comments which led to an improvement in the earlier version of the paper.

References

[1] Ahsanullah , M . (2000). Generalized order statistics from exponential distribution . *J. Statist. Plan. Infer.,* **85**, pp.85-91.

[2] Al-Hussaini, E. K. (2004). Generalized Order Statistics : Prospective and Applications. *J. Appl. Statist . Sc.* **13** (1), pp.59 – 85 .

[3] Balakrishnan, N., Malik, H. J. and Ahmed, S. E. (1988). Recurrence relations and identities for moments of order statistics,II: Specific Continuous Distributions. *Commun. Statist. – Theor. Meth.* **17**(8), pp.2657-2694 .

[4] Balakrishnan, N. and Ahsanullah, M. (1995). Relations for single and product moments of record values from exponential distribution . *J. Appl. Statist. Sc.*, **2**(1), pp.73 – 87 .

[5] Cohen, A. C. and Whitten, B. J. (1988). *Parameter estimation of reliability and life Span models*. Marcel Dekker, New York.

[6] Joshi, P. C. (1978). Recurrence relations between moments of order statistics from exponential and truncated exponential distributions. *Sankhya*, **B39**, pp.362–371.

[7] Joshi, P. C. (1982). A note on the mixed moments of order statistics from exponential and truncated exponential distributions. *J. Statist. Plan. Infer.*, **6,** pp.13–16.

[8] Kamps, U. (1995). A concept of generalized order statistics. *J. Statist. Plan. Infer.*, **48**, pp.1 – 23 .

[9] Saran, J. and Pushkarna, N. (2000). Relationships for moments of order statistics from a generalized exponential distribution. *Statistica*, **3**, pp.585–595.

In: New Developments in Applied Statistics
Editor: Mohammad Ahsanullah
ISBN: 978-1-61324-648-1

Chapter 3

DETECTION OF THE NUMBER OF SIGNALS IN THE PRESENCE OF COLORED NOISE IN DECENTRALISED PROCESSING

Madhusudan Bhandary
Mathematics Department, Columbus State University, 4225 University Avenue, Columbus, GA 31907
Wenzhen Ding
Department of Statistics, North Dakota State University, P.O. Box 5575, Waldron 201, Fargo, ND 58105

Abstract

The problem of estimating the number of signals under a colored noise model is solved when the data are received from many different places that are widely apart by using modified information criterion proposed by the authors. This criterion involves the eigenvalues of the sample canonical correlation matrix obtained from data consisting of both noise and signals as well as from the data consisting of noise only. Some simulation results are also presented.

Key words: Canonical correlation, information criterion, signal processing, white noise, colored noise.

I. Introduction

In the area of signal processing, q signals are observed at n different time points from different sources to p different sensors where ($q < p$). However, due to atmospheric interference, the signals received by the sensors do not remain undistorted. Signals are affected by a noise factor. In this area, a model often used is that the observed signal vector is the sum of a random noise vector and a linear transform of a random signal vector. One of the important problems in this area is to estimate the number of signals transmitted. The problem is equivalent to estimate the multiplicity of the smallest eigen value of the covariance matrix of observation vector.

Anderson [2], Krishnaiah [4] and Rao [6] considered the problem of testing the hypothesis of the multiplicity of the smallest eigen value of the covariance matrix. The problem of estimation of the number of signals by information theoretic criteria proposed by [1], [7] and [8] were considered in [9], [11] and [12]. In this correspondence, Bhandary [3] considered the problem of estimation of the number of signals under white noise situation when the signals are coming from the same source but received by sensors placed at many different locations.

In this paper, we consider the problem same as in Bhandary [3], but under colored noise situation. It is important to estimate the number of signals under the situation described above and under colored noise situation.

In section II, we consider the model and assumptions for signal processing and in section III, we explain the model for our problem. Section IV describes the solution of the problem and some simulation results are presented in section V.

II. Model and Assumptions

The following model [9] is used in signal processing:

$$\underset{\sim}{x}(t) = A\,\underset{\sim}{s}(t) + \underset{\sim}{n}(t) \quad (2.1)$$

where,

$\underset{\sim}{x}(t) = (x_1(t), \ldots x_p(t))'$ is the $p \times 1$ vector of observations at time t,

$\underset{\sim}{s}(t) = (s_1(t), \ldots s_q(t))'$ is the $q \times 1$ vector of unknown random signals at time t,

$\underset{\sim}{n}(t) = (n_1(t), \ldots n_p(t))'$ is the $p \times 1$ vector of random noise at time t,

and A is the $p \times q$ matrix of unknown parameters associated with the signals.

Under the assumption given in [10], we can say from model (2.1) that

$$\underset{\sim}{x}(t) \sim N_p(\underset{\sim}{0}, A\Psi A' + \Sigma_1) \quad (2.2)$$

where, Ψ = covariance matrix of $\underset{\sim}{s}(t)$ and Σ_1 = covariance matrix of $\underset{\sim}{n}(t)$ and $\underset{\sim}{s}(t)$ and $\underset{\sim}{n}(t)$ are independent.

The number of signals transmitted is q (< p), which is the rank of the matrix $A\Psi A'$.

Therefore, in this case, the estimation of the number of signals is equivalent to the estimation of the rank of the matrix $A\Psi A'$.

III. Model for the Problem

Bhandary [3] described the model in the presence of white noise in decentralized processing. The main idea of the problem is described as follows:

Suppose, q signals are going to p sensors in some location and to p sensors in some other location that is widely apart from the first location and to p sensors in some other location that is widely apart from the first two, and so on. Suppose, in this way that there are k widely separated locations, each containing p sensors.

The statistical model in this case will be described by the k equations as follows:

$$\underset{\sim}{x}_h(t) = A_h \underset{\sim}{s}(t) + \underset{\sim}{n}_h(t),\ h = 1,2,\ldots,k \tag{3.1}$$

where,

- $\underset{\sim}{x}_h(t)$ is the $p \times 1$ vector of observations at time t at the hth place,
- $\underset{\sim}{s}(t)$ is the $q \times 1$ vector of unknown random signals at time t,
- $\underset{\sim}{n}_h(t)$ is the $p \times 1$ vector of random noise at time t at the hth place,
- and A_h is the $p \times q$ matrix of unknown parameters associated with the signals at the hth place,
- $h = 1,2,\ldots,k$.

Bhandary [3] considered the above model in white noise case i.e., when $\Sigma_1 = \sigma^2 I_p$, where, Σ_1 is mentioned in (2.2) and I_p denotes identity matrix of order $p \times p$.

We will consider, in this paper, the colored noise case, i.e., when Σ_1 is arbitrary positive definite and symmetric matrix.

Under the assumption given in [10], it is trivial from model (3.1) that

$$\underset{\sim}{x}_h(t) \sim N_p(\underset{\sim}{0}, \Gamma_{hh} + \Sigma_{1h}) \tag{3.2}$$

where, $\Gamma_{hh} = A_h \Psi A'_h$ is the non-negative definite matrix of rank q (< p) and

$$\begin{aligned} Cov(\underset{\sim}{x}_h(t), \underset{\sim}{x}_{h'}(t)) &= A_h \Psi A'_{h'} \\ &= \Gamma_{hh'},\ h(\neq h') = 1,2,\ldots,k \\ &= \text{matrix of rank q (< p)}, \end{aligned} \tag{3.3}$$

and Σ_{1h} = covariance matrix of noise vector at h^{th} place,

We have observations $\underset{\sim}{x}_h(t_1), \underset{\sim}{x}_h(t_2),..., \underset{\sim}{x}_h(t_n)$ at the hth place for $h = 1,2,..., k$, and on the basis of these observations, we would like to estimate q, which is the number of signals transmitted from the source to the p sensors in k different locations that are widely apart.

IV. Solution of the Problem

We will use modified information criterion to solve the problem. Wax and Kailath [9] used Akaike's information criterion (AIC) and the Schwartz-Rissanen minimum distance length (MDL) criterion to estimate the number of signals. In [11] and [12], some information criterion that gives a consistent estimate of the integer parameter of a model was developed. Let us have a brief discussion of these approaches.

Suppose, we want to estimate q, an unknown integer parameter of a model. Let the range for q be $0 \le q \le m$. According to the criterion in [11] and [12], $\hat{q}$, the estimate of q, is chosen such that

$$I(\hat{q}, c_n) = \min \{I(0, c_n), I(1, c_n),..., I(m, c_n)\}$$

where, $I(r, c_n) = -\log L_r + c_n \gamma(r, m)$

L_r = likelihood ratio test statistic under $H_0 : q = r$

$\gamma(r, m)$ = number of free parameters to be estimated under H_0

and c_n satisfies

(i) $\lim_{n \to \infty} \frac{c_n}{n} = 0$

(ii) $\lim_{n \to \infty} \frac{c_n}{\log \log n} = \infty$ (4.1)

In our problem, it is difficult to derive exact likelihood ratio test. We propose the following modified information criterion:

Estimate q by $\hat{q}$, where $\hat{q}$ is such that

$$I'(\hat{q}, c_n) = \min_{0 \le r \le m} I'(r, c_n)$$

where, $$I'(r, c_n) = -nT_{n,r} + c_n \gamma(r, m) \tag{4.2}$$

$T_{n,r}$ = some test statistic to test $H_0 : q = r$ versus H_1 : not H_0,

$\gamma(r,m)$ is defined before, and c_n satisfies

(i) $\lim_{n\to\infty} \frac{c_n}{n} = 0$

(ii) $\lim_{n\to\infty} \frac{c_n}{\sqrt{n \log \log n}} = \infty$ (4.3)

Note:

1. For practical purposes, we can choose $c_n = \sqrt{n \log n}$, which satisfies (4.3).
2. In our case, the range for q is 1,2,…,(p-1). In this case, q can not be 0; otherwise, the off diagonals of the matrices Γ_{12}, Γ_{13}, etc. in (3.3) do not make sense.

We will solve our problem using our modified information criterion.

$$\text{Let } \underset{\sim}{y}_h(t) = \Sigma_{1h}^{-\frac{1}{2}} \underset{\sim}{x}_h(t) \ ; \ h = 1,2,\dots, k \tag{4.4}$$

Then, from (3.2), it is trivial that

$$\underset{\sim}{y}_h(t) \sim N_p(\underset{\sim}{0}, \Gamma_{hh}^* + I_p) \tag{4.5}$$

where, $\Gamma_{hh}^* = \Sigma_{1h}^{-\frac{1}{2}} \Gamma_{hh} \Sigma_{1h}^{-\frac{1}{2}}$ = non-negative definite matrix of rank q (< p).

$$\text{Let } \underset{pk\times 1}{\underset{\sim}{y}(t)} = \left(\underset{\sim}{y}_1(t), \dots, \underset{\sim}{y}_k(t) \right)' \tag{4.6}$$

Then from (3.1), (3.2), (3.3) and (4.4), we can say that $\underset{\sim}{y}(t)$ is distributed as a pk-variate normal with mean vector $\underset{\sim}{0}$ and dispersion matrix Σ_2, where Σ_2 is a $pk \times pk$ matrix as follows:

$$\Sigma_2 = \begin{pmatrix} \Gamma_{11}^* + I_p & \Gamma_{12}^* & \Gamma_{13}^* & . & . & . & \Gamma_{1k}^* \\ \Gamma_{21}^* & \Gamma_{22}^* + I_p & \Gamma_{23}^* & . & . & . & \Gamma_{2k}^* \\ \dots & \dots & \dots & \dots & \dots & \dots & \dots \\ \Gamma_{k1}^* & \Gamma_{k2}^* & . & . & . & \Gamma_{k(k-1)}^* & \Gamma_{kk}^* + I_p \end{pmatrix} \tag{4.7}$$

where, $\Gamma^*_{ji} = \Gamma^{*'}_{ij}$ = Transpose of Γ^*_{ij} , $(i \neq j) = 1,..., k$ and Γ^*_{ii} 's are symmetric matrices.

The matrix Σ_2 can also be written as follows:

$$\Sigma_2 = \begin{pmatrix} \Gamma^*_{11} + I_p & \Sigma_{22k} \\ \Sigma_{2k2} & \Sigma_{22.k} \end{pmatrix} \tag{4.8}$$

where, $\underset{p\times(k-1)p}{\Sigma_{22k}} = \left(\Gamma^*_{12} : \ \Gamma^*_{13} : \ . \ . \ . \ . \ : \Gamma^*_{1k}\right)$

$\Sigma_{2k2} = \Sigma'_{22k}$ = Transpose of the matrix Σ_{22k}

and $$\Sigma_{22.k} = \begin{pmatrix} \Gamma^*_{22} + I_p & \Gamma^*_{23} & . & . & . & \Gamma^*_{2k} \\ \Gamma^*_{32} & \Gamma^*_{33} + I_p & . & . & . & \Gamma^*_{3k} \\ \cdots & \cdots & \cdots & \cdots & \cdots & \cdots \\ \Gamma^*_{k2} & \Gamma^*_{k3} & . & . & . & \Gamma^*_{kk} + I_p \end{pmatrix} \tag{4.9}$$

The canonical correlation matrix obtained from (4.8) is

$$C^* = \Sigma_{22k} \Sigma^{-1}_{22.k} \Sigma_{2k2} (\Gamma^*_{11} + I_p)^{-1} \tag{4.10}$$

and the matrix C^* above is of rank q.

Hence, the problem of estimation of q can be solved by using the eigenvalues of C^* in (4.10) using our modified information criterion.

The procedure to estimate the eigenvalues of C^* using sample data can be described as follows:

Let $\underset{p\times 1}{\underset{\sim}{u_1}}(t_1),..., \underset{p\times 1}{\underset{\sim}{u_1}}(t_m)$ be the observations on Place 1 consisting of pure noise only, $\underset{p\times 1}{\underset{\sim}{u_2}}(t'_1),..., \underset{p\times 1}{\underset{\sim}{u_2}}(t'_m)$ be the observations on Place 2 consisting of pure noise only, and $\underset{p\times 1}{\underset{\sim}{u_k}}(t^{(k-1)}_1),..., \underset{p\times 1}{\underset{\sim}{u_k}}(t^{(k-1)}_m)$ be the observations on Place k consisting of pure noise only.

Compute $S_1 = \frac{1}{m} \sum_{i=1}^{m} \underset{\sim}{u_1}(t_i) \underset{\sim}{u_1}'(t_i)$

$$S_2 = \frac{1}{m}\sum_{i=1}^{m} \underset{\sim}{u}_2(t_i')\underset{\sim}{u}_2'(t_i')$$

$$\cdots\cdots\cdots\cdots\cdots\cdots \tag{4.11}$$

$$S_k = \frac{1}{m}\sum_{i=1}^{m} \underset{\sim}{u}_k(t_i^{(k-1)})\underset{\sim}{u}_k'(t_i^{(k-1)})$$

Let $\underset{p\times 1}{\underset{\sim}{x}_1}(t_1),\ldots,\underset{p\times 1}{\underset{\sim}{x}_1}(t_n)$ be the observations on Place 1 consisting of both signals and noise, $\underset{p\times 1}{\underset{\sim}{x}_2}(t_1'),\ldots,\underset{p\times 1}{\underset{\sim}{x}_2}(t_n')$ be the observations on Place 2 consisting of both signals and noise, and $\underset{p\times 1}{\underset{\sim}{x}_k}(t_1^{(k-1)}),\ldots,\underset{p\times 1}{\underset{\sim}{x}_k}(t_n^{(k-1)})$ be the observations on Place k consisting of both signals and noise.

Let $\underset{\sim}{y}_i(t_j^{(i-1)}) = S_i^{-\frac{1}{2}}\underset{\sim}{x}_i(t_j^{(i-1)})$ be the transformed observations where, $i = 1,2,\ldots, k$ and $j = 1,2,\ldots, n$ and $S_1, S_2,\ldots, S_k$ are obtained from (4.11).

Let $\underset{\sim}{y}_j = \left(\underset{\sim}{y}_1(t_j), \underset{\sim}{y}_2(t_j'),\ldots,\underset{\sim}{y}_k(t_j^{(k-1)})\right)'$; $j = 1,2,\ldots, n$ for convenience.

Compute
$$S_2^* = \frac{1}{n}\sum_{j=1}^{n}\underset{\sim}{y}_j\,\underset{\sim}{y}_j' \tag{4.12}$$

We can partition the matrix S_2^* as follows:

$$\underset{pk\times pk}{S_2^*} = \begin{pmatrix} \underset{p\times p}{S_{11}^*} & S_{12}^* & . & . & . & S_{1k}^* \\ S_{21}^* & \underset{p\times p}{S_{22}^*} & . & . & . & S_{2k}^* \\ \cdots & \cdots & \cdots & \cdots & \cdots & \cdots \\ S_{k1}^* & S_{k2}^* & . & . & . & \underset{p\times p}{S_{kk}^*} \end{pmatrix} = \begin{pmatrix} \underset{p\times p}{S_{11}^*} & C_{22k} \\ C_{2k2} & C_{22.k} \end{pmatrix} \tag{4.13}$$

where,
$$\underset{p\times(k-1)p}{C_{22k}} = \left(S_{12}^* : \; S_{13}^* : \; . \; . \; . \; : S_{1k}^*\right)$$

$$C_{2k2} = C_{22k}'$$

and

$$C_{22.k} = \begin{pmatrix} \underset{p\times p}{S^*_{22}} & S^*_{23} & . & . & . & S^*_{2k} \\ S^*_{32} & \underset{p\times p}{S^*_{33}} & . & . & . & S^*_{3k} \\ \cdots & \cdots & \cdots & \cdots & \cdots & \cdots \\ S^*_{k2} & S^*_{k3} & . & . & . & \underset{p\times p}{S^*_{kk}} \end{pmatrix}$$

The sample canonical correlation matrix is

$$R^* = C_{22k}(C_{22.k})^{-1} C_{2k2}(S^*_{11})^{-1} \tag{4.14}$$

Let $l'_1 \geq l'_2 \geq \ldots \geq l'_p$ be the ordered eigenvalues of the sample canonical correlation matrix R^* given by (4.14).

Define

$$I'(r, c_n) = -n\left[\sum_{i=r+1}^{p} \log l'_i - (p-r)\log\left(\frac{1}{(p-r)}\sum_{i=r+1}^{p} l'_i\right)\right] + c_n(r(2p-r)+1) \tag{4.15}$$

where, $r = 1, 2, \ldots, (p-1)$ and c_n satisfies (4.3).

Then, $\hat{q}$, the estimate of q, is such that

$$I'(\hat{q}, c_n) = \min\{I'(1, c_n), I'(2, c_n), \ldots, I'((p-1), c_n)\} \tag{4.16}$$

It can be shown, following the procedure similar to [11] and [12], that the estimator $\hat{q}$ obtained from (4.16) is a strongly consistent estimator of q.

V. Simulation Results

We generate data from multivariate normal distribution using R software for different values of k, p, q. First, we generate observations $\underset{\sim}{u_h}(t_j^{(h-1)})$ for h = 1,2,…,k and j = 1,2,…,n from multivariate normal distribution with mean vector 0 dispersion matrix Σ_{1h}. Then we calculate $S_1, S_2, \ldots, S_k$ by using (4.11). Then we generate observations $\underset{\sim}{x_h}(t_j^{(h-1)})$ for h = 1,2,…,k and j = 1,2,…,n from multivariate normal distribution with mean vector 0 and dispersion matrix $\Gamma_{hh} + \Sigma_{1h}$. The sample covariance matrix S_2^* given by (4.13) is calculated and then we calculate the sample canonical correlation matrix R^* by using (4.14), and then $I'(r, c_n)$ is calculated by using (4.15) for $c_n = \sqrt{n \log n}$. We present our results by Tables 1-12 and see that our method estimates the number of signals very close to the true number of signals and hence we recommend this method for use.

Table 1. k=2, p=3, q=2, number of simulations=3000, n=10 and 30

$$\Gamma_{11} = \begin{pmatrix} 1.09 & 1.22 & 0.35 \\ 1.22 & 2.60 & 0.86 \\ 0.35 & 0.86 & 0.29 \end{pmatrix}, \quad \Gamma_{22} = \begin{pmatrix} 1.85 & 1.95 & 0.86 \\ 1.95 & 0.50 & 1.29 \\ 0.86 & 1.29 & 2.29 \end{pmatrix}, \Sigma_{11} = \begin{pmatrix} 2.5 & 1.50 & 0.2 \\ 1.5 & 3.75 & 0.5 \\ 0.2 & 0.50 & 2.0 \end{pmatrix}, \quad \Sigma_{12} = \begin{pmatrix} 2.09 & 1.22 & 0.35 \\ 1.22 & 3.60 & 0.86 \\ 0.35 & 0.86 & 1.29 \end{pmatrix}$$

Sample size	$I^*(r,c_n)$	Minimum	Estimation
n=10	r $I^*(r, c_n)$ 1 44.94590 2 43.18673	$\min_r I^*(r,c_n) = I^*(2,c_n)$	$\hat{q} = 2$
n=30	r $I^*(r, c_n)$ 1 131.22140 2 90.91155	$\min_r I^*(r,c_n) = I^*(2,c_n)$	$\hat{q} = 2$

Table 2. k=2, p=4, q=3, number of simulations=3000, n=10 and 30

$$\Gamma_{11} = \begin{pmatrix} 5.29 & 2.93 & 2.25 & 2.05 \\ 2.93 & 5.31 & 2.67 & 4.39 \\ 2.25 & 2.67 & 2.33 & 3.29 \\ 2.05 & 4.39 & 3.29 & 5.29 \end{pmatrix}, \quad \Gamma_{22} = \begin{pmatrix} 6.29 & 1.06 & 2.29 & 5.64 \\ 1.06 & 2.34 & 1.46 & 3.59 \\ 2.29 & 1.46 & 1.73 & 3.30 \\ 5.64 & 3.59 & 3.30 & 8.09 \end{pmatrix}, \Sigma_{11} = \begin{pmatrix} 6.39 & 2.93 & 2.25 & 2.05 \\ 2.93 & 6.41 & 2.67 & 4.39 \\ 2.25 & 2.67 & 3.43 & 3.29 \\ 2.05 & 4.39 & 3.29 & 6.39 \end{pmatrix}, \quad \Sigma_{12} = \begin{pmatrix} 7.69 & 1.06 & 2.29 & 5.64 \\ 1.06 & 3.74 & 1.46 & 3.51 \\ 2.29 & 1.46 & 3.13 & 3.30 \\ 5.64 & 3.51 & 3.30 & 9.49 \end{pmatrix}$$

Sample size	$I^*(r,c_n)$	Minimum	Estimation
n=10	r $I^*(r, c_n)$ 1 86.83221 2 99.45533 3 76.77641	$\min_r I^*(r,c_n) = I^*(3,c_n)$	$\hat{q} = 3$
n=30	r $I^*(r, c_n)$ 1 184.0369 2 201.4410 3 161.6205	$\min_r I^*(r,c_n) = I^*(3,c_n)$	$\hat{q} = 3$

Table 3. k=2, p=4, q=2, number of simulations=3000, n=10 and 30

$$\Gamma_{11} = \begin{pmatrix} 8.7921 & 6.5220 & 4.0794 & 2.8227 \\ 6.5220 & 8.3924 & 2.6738 & 3.8230 \\ 4.0794 & 2.6738 & 1.9277 & 1.1383 \\ 2.8227 & 3.8230 & 1.1383 & 1.7474 \end{pmatrix}, \Gamma_{22} = \begin{pmatrix} 2.2114 & 2.6193 & 2.7159 & 3.9824 \\ 2.6193 & 3.1585 & 3.0773 & 4.7690 \\ 2.7159 & 3.0773 & 3.6829 & 4.7614 \\ 3.9824 & 4.7690 & 4.7614 & 7.2200 \end{pmatrix},$$

$$\Sigma_{11} = \begin{pmatrix} 6.39 & 2.93 & 2.25 & 2.05 \\ 2.93 & 6.41 & 2.67 & 4.39 \\ 2.25 & 2.67 & 3.43 & 3.29 \\ 2.05 & 4.39 & 3.29 & 6.39 \end{pmatrix}, \quad \Sigma_{12} = \begin{pmatrix} 7.69 & 1.06 & 2.29 & 5.64 \\ 1.06 & 3.74 & 1.46 & 3.51 \\ 2.29 & 1.46 & 3.13 & 3.30 \\ 5.64 & 3.51 & 3.30 & 9.49 \end{pmatrix}$$

Sample size	$I^*(r,c_n)$	Minimum	Estimation
n=10	r $I^*(r, c_n)$ 1 83.12055 2 70.60011 3 76.77641	$\min_r I^*(r,c_n) = I^*(2,c_n)$	$\hat{q} = 2$
n=30	r $I^*(r, c_n)$ 1 164.3040 2 144.0954 3 161.6205	$\min_r I^*(r,c_n) = I^*(2,c_n)$	$\hat{q} = 2$

Table 4. k=2, p=4, q=1, number of simulations=3000, n=10 and 30

$$\Gamma_{11} = \begin{pmatrix} 6.9696 & 8.8440 & 10.824 & 14.9952 \\ 8.8440 & 11.2225 & 13.735 & 19.0280 \\ 10.8240 & 13.7350 & 16.810 & 23.2880 \\ 14.9952 & 19.0280 & 23.288 & 32.2624 \end{pmatrix}, \Gamma_{22} = \begin{pmatrix} 18.7489 & 14.9385 & 9.3528 & 24.9841 \\ 14.9385 & 11.9025 & 7.4520 & 19.9065 \\ 9.3528 & 7.4520 & 4.6656 & 12.4632 \\ 24.9841 & 19.9065 & 12.4632 & 33.2929 \end{pmatrix},$$

$$\Sigma_{11} = \begin{pmatrix} 6.39 & 2.93 & 2.25 & 2.05 \\ 2.93 & 6.41 & 2.67 & 4.39 \\ 2.25 & 2.67 & 3.43 & 3.29 \\ 2.05 & 4.39 & 3.29 & 6.39 \end{pmatrix}, \quad \Sigma_{12} = \begin{pmatrix} 7.69 & 1.06 & 2.29 & 5.64 \\ 1.06 & 3.74 & 1.46 & 3.51 \\ 2.29 & 1.46 & 3.13 & 3.30 \\ 5.64 & 3.51 & 3.30 & 9.49 \end{pmatrix}$$

Table 4. Continued

Sample size	$I^*(r,c_n)$	Minimum	Estimation
n=10	r $I^*(r, c_n)$ 1 62.25367 2 75.62658 3 76.77641	$\min_r I^*(r,c_n) = I^*(1,c_n)$	$\hat{q} = 1$
n=30	r $I^*(r, c_n)$ 1 117.0428 2 147.7876 3 161.6205	$\min_r I^*(r,c_n) = I^*(1,c_n)$	$\hat{q} = 1$

Table 5. k=3, p=3, q=2, number of simulations=3000, n=10 and 30

$$\Gamma_{11} = \begin{pmatrix} 1.09 & 1.22 & 0.35 \\ 1.22 & 2.60 & 0.86 \\ 0.35 & 0.86 & 0.29 \end{pmatrix}, \quad \Sigma_{11} = \begin{pmatrix} 2.5 & 1.50 & 0.2 \\ 1.5 & 3.75 & 0.5 \\ 0.2 & 0.50 & 2.0 \end{pmatrix},$$

$$\Gamma_{22} = \begin{pmatrix} 1.85 & 1.95 & 0.86 \\ 1.95 & 0.50 & 1.29 \\ 0.86 & 1.29 & 2.29 \end{pmatrix}, \quad \Sigma_{12} = \begin{pmatrix} 2.09 & 1.22 & 0.35 \\ 1.22 & 3.60 & 0.86 \\ 0.35 & 0.86 & 1.29 \end{pmatrix},$$

$$\Gamma_{33} = \begin{pmatrix} 1.69 & 3.41 & 2.79 \\ 3.41 & 6.98 & 5.86 \\ 2.79 & 5.86 & 5.14 \end{pmatrix}, \quad \Sigma_{13} = \begin{pmatrix} 3.66 & 1.92 & 2.15 \\ 1.92 & 1.80 & 1.61 \\ 2.15 & 1.61 & 2.19 \end{pmatrix}.$$

Sample size	$I^*(r,c_n)$	Minimum	Estimation
n=10	r $I^*(r, c_n)$ 1 57.46849 2 43.18673	$\min_r I^*(r,c_n) = I^*(2,c_n)$	$\hat{q} = 2$
n=30	r $I^*(r, c_n)$ 1 120.85963 2 90.91155	$\min_r I^*(r,c_n) = I^*(2,c_n)$	$\hat{q} = 2$

Table 6. k=3, p=3, q=1, number of simulations=3000, n=10 and 30

$$\Gamma_{11} = \begin{pmatrix} 0.16 & 0.52 & 0.84 \\ 0.52 & 1.69 & 2.73 \\ 0.84 & 2.73 & 4.41 \end{pmatrix}, \quad \Sigma_{11} = \begin{pmatrix} 1.46 & 0.52 & 0.84 \\ 0.52 & 2.99 & 2.73 \\ 0.84 & 2.73 & 5.71 \end{pmatrix},$$

$$\Gamma_{22} = \begin{pmatrix} 2.89 & 1.02 & 0.34 \\ 1.02 & 0.36 & 0.12 \\ 0.34 & 0.12 & 0.04 \end{pmatrix}, \quad \Sigma_{12} = \begin{pmatrix} 4.59 & 1.02 & 0.34 \\ 1.02 & 2.06 & 0.12 \\ 0.34 & 0.12 & 1.74 \end{pmatrix},$$

$$\Gamma_{33} = \begin{pmatrix} 1.44 & 0.60 & 2.76 \\ 0.60 & 0.25 & 1.15 \\ 2.76 & 1.15 & 5.29 \end{pmatrix}, \quad \Sigma_{13} = \begin{pmatrix} 3.66 & 1.92 & 2.15 \\ 1.92 & 1.80 & 1.61 \\ 2.15 & 1.61 & 2.19 \end{pmatrix}.$$

Sample size	**$I^*(r,c_n)$**	**Minimum**	**Estimation**
n=10	r $I^*(r, c_n)$ 1 29.47843 2 43.18673	$\min_r I^*(r,c_n) = I^*(1,c_n)$	$\hat{q} = 1$
n=30	r $I^*(r, c_n)$ 1 70.90432 2 90.91155	$\min_r I^*(r,c_n) = I^*(1,c_n)$	$\hat{q} = 1$

Table 7. k=3, p=4, q=3, number of simulations=3000, n=10 and 30

$$\Gamma_{11} = \begin{pmatrix} 5.29 & 2.93 & 2.25 & 2.05 \\ 2.93 & 5.31 & 2.67 & 4.39 \\ 2.25 & 2.67 & 2.33 & 3.29 \\ 2.05 & 4.39 & 3.29 & 5.29 \end{pmatrix}, \quad \Sigma_{11} = \begin{pmatrix} 6.39 & 2.93 & 2.25 & 2.05 \\ 2.93 & 6.41 & 2.67 & 4.39 \\ 2.25 & 2.67 & 3.43 & 3.29 \\ 2.05 & 4.39 & 3.29 & 6.39 \end{pmatrix},$$

$$\Gamma_{22} = \begin{pmatrix} 6.29 & 1.06 & 2.29 & 5.64 \\ 1.06 & 2.34 & 1.46 & 3.59 \\ 2.29 & 1.46 & 1.73 & 3.30 \\ 5.64 & 3.59 & 3.30 & 8.09 \end{pmatrix}, \quad \Sigma_{12} = \begin{pmatrix} 7.69 & 1.06 & 2.29 & 5.64 \\ 1.06 & 3.74 & 1.46 & 3.51 \\ 2.29 & 1.46 & 3.13 & 3.30 \\ 5.64 & 3.51 & 3.30 & 9.49 \end{pmatrix},$$

$$\Gamma_{33} = \begin{pmatrix} 1.87 & 2.22 & 1.44 & 1.21 \\ 2.22 & 4.58 & 3.50 & 2.71 \\ 1.44 & 3.50 & 2.96 & 2.62 \\ 1.21 & 2.71 & 2.62 & 2.93 \end{pmatrix}, \quad \Sigma_{13} = \begin{pmatrix} 3.65 & 3.63 & 2.03 & 3.61 \\ 3.63 & 5.75 & 3.32 & 3.53 \\ 2.03 & 3.32 & 2.94 & 2.50 \\ 3.61 & 3.53 & 2.50 & 4.04 \end{pmatrix}.$$

Table 7. Continued

Sample size	$I^*(r,c_n)$	Minimum	Estimation
n=10	r $I^*(r, c_n)$ 1 82.38802 2 96.45167 3 76.77641	$\min_r I^*(r,c_n) = I^*(3,c_n)$	$\hat{q} = 3$
n=30	r $I^*(r, c_n)$ 1 181.4265 2 167.9350 3 161.6205	$\min_r I^*(r,c_n) = I^*(3,c_n)$	$\hat{q} = 3$

Table 8. k=3, p=4, q=2, number of simulations=3000, n=10 and 30

$$\Gamma_{11} = \begin{pmatrix} 9.01 & 7.06 & 4.28 & 2.84 \\ 7.06 & 9.05 & 3.11 & 3.80 \\ 4.28 & 3.11 & 2.05 & 1.24 \\ 2.84 & 3.80 & 1.24 & 1.60 \end{pmatrix}, \quad \Sigma_{11} = \begin{pmatrix} 6.39 & 2.93 & 2.25 & 2.05 \\ 2.93 & 6.41 & 2.67 & 4.39 \\ 2.25 & 2.67 & 3.43 & 3.29 \\ 2.05 & 4.39 & 3.29 & 6.39 \end{pmatrix},$$

$$\Gamma_{22} = \begin{pmatrix} 2.34 & 2.73 & 2.67 & 4.14 \\ 2.73 & 3.25 & 3.26 & 4.90 \\ 2.67 & 3.26 & 3.37 & 4.88 \\ 4.14 & 4.90 & 4.88 & 7.40 \end{pmatrix}, \quad \Sigma_{12} = \begin{pmatrix} 7.69 & 1.06 & 2.29 & 5.64 \\ 1.06 & 3.74 & 1.46 & 3.51 \\ 2.29 & 1.46 & 3.13 & 3.30 \\ 5.64 & 3.51 & 3.30 & 9.49 \end{pmatrix},$$

$$\Gamma_{33} = \begin{pmatrix} 6.98 & 3.17 & 4.00 & 4.89 \\ 3.17 & 1.93 & 1.70 & 1.81 \\ 4.00 & 1.70 & 2.32 & 2.90 \\ 4.89 & 1.81 & 2.90 & 3.77 \end{pmatrix}, \quad \Sigma_{13} = \begin{pmatrix} 3.65 & 3.63 & 2.03 & 3.61 \\ 3.63 & 5.75 & 3.32 & 3.53 \\ 2.03 & 3.32 & 2.94 & 2.50 \\ 3.61 & 3.53 & 2.50 & 4.04 \end{pmatrix}.$$

Table 8. Continued

Sample size	$I^*(r,c_n)$	Minimum	Estimation
n=10	r $I^*(r, c_n)$ 1 41.59592 2 64.69191 3 76.77641	$\min_r I^*(r,c_n) = I^*(1,c_n)$	$\hat{q}=1$
n=30	r $I^*(r, c_n)$ 1 161.1329 2 154.4320 3 161.6205	$\min_r I^*(r,c_n) = I^*(2,c_n)$	$\hat{q}=2$

References

[1] H. Akaike, " Information theory and extension of the maximum likelihood principle," in *Proc. Second Int. Symp. Inform. Theory, Supp. Problems Contr. Inform. Theory*, 1972, pp.267-281.

[2] T.W. Anderson, " Asymptotic theory for principal component analysis," Ann. Math. Statist., Vol. 34, pp. 122-138,1963.

[3] M. Bhandary, " Detection of the number of signals in the presence of white noise in decentralized processing," *IEEE Trans. on signal processing*, Vol. 46, No. 3, 800-803,1998.

[4] P.R. Krishnaiah, " Some development on real multivariate distributions," in *Developments in Statistics*, Vol. 1, P.R. Krishnaiah, Ed. New York : Academic, 1978, pp. 135-169.

[5] C.R. Rao, *Linear Statistical Inference and Its Applications*. New York : Wiley, 1973.

[6] ________," Likelihood ratio tests for relationships between two covariance matrices," in *Studies in Econometrics*, Time Series and Multivariate Statistics, T. Amemiya, S. Karlin, and L. Goodman, Eds. New York : Academic, 1983.

[7] J. Rissanen," Modeling by shortest data description," *Automatica*, Vol. 14, pp. 465-471, 1978.

[8] G. Schwartz,' Estimating the dimension of a model," *Ann. Statist.*, Vol. 6, pp. 461-464, 1978.

[9] M. Wax and T. Kailath," Detection of the number of signals by information theoretic criteria," *IEEE Trans. Acoust., Speech, Signal Processing*, Vol. ASSP-33, 387-392, 1985.

[10] M. Wax, T.J. Shan, and T. Kailath," Spatio temporal spectral analysis by eigen structure methods," IEEE Trans. Acoust., Speech, Signal Processing, Vol. ASSP-32, 817-827, 1984.

[11] L.C. Zhao, P.R. Krishnaiah, and Z.D. Bai," On detection of number of signals in presence of white noise," *J. Multivariate Anal.*, Vol. 20, pp. 1-25, 1986.

[12] ______________," On detection of number of signals when the noise covariance matrix is arbitrary," *J. Multivariate Anal.*, Vol. 20, pp. 26-49, 1986.

In: New Developments in Applied Statistics
Editor: Mohammad Ahsanullah

ISBN: 978-1-61324-648-1

Chapter 4

BALANCED GROUPS RANKED SET SAMPLES FOR ESTIMATING THE POPULATION MEDIAN

Abdul Aziz Jemain[1,*]***, Amer Al-Omari***[2,♣] ***and Kamarulzaman Ibrahim***[1,°]

[1] School of Mathematical Sciences, Faculty of Science and Technology, University Kebangsaan Malaysia, 43600 UKM Bangi, Selangor, Malaysia

[2]Department of Mathematics, Jerash Private University, Jordan

Abstract

Balanced groups ranked set samples method (BGRSS) is suggested for estimating the population median with samples of size $m = 3k$ where $(k = 1, 2, \ldots)$. The BGRSS sample median is considered as estimator of the population median. It is found that, the BGRSS produces unbiased estimators with smaller variance than the commonly used simple random sampling (SRS) when the underlying distribution is symmetric. For asymmetric distributions, the BGRSS estimators have a small bias. A real data is used to illustrate the BGRSS method.

Keywords: Balanced groups ranked set sampling; ranked set sampling; simple random sampling.

1. Introduction

The RSS suggested by McIntyre (1952) for estimating mean pasture yields was found to have greater efficiency than SRS. He also suggested that this method is particularly suitable where the experimental or sampling units in a study can be more easily ranked than quantified. To obtain a sample of size m using RSS, randomly selected m simple random samples each of size m from the target population and rank the units within each sample with

* Email address: azizj@pkrisc.cc.ukm.my
♣ Email address: alomari_amer@yahoo.com
° Email address: kamarulz@pkrisc.cc.ukm.my

respect to a variable of interest. The ith smallest rank unit of the ith sample $(i = 1, 2, ..., m)$ is drawn and measured. This method is repeated n times if needed to obtain a sample of size mn out of nm^2 units. Takahasi and Wakimoto (1968) proposed the same method with the mathematical theory of ranked set sampling. Dell and Clutter (1972) showed that the mean of the RSS is an unbiased estimator of the population mean, whether there are errors in ranking or not. Samawi et al. (1996) investigated the extreme ranked set samples (ERSS) for estimating a population mean. Muttlak (1997) suggested using median ranked set sampling (MRSS) to estimate the population mean. Muttlak (2003a, 2003b) suggested quartile ranked set sampling (QRSS) and percentile ranked set sampling (PRSS) for estimating the population mean and showed that PRSS and QRSS produced unbiased estimators of the population mean when the underlying distribution is symmetric. Jemain and Al-Omari (2006a) suggested double quartile ranked set sampling (DQRSS) for estimating the population mean and showed that the mean based on DQRSS is an unbiased estimator and more efficient than those based on SRS, RSS and QRSS if the underlying distribution is symmetric. Details a bout RSS can be found in several works (see Al-Saleh and Al-Omari (2002), Jemain and Al-Omari (2006b), and Ozturk and Deshpande (2006)).

This paper is presented as follows: in Section 2, we describe the BGRSS and illustrate two cases as examples. In Section 3, Estimation of the population median using BGRSS is considered. In Section 4, a simulation study based on BGRSS for estimation of the median is considered. This is followed by a real data set to illustrate the BGRSS, as given in section 5. Finally, we summarized our results in Section 6.

2. Balanced Groups Ranked Set Samples

The balanced groups ranked set sampling (BGRSS) can be described as follows:

Step 1: Randomly select $m = 3k$ $(k = 1, 2, ...)$ sets each of size m from the target population, and rank the unit within each set with respect to the variable of interest.

Step 2: Allocate the $3k$ selected sets randomly into three groups, each of size k sets.

Step 3: For each group in step (2), select for measurement the lowest rank unit from each set in the first group, and the median unit from each set in the second group, and the largest rank unit from each set in the third group.

By this way we have a measured sample of size $m = 3k$ units in one cycle. The Steps 1-3 can be repeated n times to increase the sample size to $3kn$ out of $9k^2n$ units.

It is clear that the BGRSS method differs from the usual RSS with n cycles, since in the usual RSS with n cycles, we identify and measure the ith smallest of the ith sample $(i = 1, 2, ..., m)$ and repeat the process n times. But in the BGRSS method the measured units consist only from the same number of each of minimums, medians and maximums. Also, BGRSS is different from the ERSS, QRSS and PRSS methods, where in the case of odd sample size for the last three methods we identify and measure the median from one sample and in the case of even sample size the median did not selected from any sample.

Indeed, the BGRSS method is easy to be applied since we only need to identify and measure the lowest rank units of the first k sets, and the medians of the second k sets, and the largest rank units from the last k sets.

Let us consider the following example to illustrate BGRSS for estimating the population median.

Example

Case 1: let $k=1$, so $m=3$. Then we may have 3 sets of SRS each of size 3, as follows:

$$\{X_{11}, X_{12}, X_{13}\},\ \{X_{21}, X_{22}, X_{23}\},\ \{X_{31}, X_{32}, X_{33}\}.$$

After ranking the units with respect to a variable of interest allocate them into three groups where each contains one set of size three units as shown below:

First group, Second group, Third group

$$A_1 = \{X_{1(1:3)}, X_{1(2:3)}, X_{1(3:3)}\},\ A_2 = \{X_{2(1:3)}, X_{2(2:3)}, X_{2(3:3)}\},\ A_3 = \{X_{3(1:3)}, X_{3(2:3)}, X_{3(3:3)}\}.$$

Now, select the smallest rank unit form the first group, the median from the second group, and the largest rank unit from the third group as:

$$X_{1(1:3)} = \min(A_1),\ X_{2(2:3)} = \text{median}(A_2),\ X_{3(3:3)} = \max(A_3).$$

The final set $\{X_{1(1:3)}, X_{2(2:3)}, X_{3(3:3)}\}$ is the BGRSS of size 3. These units are measured for estimating the median η of the variable of interest as:

$$\hat{\eta}_{BGRSSO} = \text{med}\{X_{1(1:3)}, X_{2(2:3)}, X_{3(3:3)}\}.$$

It is of interest to note that if $k=1$, then the BGRSSO is the same as the usual RSS method in the case of estimating the population median.

Case 2: if $k=2$, then $m=6$. So, we have six SRS sets each of size six. We first rank the unit within each set with respect to a variable of interest, and then allocate them into 3 groups where each contains two sets each of size six. After ranking and applying the BGRSS method, the sets appear as shown below:

$$\text{First group}\begin{cases} A_1 = \{\boxed{X_{1(1:6)}}, X_{1(2:6)}, X_{1(3:6)}, X_{1(4:6)}, X_{1(5:6)}, X_{1(6:6)}\} \\ A_2 = \{\boxed{X_{2(1:6)}}, X_{2(2:6)}, X_{2(3:6)}, X_{2(4:6)}, X_{2(5:6)}, X_{2(6:6)}\} \end{cases},$$

$$\text{Second group}\begin{cases} A_3 = \left\{X_{3(1:6)}, X_{3(2:6)}, \boxed{X_{3(3:6)}}, \boxed{X_{3(4:6)}}, X_{3(5:6)}, X_{3(6:6)}\right\} \\ A_4 = \left\{X_{4(1:6)}, X_{4(2:6)}, \boxed{X_{4(3:6)}}, \boxed{X_{4(4:6)}}, X_{4(5:6)}, X_{4(6:6)}\right\} \end{cases},$$

$$\text{Third group}\begin{cases} A_5 = \left\{X_{5(1:6)}, X_{5(2:6)}, X_{5(3:6)}, X_{5(4:6)}, X_{5(5:6)}, \boxed{X_{5(6:6)}}\right\} \\ A_6 = \left\{X_{6(1:6)}, X_{6(2:6)}, X_{6(3:6)}, X_{6(4:6)}, X_{6(5:6)}, \boxed{X_{6(6:6)}}\right\} \end{cases}.$$

Finally, the set

$\left\{X_{1(1:6)}, X_{2(1:6)}, \frac{1}{2}\left(X_{3(3:6)} + X_{3(4:6)}\right), \frac{1}{2}\left(X_{4(3:6)} + X_{4(4:6)}\right), X_{5(6:6)}, X_{6(6:6)}\right\}$ is a BGRSS of size 6, which can be used for estimating the population median as:

$$\hat{\eta}_{BGRSSE} = \text{med}\left\{X_{1(1:6)}, X_{2(1:6)}, \frac{1}{2}\left(X_{3(3:6)} + X_{3(4:6)}\right), \frac{1}{2}\left(X_{4(3:6)} + X_{4(4:6)}\right), X_{5(6:6)}, X_{6(6:6)}\right\}.$$

In the following section we shall give some notations and introduce an estimator of the population median using BGRSS.

3. Estimation of the Population Median Using BGRSS

Let $X_1, X_2, \ldots, X_m$ be a random sample with probability density function $f(x)$ with mean μ and variance σ^2. Let $X_{11}, X_{12}, \ldots, X_{1m}$; $X_{21}, X_{22}, \ldots, X_{2m}; \ldots, X_{m1}, X_{m2}, \ldots,$ X_{mm} be independent random variables all with the same cumulative distribution function $F(x)$.

The SRS estimator of the population median η is defined as:

$$\hat{\eta}_{SRS} = \begin{cases} X_{\left(\frac{m+1}{2}:m\right)} & , \text{if } m \text{ is odd} \\ \frac{1}{2}\left(X_{\left(\frac{m}{2}:m\right)} + X_{\left(\frac{m+2}{2}:m\right)}\right) & , \text{if } m \text{ is even} \end{cases} \quad (3.1)$$

The RSS estimator of the population median η from a sample of size *m* is given by

$$\hat{\eta}_{RSS} = \text{med}\left\{X_{i(i:i)}, i = 1, 2, \ldots, m\right\}. \quad (3.2)$$

If m is odd, let $X_{i(1:m)}$ be the lowest rank unit of the ith sample $(i = 1,2,...,k)$, and let $X_{i\left(\frac{m+1}{2}:m\right)}$ be the median of the ith sample $(i = k+1, k+2,...,2k)$, and let $X_{i(m:m)}$ be the largest rank unit of the ith sample $(i = 2k+1, 2k+2,...,3k)$. So that $X_{1(1:m)},..., X_{k(1:m)}, X_{k+1\left(\frac{m+1}{2}:m\right)},...,\ X_{2k\left(\frac{m+1}{2}:m\right)}, X_{2k+1(m:m)},..., X_{3k(m:m)}$ denote the measured BGRSSO which are iid. In the case of even sample size, let $\frac{1}{2}\left(X_{i\left(\frac{m}{2}:m\right)} + X_{i\left(\frac{m+2}{2}:m\right)}\right)$ be the median of the ith sample $(i = k+1, k+2,...,2k)$. The measured units $X_{1(1:m)},..., X_{k(1:m)}, \frac{1}{2}\left(X_{k+1\left(\frac{m}{2}:m\right)} + X_{k+1\left(\frac{m+2}{2}:m\right)}\right),..., \frac{1}{2}\left(X_{2k\left(\frac{m}{2}:m\right)} + X_{2k\left(\frac{m+2}{2}:m\right)}\right)$, $X_{2k+1(m:m)},..., X_{3k(m:m)}$ denotes the BGRSSE. The BGRSS estimator of the population median η if m is odd is given by:

$$\hat{\eta}_{BGRSSO} = \text{median}\left\{\begin{array}{ll} X_{i(1:m)} & ,i = 1,...,k \\ X_{i\left(\frac{m+1}{2}:m\right)} & ,i = k+1,...,2k \\ X_{i(m:m)} & ,i = 2k+1,...,3k \end{array}\right\}. \tag{3.3}$$

If m is even the estimator is given by

$$\hat{\eta}_{BGRSSE} = \text{median}\left\{\begin{array}{ll} X_{i(1:m)} & ,i = 1,...,k \\ \frac{1}{2}\left(X_{i\left(\frac{m}{2}:m\right)} + X_{i\left(\frac{m+2}{2}:m\right)}\right) & ,i = k+1,...,2k \\ X_{i(m:m)} & ,i = 2k+1,...,3k \end{array}\right\}. \tag{3.4}$$

The properties of $\hat{\eta}_{BGRSS}$ are:

1) If the underlying distribution is symmetric about the population mean μ then
   a) $\hat{\eta}_{BGRSS}$ is an unbiased estimator of the population median η.
   b) $\text{Var}(\hat{\eta}_{BGRSS}) < \text{Var}(\hat{\eta}_{SRS})$.

2) If the underlying distribution is asymmetric about μ then

a) $\hat{\eta}_{BGRSS}$ has a small bias.

b) $\text{MSE}(\hat{\eta}_{BGRSS}) < \text{MSE}(\hat{\eta}_{SRS})$.

In the following section, a simulation study is conducted to investigate the performance of BGRSS in estimating the population median.

4. Simulation Study

In this section, we compare the proposed estimators of the population median using BGRSS relative to SRS method. Six probability distribution functions were considered for the populations: uniform U(0,1), normal N(0,1), logistic L(0,1), exponential Exp(1), beta B(7,4) and gamma (2,1). If the underlying distribution is symmetric the efficiency of the BGRSS relative to SRS is defined as

$$eff(\hat{\eta}_{SRS}, \hat{\eta}_{BGRSS}) = \frac{\text{Var}(\hat{\eta}_{SRS})}{\text{Var}(\hat{\eta}_{BGRSS})}. \tag{4.1}$$

But if the distribution is asymmetric the efficiency is defined as

$$eff(\hat{\eta}_{SRS}, \hat{\eta}_{BGRSS}) = \frac{\text{MSE}(\hat{\eta}_{SRS})}{\text{MSE}(\hat{\eta}_{BGRSS})}. \tag{4.2}$$

Results of the efficiency and bias obtained using BGRSS are provided in Table 1.

Table 1. The efficiency of median estimation using BGRSS with respect to SRS estimator with $m = 3, 6, ..., 21$.

Distribution		$k=1$	$k=2$	$k=3$	$k=4$	$k=5$	$k=6$	$k=7$
		$m=3$	$m=6$	$m=9$	$m=12$	$m=15$	$m=18$	$m=21$
U (0,1)	*Eff*	1.454	2.131	2.343	3.122	3.187	4.386	4.418
N (0,1)	*Eff*	1.614	2.352	2.482	3.430	3.482	4.657	4.723
L (0,1)	*Eff*	1.704	2.363	2.479	3.430	3.522	4.599	4.801
Exp (1)	*Eff*	1.779	2.337	2.478	3.443	3.602	4.775	4.894
	Bias	0.085	0.074	0.022	0.022	0.010	0.010	0.005
B (7,4)	*Eff*	1.590	2.359	2.384	3.373	3.498	4.670	4.732
	Bias	0.003	0.001	0.001	0.001	0.000	0.000	0.000
G (2,1)	*Eff*	1.738	2.447	2.461	3.460	3.551	4.624	4.812
	Bias	0.090	0.073	0.022	0.021	0.007	0.008	0.002

Based on Table 1, we may conclude the followings:

1. A gain in efficiency is obtained by using BGRSS for estimating the population median for all distributions considered in this study. As an example, for $m = 6$, the

efficiency of the BGRSS is 2.363 for estimating the population median of a logistic distribution with parameters 0 and 1. For asymmetric distributions the BGRSS median estimator has a small bias which is decreasing to zero as m increase. As an example, for gamma distribution with parameters 2 and 1, and with $m=9$, we have $eff\left(\hat{\eta}_{SRS},\hat{\eta}_{BGRSS}\right)=2.461$ with bias 0.022.

2. The efficiency of BGRSS is increases in the sample size. As an example, for $m=3$ the efficiency of BGRSS is 1.704 and with $m=9$ the 2.479 for estimating the population median of logistic distribution with parameters 0 and 1.

5. Example for Real Data

In this section, a collection of a real data set is used to illustrate the BGRSS for estimating the population median. These data consists of height (H), and the diameter (D) at breast height of 399 trees. For more details, see Platt et al. [10]. Table 2 contains the summary statistics of the data. We will show how to use the RSS and BGRSS to estimate the median for the diameter of 399 trees based on height.

Table 2. Summary statistics of 399 tree data.

Variable	Mean	Variance	Skewness	Kurtosis	Median
Diameter (D) in cm	20.84	310.11	0.884	-0.423	14.5
Height (H) in feet.	52.36	325.14	1.619	1.776	29
Correlation coefficient ρ			0.908		

It is known that the coefficient of skewness should be close to zero for symmetrically distributed data. Since the coefficient of skewness of the diameter is 0.884, these data are asymmetrically distributed.

To illustrate RSS and BGRSS, we first fix the values of height and do the ranking based on the values of diameters. Let $k=2$, then $m=6$. For estimating the median diameter (D) using RSS and BGRSS methods consider the following steps:

Step 1: Randomly select 6 independent simple random samples of size $m=6$ ordered pairs *(D,H)* as:

Set 1: {{4.2,5},{9.6,16},{8.2,20},{35.7,66},{6,12},{6.8,18}}
Set 2: {{5.1,12},{7.5,18},{75.4,244},{3,4},{7.2,26},{2.2,3}}
Set 3: {{6,12},{41.1,105},{4.5,12},{58.5,192},{20.2,38},{7.5,19}}
Set 4: {{60.6,180},{3.6,5},{4.7,14},{25.5,40},{14.7,33},{3.3,5}}
Set 5: {{50.9,84},{8.6,22},{32.7,53},{6.2,7},{8.2,22},{47.6,154}}
Set 6: {{11.5,21},{6,16},{39,116},{27,29},{5.2,6},{19.6,40}}.

Step 2: For each set in Step 1, rank the pairs within each set based on their height (in bold) from lowest to highest as shown below:

Set 1: $\{\{4.2,\mathbf{5}\},\{6,\mathbf{12}\},\{6.8,\mathbf{18}\},\{8.2,\mathbf{20}\},\{9.6,\mathbf{16}\},\{35.7,\mathbf{66}\}\}$
Set 2: $\{\{2.2,\mathbf{3}\},\{3,\mathbf{4}\},\{5.1,\mathbf{12}\},\{7.2,\mathbf{26}\},\{7.5,\mathbf{18}\},\{75.4,\mathbf{244}\}\}$
Set 3: $\{\{4.5,\mathbf{12}\},\{6,\mathbf{12}\},\{7.5,\mathbf{19}\},\{20.2,\mathbf{38}\},\{41.1,\mathbf{105}\},\{58.5,\mathbf{192}\}\}$
Set 4: $\{\{3.3,\mathbf{5}\},\{3.6,\mathbf{5}\},\{4.7,\mathbf{14}\},\{14.7,\mathbf{33}\},\{25.5,\mathbf{40}\},\{60.6,\mathbf{180}\}\}$
Set 5: $\{\{6.2,\mathbf{7}\},\{8.2,\mathbf{22}\},\{8.6,\mathbf{22}\},\{32.7,\mathbf{53}\},\{47.6,\mathbf{154}\},\{50.9,\mathbf{84}\}\}$
Set 6: $\{\{5.2,\mathbf{6}\},\{6,\mathbf{16}\},\{11.5,\mathbf{21}\},\{19.6,\mathbf{40}\},\{27,\mathbf{29}\},\{39,\mathbf{116}\}\}$

Step 3: Now, we consider RSS, BGRSS and SRS:

1. Under RSS method, from the *i*th subset, select and measure the diameter corresponding to the *i*th ordered height values. The six RSS diameters are: 4.2, 3, 7.5, 14.7, 47.6, and 39. Estimator of the median for the diameter using RSS, is given by: $\hat{\eta}_{D,RSS} = \text{median}\{4.2,3,7.5,14.7,47.6,39\} = \dfrac{7.5+14.7}{2} = 11.1\text{cm/feet}$.
2. Under the BGRSS, the lowest rank units are measured from the first two subsets, the median is measured from the third and fourth subsets and largest rank units are measured from the last two subsets. Thus, diameters considered are: 4.2, 2.2, 13.85, 9.7, 50.9 and 39. The median diameter can be estimated using BGRSS as: $\hat{\eta}_{D,BGRSS} = \text{median}\{4.2, 2.2, 13.85, 9.7, 50.9, 39\} = \dfrac{9.7+13.85}{2} = 11.775\text{cm/feet}$
3. Under SRS we have 6 estimates of median. Let $\hat{\eta}_{D,SRSi}$ be the median of the *i*th set $(i = 1,2,...,6)$. So we have: $\hat{\eta}_{D,SRS1} = 7.5\,\text{cm/feet}$, $\hat{\eta}_{D,SRS2} = 6.15\,\text{cm/feet}$, $\hat{\eta}_{D,SRS3} = 13.85\,\text{cm/feet}$, $\hat{\eta}_{D,SRS4} = 9.7\,\text{cm/feet}$, $\hat{\eta}_{D,SRS5} = 20.65\,\text{cm/feet}$, $\hat{\eta}_{D,SRS6} = 15.55\,\text{cm/feet}$.

Table 3 provide the means, variances and the efficiencies based on 50,000 simulated values of $\hat{\eta}_{D,SRS}$ and $\hat{\eta}_{D,BGRSS}$.

Table 3. The efficiency values of median estimation using BGRSS with respect to SRS estimator when $m = 3,6,...,15$

Size	**SRS**		**BGRSS**		
m	**Median**	**MSE**	**Median**	**MSE**	**Efficiency**
3	18.2240	186.8200	17.0205	120.3330	1.5525
6	17.1245	93.2408	16.6486	40.8479	2.2826
9	16.0511	71.1245	14.9031	25.3861	2.8017
12	15.2209	38.2334	14.4382	9.77257	3.9123
15	15.6588	47.4886	14.9562	12.1199	3.9182

From Table 3 the estimated median is found to be close to the real value of the population median. Also, we can see that the BGRSS method is more efficient than the SRS for estimating the population median of the diameter.

6. Summary

A gain in efficiency is obtained using BGRSS for estimating the population median. Thus, it is recommended to use BGRSS for estimating the population median of symmetric and asymmetric distributions.

References

Al-Saleh, M. F. and Al-Omari, A. I. (2002). Multistage ranked set sampling. *Journal of Statistical Planning and Inference*, **102**, 273-286.

Dell, T. R. and Clutter, J. L. (1972). Ranked set sampling theory with order statistics background. *Biometrika* **28**, 545-555.

Jemain, A. A. and Al-Omari, A. I. (2006a). Double percentile ranked set samples for estimating the population mean. *Advances and Applications in Statistics*. **6**(3) 261-276.

Jemain, A. A. and Al-Omari, A. I. (2006b). Double quartile ranked set samples. *Pakistan Journal of Statistics*. **22**(3): 217-228.

McIntyre, G. A. (1952). A method for unbiased selective sampling using ranked sets. *Australian Journal of Agricultural Research*, **3**, 385-390.

Muttlak, H. A. (2003). Investigating the use of quartile ranked set samples for estimating the population mean. *Journal of Applied Mathematics and Computation*, **146**, 437- 443.

Muttlak, H. A. (2003) Modified ranked set sampling methods. *Pakistan Journal of Statistics*. **19**(3) 315-323.

[8] Muttlak, H. A. (1997). Median ranked set sampling, *Journal of Applied Statistical Sciences*. **6**(4) 577-586.

Ozturk, O. and Deshpande, J.V. (2006). Ranked Set Sample nonparametric quantile confidence intervals. *Journal of Statistical Planning and Inference*. **136**: 570-577.

Platt, W. J., Evans, G. W. and Rathbun, S. L. (1988). The population dynamics of a long lived conifer. *The American Naturalis*, **131**:391-525.

Samawi, H., Abu-Dayyeh, W. and Ahmed, S. (1996). Extreme ranked set sampling, *The Biometrical Journal*. **30**, 577-586.

Takahasi, K. and Wakimoto, K. (1968). On the unbiased estimates of the opulation mean based on the sample stratified by means of ordering. *Annals of the Institute of Statistical Mathematics*. **20**, 1-31.

In: New Developments in Applied Statistics
Editor: Mohammad Ahsanullah

ISBN: 978-1-61324-648-1

Chapter 5

A Finite Population Loss and Delay Queueing System with Interdependent Interarrival and Service Times, Controllable Arrival Rates and No Passing

***A. Srinivasan*[1,*] *and M. Thiagarajan*[2,♣]**

[1] Bishop Heber College, Tiruchirappalli , Tamil Nadu ,India
[2] St. Joseph's College, Tiruchirappalli , Tamil Nadu ,India

Abstract

Many models on loss and delay queueing system with no passing have been studied in the past. Due to restriction of no passing the customers are allowed to depart from the system in the chronological order of their arrival and the customers are classified into two classes (elective or emergency) depending upon whether they can wait or are lost when all servers are busy. The arrival and service processes are taken to be independent in most of the models. In this paper, we consider a finite population , loss and delay queueing system with no passing where arrival and service process are interdependent. In addition to this interdependence, an operating strategy , called controllable arrival rates , is incorporated in the model. Such a model is much useful in controlling congestion in buffers in the computer communication system. The steady state solutions and the analytical expressions for the expected waiting time of a customer in the system are derived explicitly. The numerical results for the expected waiting times of the two classes of customers for varying arrival rates(faster and slower) , the effect of dependence parameter between inter arrival and service times and the effect of switching over to varying arrival rates have been provided in the tables.

Key words: Interdependent arrival and service processes, controllable arrival rates, loss and delay, no passing, parallel channels, finite population.

Primary Subject: 60K25, Secondary Subjects : 68M20, 90B22

---

[*] Email address: mathsrinivas@yahoo.com
[♣] Email address: mathsthiags@yahoo.com

1. Introduction

Many authors investigated the queue situation where customers leave the system in the same order in which they arrive , because of a physical restriction that no passing is allowed . Washburn [22] introduced no passing concept for M/M/m queueing model with infinite population. Sharma and Kumar [13] developed the same problem for a finite model.

In the loss and delay queueing system , the customers are classified into two classes. They are (i) Elective customers and (ii) Emergency customers. The elective customers have patience to form a queue and wait while the emergency customers find the serves busy on their arrival, leave the system and are lost. Shonic [15] analysed combination of loss and delay system and gave its utilization for hospital administration concerning with occupancy of hospital beds. Akinpelu and Shoog [2] studied problem of controlling overloads in common channel signaling networks and showed how there was trade-off between loss and delay. Shore [16] obtained inverse distribution function and loss and delay function in GI/G/c queueing model.

Jain et al. [8] introduced no passing multiserver queueing model with two type of customers and discouragement and obtained various traffic measures numerically. Jain [9] analysed the finite population loss and delay queueing system with no passing concept.

In general, it is assumed that the arrival and service processes are independent. But there are many practical situations in which they are interdependent. Much work has been reported in literature regarding interdependent queueing models. Conolly [4] and Conolly and Hadidi [6] have discussed the single server correlated queueing model. Conolly and Choo [5] have studied the correlated queue with exponential demand M/M/1 queueing system with interdependent arrival and service process. Borst and Combe [3] have analysed the busy period of the correlated queues. Srinivasa Rao et.al [17] have considered M/M/1/∞ interdependent queueing model with controllable arrival rates. Aftab Begum and Maheswari [1] have analysed M/M/c/∞ interdependent queueing model with controllable arrival rates. Srinivasan and Thiagarajan [18] have obtained the expected waiting time and system size for the M/M/1/K interdependent queueing model with controllable arrival rates. Srinivasan and Thiagarajan [19] have generalized their results for the M/M/c/K interdependent queueing model with controllable arrival rates.

In order to have optimal policies and to approximate the practical situations more close to the reality in data voice transmission, computer communication, production process etc., controlling the congestion in buffers plays an important role. This valid point is taken into account in our queueing model, which has not been studied in all the earlier models of mixed type(loss and delay) of customers.

In this model , the system has three arrival rates, λ_{01}- mean arrival rate of elective customers when the system is in the faster rate of arrivals, λ_{02}- mean arrival rate of emergency customers when the system is in the faster rate of arrivals, λ_{11}- mean arrival rate of elective customers when the system is in the slower rate of arrivals, which control the arrivals. It is also assumed that whenever the system size reaches c (number of parallel servers), the arrival rate changes from λ_{02} to λ_{01} for emergency customers and it continues with that rate as long as the content in the system reaches a certain prescribed integer R. When the system size reaches R, the arrival rate changes from λ_{01} to λ_{11} and it continues with that rate as long as the content in the system is greater than some other prescribed integer r ($r \geq 0$, $c<r<R<N$). When the system size reaches r, the arrival rate changes back to λ_{01} and the process is repeated. From the service point of view the customers can be classified into two classes A and B. The customers who have zero service time are called of type A. The (1–p) proportion of the total customers are of type A. The remaining p proportion of the customers are of type B having exponential service times with rate μ.

2. Description of the Model

Consider a c-server, finite capacity, loss and delay and no passing queueing system with the following assumptions:

The arrival process $\{X_1(t)\}$ and the service completion process $\{X_2(t)\}$ of the system are correlated and follow a bivariate Poisson process [23] given by

$$P\{X_1(t) = x_1, X_2(t) = x_2\}$$
$$= \frac{e^{-(\lambda_{ij}+\mu_n-\in)} \sum_{k=0}^{\min(x_1,x_2)} (\in t)^k \left[(\lambda_{ij}-\in)\, t\right]^{x_1-k} \left[(\mu_n-\in)\, t\right]^{x_2-k}}{k!\,(x_1-k)!\,(x_2-k)!} \tag{2.1}$$

$$x_1, x_2 = 0,1,2,;\ \lambda_{01}, \lambda_{02}, \lambda_{11}, \mu_n > 0,$$
$$n = 0,1,2,\ldots c-1,c,c+1,\ldots r-1,r,r+1,\ldots R-1,R,R+1,\ \ldots\ ,N-1, N\ ;\ 0 \leq \in < \min(\lambda_{ij}, \mu_n)$$

with parameters λ_{01}, λ_{02}, λ_{11}, μ_n and $\in$ as mean arrival rate of elective customers when the system is in the faster rate of arrivals, mean arrival rate of emergency customers when the system is in the faster rate of arrivals, mean arrival rate of elective customers when the system is in the slower rate of arrivals, mean service rate of customers of type B and the co-variance between arrival and service processes respectively. Also the mean arrival rate and mean service rate when the system size is n is defined as

$$\lambda_n = \begin{cases} (N-n)\lambda_{0j}\,p & ; \quad 0 \leq n < c, \quad j = 1,2 \\ (N-n)\lambda_{01}\,p & ; \quad c \leq n \leq R-1 \\ (N-n)\lambda_{11}\,p & ; \quad r+1 \leq n \leq N \end{cases} \tag{2.2}$$

$$\mu_n = \begin{cases} n\mu & ; \quad 0 \leq n < c \\ c\mu & ; \quad c \leq n \leq N \end{cases} \tag{2.3}$$

3. Steady State Equations

Let $P_n(0)$ be the steady state probability that there are n customers in the system when the system is in the faster rate of arrivals. Let $P_n(1)$ be the steady state probability that there are n customers in the system when the system is in the slower rate of arrivals.

We observe that only $P_n(0)$ exists when n = 0,1,2,....c–1, c, c+1, ... r–1, r ; both $P_n(0)$ and $P_n(1)$ exist when n = r+1, r+2, R–2, R–1 ; only $P_n(1)$ exists when n = R, R+1,....,N. Further $P_n(0) = P_n(1)= 0$ when n > N.

Then the steady state probability difference equations are given by

$$-N(\lambda_0 - 2\in)p\, P_0(0) + (\mu - \in)P_1(0) = 0 \tag{3.1}$$

$$-[(N-n)(\lambda_0 - 2\in)p + n(\mu - \in)]P_n(0) + (N-n+1)(\lambda_0 - 2\in)p\, P_{n-1}(0) + (n+1)(\mu - \in)P_{n+1}(0) = 0 \quad (1 \le n \le c-1) \tag{3.2}$$

$$-[(N-c)(\lambda_{01} - \in)p + c(\mu - \in)]P_c(0) + (N-c+1)(\lambda_0 - 2\in)pP_{c-1}(0) + c(\mu - \in)P_{c+1}(0) = 0 \tag{3.3}$$

$$-[(N-n)(\lambda_{01} - \in)p + c(\mu - \in)]P_n(0) + (N-n+1)(\lambda_{01} - \in)pP_{n-1}(0) + c(\mu - \in)P_{n+1}(0) = 0 \quad (c+1 \le n \le r-1) \tag{3.4}$$

$$-[(N-r)(\lambda_{01} - \in)p + c(\mu - \in)]P_r(0) + (N-r+1)(\lambda_{01} - \in)pP_{r-1}(0) + c(\mu - \in)P_{r+1}(0) + c(\mu - \in)P_{r+1}(1) = 0 \tag{3.5}$$

$$-[(N-n)(\lambda_{01} - \in)p + c(\mu - \in)]P_n(0) + (N-n+1)(\lambda_{01} - \in)pP_{n-1}(0) + c(\mu - \in)P_{n+1}(0) = 0 \quad (r+1 \le n \le R-2) \tag{3.6}$$

$$-[(N-R+1)(\lambda_{01} - \in)p + c(\mu - \in)]P_{R-1}(0) + (N-R+2)(\lambda_{01} - \in)p\, P_{R-2}(0) = 0 \tag{3.7}$$

$$-[(N-r-1)(\lambda_{11} - \in)p + c(\mu - \in)]P_{r+1}(1) + c(\mu - \in)P_{r+2}(1) = 0 \tag{3.8}$$

$$-[(N-n)(\lambda_{11} - \in)p + c(\mu - \in)]P_n(1) + (N-n+1)(\lambda_{11} - \in)pP_{n-1}(1) + c(\mu - \in)P_{n+1}(1) = 0 \quad (r+2 \le n \le R-1) \tag{3.9}$$

$$-[(N-R)(\lambda_{11} - \in)p + c(\mu - \in)]P_R(1) + (N-R+1)(\lambda_{11} - \in)p\, P_{R-1}(1) + (N-R+1)(\lambda_{01} - \in)p\, P_{R-1}(0) + c(\mu - \in)P_{R+1}(1) = 0 \tag{3.10}$$

$$-\left[(N-n)(\lambda_{11}-\in)p+c(\mu-\in)\right]P_n(1)+(N-n+1)(\lambda_{11}-\in)p\,P_{n-1}(1)$$

$$+c(\mu-\in)P_{n+1}(1)=0\quad(R+1\le n\le N-1) \tag{3.11}$$

$$-c(\mu-\in)P_N(1)+(\lambda_{11}-\in)p\,P_{N-1}(1)=0 \tag{3.12}$$

Write $s_1=\dfrac{(\lambda_{01}-\in)p}{\mu-\in}$, $s_2=\dfrac{(\lambda_{02}-\in)p}{\mu-\in}$, $t_1=\dfrac{(\lambda_{11}-\in)p}{\mu-\in}$ and

$$s_0=s_1+s_2=\frac{(\lambda_0-2\in)p}{\mu-\in}\quad\text{where}\quad \lambda_0=\lambda_{01}+\lambda_{02}$$

From (3.1) and (3.2),

$$P_n(0)=\frac{(N)_n(s_0)^n}{n!}P_0(0),\qquad n=0,1,2,\ldots\ldots,c-1$$

where $(N)_n=N(N-1)(N-2)\ldots\ldots(N-n+1)$.(3.13)

From (3.3) and (3.4),

$$P_n(0)=\frac{(N)_c(s_0)^c}{c!}(N-c)_{n-c}\left(\frac{s_1}{c}\right)^{n-c}P_0(0),\ n=c,c+1,\ldots\ldots,r \tag{3.14}$$

From (3.5) and (3.6),

$$P_n(0)=\frac{(N)_c(s_0)^c}{c!}(N-c)_{n-c}\left(\frac{s_1}{c}\right)^{n-c}P_0(0)$$

$$-\left[(N-r-1)_{n-r-1}\left(\frac{s_1}{c}\right)^{n-r-1}+(N-r-2)_{n-r-2}\left(\frac{s_1}{c}\right)^{n-r-2}+\ldots..+1\right]P_{r+1}(1),$$

$$n=r+1,r+2,\ldots..,R-1. \tag{3.15}$$

From (3.5), (3.7) and (3.15),

$$P_{r+1}(1)=\frac{A_1}{A_2}P_0(0)\ ,$$

where $A_1=\left[(N-R+1)^2\left(\dfrac{s_1}{c}\right)-1\right](N-c)_{R-2-c}\left(\dfrac{s_1}{c}\right)^{R-1-c}\dfrac{(N)_c(s_0)^c}{c!}$,

$$A_2 = (N-r-1)_{R-r-1}\left(\frac{s_1}{c}\right)^{R-r-1} + (N-r-2)_{R-r-2}\left(\frac{s_1}{c}\right)^{R-r-2} + \dots\dots + (N-R+1)\left(\frac{s_1}{c}\right) + 1 \qquad (3.16)$$

From (3.8) and (3.9),

$$P_n(1) = \left[(N-r-1)_{n-r-1}\left(\frac{t_1}{c}\right)^{n-r-1} + (N-r-2)_{n-r-2}\left(\frac{t_1}{c}\right)^{n-r-2} + \dots + 1\right]P_{r+1}(1),$$
$$n = r+1, r+2, \dots R-1, R \qquad (3.17)$$

where $P_{r+1}(1)$ is given by (3.16)

From (3.10),(3.11) and (3.12),

$$P_n(1) = \left[(N-r-1)_{n-r-1}\left(\frac{t_1}{c}\right)^{n-r-1} + (N-r-2)_{n-r-2}\left(\frac{t_1}{c}\right)^{n-r-1} + \dots\dots + 1\right]P_{r+1}(1)$$
$$- A_3\left[1 + (N-R)\left(\frac{t_1}{c}\right) + \dots + (N-R)_{n-R-1}\left(\frac{t_1}{c}\right)^{n-R-1}\right]P_0(0)$$
$$+ A_1\left[1 + (N-R-1)\left(\frac{t_1}{c}\right) + \dots + (N-R-1)_{n-R-1}\left(\frac{t_1}{c}\right)^{n-R-1}\right]P_0(0)$$
$$n = R+1, R+2\dots\dots\dots, N. \qquad (3.18)$$

where $A_3 = (N-c)_{R-c}\left(\frac{s_1}{c}\right)^{R-c}\frac{(N)_c(s_0)^c}{c!}$, A_1 and $P_{r+1}(1)$ are given by (3.16)

4. System Characteristics

The probability that the system is in state i is $P(i) = \sum_{n=0}^{N} P_n(i)$, $i = 0,1$

The probability that the system is in faster rate of arrivals is

$$P(0) = \sum_{n=0}^{c-1} P_n(0) + \sum_{n=c}^{r} P_n(0) + \sum_{n=r+1}^{R-1} P_n(0) \qquad (4.1)$$

From (3.13),

$$\sum_{n=0}^{c-1} P_n(0) = \sum_{n=0}^{c-1} \frac{(N)_n (s_0)^n}{n!} P_0(0) \tag{4.2}$$

From (3.14),

$$\sum_{n=c}^{r} P_n(0) = \frac{(N)_c (s_0)^c}{c!} \, {}_2F_1\left(1, c-N;-;-\frac{s_1}{c}\right) P_0(0) \tag{4.3}$$

where $(1)_n = 0$ or $(c\text{-}N)_n = 0$ when $n > r\text{-}c$ in the Hyper Geometric function.

From (3.15),

$$\sum_{N=r+1}^{R-1} P_n(0) = \left[\frac{(N)_c (s_0)^c}{c!}\left(\frac{s_1}{c}\right)^{r+1-c} (N-c)_{r+1-c} - \frac{A_1}{A_2}\right] {}_2F_1\left(1, 1+r-N;-;-\frac{s_1}{c}\right) P_0(0)$$
$$-\left[{}_2F_1\left(1,2+r-N;-;-\frac{s_1}{c}\right) + {}_2F_1\left(1,3+r-N;-;-\frac{s_1}{c}\right) + ... + 1\right]\frac{A_1}{A_2} P_0(0)$$

where A_1 and A_2 are given by (3.16), ... (4.4)

$(1)_n$=0 or $(1+r\text{-}N)_n$=0 when $n > R\text{-}r\text{-}2$ in the first Hyper Geometric function,

$(1)_n$=0 or $(2+r\text{-}N)_n$=0 when $n > R\text{-}r\text{-}3$ in the second Hyper Geometric function,

$(1)_n$=0 or $(3+r\text{-}N)_n$=0 when $n > R\text{-}r\text{-}4$ in the third Hyper Geometric function and so on.

From (4.1), (4.2),(4.3) and (4.4),

$$P(0) = \sum_{n=0}^{c-1} \frac{(N)_n (s_0)^n}{n!} P_0(0) + \sum_{n=c}^{r} \frac{(N)_c (s_0)^c}{c!} \, {}_2F_1\left(1, c-N;-;-\frac{s_1}{c}\right) P_0(0)$$
$$+\left[\frac{(N)_c (s_0)^c}{c!}\left(\frac{s_1}{c}\right)^{r+1-c} (N-c)_{r+1-c} - \frac{A_1}{A_2}\right] {}_2F_1\left(1, 1+r-N;-;-\frac{s_1}{c}\right) P_0(0)$$
$$-\left[{}_2F_1\left(1,2+r-N;-;-\frac{s_1}{c}\right) + {}_2F_1\left(1,3+r-N;-;-\frac{s_1}{c}\right) + ... + 1\right]\frac{A_1}{A_2} P_0(0) \tag{4.5}$$

The probability that the system is in slower rate of arrivals is

$$P(1) = \sum_{n=r+1}^{R-1} P_n(1) + \sum_{n=R}^{N} P_n(1) \tag{4.6}$$

From (3.16) and (3.17),

$$\sum_{n=r+1}^{R-1} P_n(1) = \left[{}_2F_1\left(1, 1+r-N; -; -\frac{t_1}{c}\right) + {}_2F_1\left(1, 2+r-N; -; -\frac{t_1}{c}\right) + ... + 1 \right] \frac{A_1}{A_2} P_0(0)$$

where A_1,A_2 are given by (3.16), (4.7)

$(1)_n = 0$ or $(1+r-N)_n = 0$ when n > R-r-2 in the first Hyper Geometric function,

$(1)_n = 0$ or $(2+r-N)_n = 0$ when n > R-r-3 in the second Hyper Geometric function and so on.

From (3.16) and (3.18),

$$\sum_{n=R}^{N} P_n(1) =$$

$$= \sum_{n=R}^{N} \left[(N-r-1)_{n-r-1} \left(\frac{t_1}{c}\right)^{n-r-1} + (N-r-2)_{n-r-2} \left(\frac{t_1}{c}\right)^{n-r-2} + ... + 1 \right] \frac{A_1}{A_2} P_0(0)$$

$$- \sum_{n=R}^{N} \left[(N-R)_{n-R-1} \left(\frac{t_1}{c}\right)^{n-R-1} + (N-R)_{n-R-2} \left(\frac{t_1}{c}\right)^{n-R-2} + ... + 1 \right] A_3 P_0(0)$$

$$+ \sum_{n=R}^{N} \left[(N-R-1)_{n-R-1} \left(\frac{t_1}{c}\right)^{n-R-1} + (N-R-1)_{n-R-2} \left(\frac{t_1}{c}\right)^{n-R-2} + ... + 1 \right] A_1 P_0(0)$$

where A_1,A_2 are given by (3.16) and A_3 is given by (3.18) (4.8)

From (4.8),

$$\sum_{n=R}^{N} \left[(N-r-1)_{n-r-1} \left(\frac{t_1}{c}\right)^{n-r-1} + (N-r-2)_{n-r-2} \left(\frac{t_1}{c}\right)^{n-r-2} + ... + 1 \right] \frac{A_1}{A_2} P_0(0)$$

$$= \frac{A_1 A_4}{A_2} {}_2F_1\left(1, R-N; -; -\frac{t_1}{c}\right) P_0(0) \tag{4.9}$$

where $A_4 = (N-r-1)_{R-r-1} \left(\frac{t_1}{c}\right)^{R-r-1} + (N-r-2)_{R-r-2} \left(\frac{t_1}{c}\right)^{R-r-2} + ... + (N-R+1)$,

$(1)_n = 0$ or $(R-N)_n = 0$ when $n > N-R$ in the Hyper Geometric function.

From (4.8),

$$\sum_{n=R}^{N}\left[(N-R)_{n-R-1}\left(\frac{t_1}{c}\right)^{n-R-1}+(N-R)_{n-R-2}\left(\frac{t_1}{c}\right)^{n-R-2}+\ldots\ldots+1\right]A_3P_0(0)$$

$$=\left[{}_2F_1\left(1,R-N;-;-\frac{t_1}{c}\right)+{}_2F_1\left(1,R-N;-;-\frac{t_1}{c}\right)+\ldots.+1\right]A_3P_0(0) \qquad (4.10)$$

where $(1)_n = 0$ or $(R\text{-}N)_n = 0$ when $n>N\text{-}R\text{-}1$ in the first Hyper Geometric function, $(1)_n = 0$ or $(R\text{-}N)_n = 0$ when $n>N\text{-}R\text{-}2$ in the Second Hyper Geometric function and so on.

From (4.8),

$$\sum_{n=R}^{N}\left[(N-R-1)_{n-R-1}\left(\frac{t_1}{c}\right)^{n-R-1}+(N-R-1)_{n-R-2}\left(\frac{t_1}{c}\right)^{n-R-2}+\ldots\ldots+1\right]A_1P_0(0)$$

$$=\left[{}_2F_1\left(1,1+R-N;-;-\frac{t_1}{c}\right)+{}_2F_1\left(1,1+R-N;-;-\frac{t_1}{c}\right)+\ldots.+1\right]A_1P_0(0) \qquad (4.11)$$

where $(1)_n = 0$ or $(1\text{+}R\text{-}N)_n=0$ when $n>N\text{-}R\text{-}1$ in the first Hyper Geometric function,

$(1)_n =0$ or $(1\text{+}R\text{-}N)_n=0$ when $n>N\text{-}R\text{-}2$ in the Second Hyper Geometric function and so on.

From (4.6), (4.7), (4.8), (4.9), (4.10) and (4.11),

$$P(1)=\left[{}_2F_1\left(1,1+r-N;-;-\frac{t_1}{c}\right)+{}_2F_1\left(1,2+r-N;-;-\frac{t_1}{c}\right)+\ldots\ldots\ldots+1\right]\frac{A_1}{A_2}P_0(0)$$
$$+{}_2F_1\left(1,R-N;-;-\frac{t_1}{c}\right)\frac{A_1A_4}{A_2}P_0(0)$$
$$-\left[{}_2F_1\left(1,R-N;-;-\frac{t_1}{c}\right)+{}_2F_1\left(1,R-N;-;-\frac{t_1}{c}\right)+\ldots\ldots..+1\right]A_3P_0(0)$$
$$+\left[{}_2F_1\left(1,1+R-N;-;-\frac{t_1}{c}\right)+{}_2F_1\left(1,1+R-N;-;-\frac{t_1}{c}\right)+\ldots\ldots..+1\right]A_1P_0(0) \qquad (4.12)$$

The probability $[P_0(0)]$ that the system is empty can be calculated from the normalizing condition

$$P(0)+P(1) = 1 \qquad (4.13)$$

From (4.5), (4.12) and (4.13),

$$P_0(0) = \left\{ \sum_{n=0}^{c-1} \frac{(N)_n (s_0)^n}{n!} + \frac{(N)_c (s_0)^c}{c!} {}_2F_1\left(1, c-N; -, -\frac{s_1}{c}\right) \right.$$

$$+ \left[\frac{(N)_c (s_0)^c}{c!} \left(\frac{s_1}{c}\right)^{r+1-c} (N-c)_{r+1-c} - \frac{A_1}{A2} \right] {}_2F_1\left(1, 1+r-N; -; -\frac{s_1}{c}\right)$$

$$- \left[{}_2F_1\left(1, 2+r-N; -; -\frac{s_1}{c}\right) + {}_2F_1\left(1, 3+r-N; -; -\frac{s_1}{c}\right) + \dots\dots + 1 \right] \frac{A_1}{A_2}$$

$$+ \left[{}_2F_1\left(1, 1+r-N; -; -\frac{t_1}{c}\right) + {}_2F_1\left(1, 2+r-N; -; -\frac{t_1}{c}\right) + \dots\dots + 1 \right] \frac{A_1}{A_2}$$

$$+ {}_2F_1\left(1, R-N; -; -\frac{t_1}{c}\right) \frac{A_1 A_4}{A_2}$$

$$- \left[{}_2F_1\left(1, R-N; -; -\frac{t_1}{c}\right) + {}_2F_1\left(1, R-N; -; -\frac{t_1}{c}\right) + \dots\dots + 1 \right] A_3$$

$$\left. + \left[{}_2F_1\left(1, 1+R-N; -; -\frac{t_1}{c}\right) + {}_2F_1\left(1, 1+R-N; -; -\frac{t_1}{c}\right) + \dots\dots + 1 \right] A_1 \right\}^{-1} \quad (4.14)$$

The expected waiting time $E[W_{A_0}]$ for the customer of type A when the system is in the faster rate of arrivals is given by [9]

$$E\left[W_{A_0}\right] = \frac{1}{\mu - \in} \left\{ \sum_{n=0}^{c-1} a_n P_n(0) + \left(\sum_{n=c}^{r} P_n(0) + \sum_{n=r+1}^{R-1} P_n(0) \right) \left(\frac{n-c+1}{c} + a_{c-1} \right) \right\}$$

$$\text{where } a_n = \begin{cases} 0 & ; \quad n = 0 \\ \sum_{j=1}^{n} \frac{1}{j} & ; \quad n = 1, 2 \dots, N \end{cases} \quad (4.15)$$

Using (4.2), (4.3) and (4.4) in (4.15), we get

$$(\mu - \in) E\left[W_{A_0}\right] = \sum_{n=0}^{c-1} \left(\sum_{j=1}^{n} \frac{1}{j} \right) \frac{(N)_n (s_0)^n}{n!} P_0(0)$$

$$+ \left(\sum_{j=1}^{c-1} \frac{1}{j} \right) \left[\frac{(N)_c (s_0)^c}{c!} {}_2F_1\left(1, c-N; -; -\frac{s_1}{c}\right) P_0(0) \right.$$

$$+ \frac{(N)_c (s_0)^c}{c!} \left(\frac{s_1}{c}\right)^{r+1-c} (N-c)_{r+1-c} \; {}_2F_1\left(1, 1+r-N; -; -\frac{s_1}{c}\right) P_0(0)$$

$$-\left({}_2F_1\left(1,1+r-N;-\frac{s_1}{c}\right)+{}_2F_1\left(1,2+r-N;-;-\frac{s_1}{c}\right)+\ldots\ldots+1\right)\frac{A_1}{A_2}P_0(0)\Bigg]$$

$$+\frac{(N)_c(s_0)^c}{c!}\frac{\left[1-(r-c+2)(N-c)_{r-c+1}\left(\frac{s_1}{c}\right)^{r-c+1}+(r-c+1)(N-c)_{r-c+2}\left(\frac{s_1}{c}\right)^{r-c+2}\right]}{\left[1-(N-c)\left(\frac{s_1}{c}\right)\right]^2}P_0(0)$$

$$+\frac{(N)_c(s_0)^c}{cc!}\left(\frac{s_1}{c}\right)^{r+1-c}(N-c)_{r+1-c}\Bigg[(r+1-c)(R-r+2)$$

$$+\frac{\left(1-(R-r)(N-r-1)_{R-r-1}\left(\frac{s_1}{c}\right)^{R-r-1}+(R-r-1)(N-r-1)_{R-r}\left(\frac{s_1}{c}\right)^{R-r}\right)}{\left(1-(N-r-1)\left(\frac{s_1}{c}\right)\right)^2}\Bigg]P_0(0)$$

$$-\frac{A_1}{cA_2}\Bigg[(r-c+2)+(r-c+3)\left(1+(N-r-1)\left(\frac{s_1}{c}\right)\right)+(r-c+4)\left(1+(N-r-1)\left(\frac{s_1}{c}\right)+(N-r-1)_2\left(\frac{s_1}{c}\right)^2\right)$$

$$+\ldots\ldots+(R-c)\left(1+(N-r-1)\left(\frac{s_1}{c}\right)+(N-r-1)_2\left(\frac{s_1}{c}\right)^2+\ldots\ldots+(N-r-1)_{R-r-2}\left(\frac{s_1}{c}\right)^{R-r-2}\right)\Bigg]P_0(0)$$

where $P_0(0)$ is given by (4.14) (4.16)

The expected waiting time $E[W_{B_0}]$ for the customers of typeB where the system is in the faster rate of arrivals is given by [9]

$$E\left[W_{B_0}\right]=\frac{1}{\mu-\in}\left\{\sum_{n=0}^{c-1}a_{n+1}P_n(0)+\left(\sum_{n=c}^{r}P_n(0)+\sum_{n=r+1}^{R-1}P_n(0)\right)\left(\frac{n-c+1}{c}+a_c\right)\right\}\qquad(4.17)$$

where a_{n+1} and a_c are given by (4.15).

Using (4.2), (4.3) and (4.4) in (4.17), we get

$$(\mu-\in)E\left[W_{B_0}\right]=\sum_{n=0}^{c-1}\left(\sum_{j=1}^{n+1}\frac{1}{j}\right)\frac{(N)_n(s_0)^n}{n!}P_0(0)$$

$$+\left(\sum_{j=1}^{c}\frac{1}{j}\right)\left[\frac{(N)_c(s_0)^c}{c!}\,{}_2F_1\left(1,c-N;-;-\frac{s_1}{c}\right)P_0(0)\right.$$

$$+\frac{(N)_c(s_0)^c}{c!}\left(\frac{s_1}{c}\right)^{r+1-c}(N-c)_{r+1-c}\,{}_2F_1\left(1,1+r-N;-;-\frac{s_1}{c}\right)P_0(0)$$

$$\left.-\left({}_2F_1\left(1,1+r-N;-\frac{s_1}{c}\right)+{}_2F_1\left(1,2+r-N;-;-\frac{s_1}{c}\right)+\ldots\ldots+1\right)\frac{A_1}{A_2}P_0(0)\right]$$

$$+\frac{(N)_c(s_0)^c}{c!}\frac{\left[1-(r-c+2)(N-c)_{r-c+1}\left(\frac{s_1}{c}\right)^{r-c+1}+(r-c+1)(N-c)_{r-c+2}\left(\frac{s_1}{c}\right)^{r-c+2}\right]}{\left[1-(N-c)\left(\frac{s_1}{c}\right)\right]^2}P_0(0)$$

$$+\frac{(N)_c(s_0)^c}{cc!}\left(\frac{s_1}{c}\right)^{r+1-c}(N-c)_{r+1-c}\left[(r+1-c)(R-r+2)\right.$$

$$\left.+\frac{\left(1-(R-r)(N-r-1)_{R-r-1}\left(\frac{s_1}{c}\right)^{R-r-1}+(R-r-1)(N-r-1)_{R-r}\left(\frac{s_1}{c}\right)^{R-r}\right)}{\left(1-(N-r-1)\left(\frac{s_1}{c}\right)\right)^2}\right]P_0(0)$$

$$-\frac{A_1}{cA_2}\left[(r-c+2)+(r-c+3)\left(1+(N-r-1)\left(\frac{s_1}{c}\right)\right)+(r-c+4)\left(1+(N-r-1)\left(\frac{s_1}{c}\right)+(N-r-1)_2\left(\frac{s_1}{c}\right)^2\right)\right.$$

$$\left.+\ldots\ldots+(R-c)\left(1+(N-r-1)\left(\frac{s_1}{c}\right)+(N-r-1)_2\left(\frac{s_1}{c}\right)^2+\ldots\ldots+(N-r-1)_{R-r-2}\left(\frac{s_1}{c}\right)^{R-r-2}\right)\right]P_0(0)$$

where $P_0(0)$ is given by (4.14) (4.18)

The expected waiting time $E[W_{A_1}]$ for the customer of type A when the system is in the slower rate of arrivals is given by [9]

$$E\left[W_{A_1}\right]=\frac{1}{\mu-\in}\left(\sum_{n=r+1}^{R-1}P_n(1)+\sum_{n=R}^{N}P_n(1)\right)\left(\frac{n-c+1}{c}+a_{c-1}\right) \tag{4.19}$$

where a_{c-1} is given by (4.15)

Using (4.7), (4.8), (4.9) , (4.10) and (4.11) in (4.19) , we get

$$(\mu-\in)E\left[W_{A_1}\right]=\left(\sum_{j=1}^{c-1}\frac{1}{j}\right)\left[\left({}_2F_1\left(1,1+r-N;-;-\frac{t_1}{c}\right)+{}_2F_1\left(1,2+r-N;-;-\frac{t_1}{c}\right)+\ldots\ldots+1\right)\right.$$

$$\frac{A_1}{A_2}P_0(0)+\left((N-r-1)_{R-r-1}\left(\frac{t_1}{c}\right)^{R-r-1}+(N-r-2)_{R-r-2}\left(\frac{t_1}{c}\right)^{R-r-2}+\ldots\ldots+(N-R+1)\right)$$

$${}_2F_1\left(1,R-N;-;-\frac{t_1}{c}\right)\frac{A_1}{A_2}P_0(0)-\left({}_2F_1\left(1,R-N;-;-\frac{t_1}{c}\right)+\right.$$

$$\left.+{}_2F_1\left(1,R-N;-;-\frac{t_1}{c}\right)+\ldots\ldots\ldots+1\right)A_3P_0(0)+\left({}_2F_1\left(1,1+R-N;-;-\frac{t_1}{c}\right)\right.$$

$$\left.\left.+{}_2F_1\left(1,1+R-N;-;-\frac{t_1}{c}\right)+\ldots\ldots+1\right)A_1P_0(0)\right]+\left[(r-c+2)+\right.$$

$$+(r-c+3)\left(1+(N-r-1)\left(\frac{t_1}{c}\right)\right)+(r-c+4)\left(1+(N-r-1)\left(\frac{t_1}{c}\right)+(N-r-1)_2\left(\frac{t_1}{c}\right)^2\right)$$

$$\left.+\ldots+(R-c)\left(1+(N-r-1)\left(\frac{t_1}{c}\right)+(N-r-1)_2\left(\frac{t_1}{c}\right)^2+\ldots+(N-r-1)_{R-r-2}\left(\frac{t_1}{c}\right)^{R-r-2}\right)\right]$$

$$\frac{A_1}{cA_2}P_0(0)+\left[(R-c-1)\left((N-r-1)_{R-r-1}\left(\frac{t_1}{c}\right)^{R-r-1}+(N-r-2)_{R-r-2}\left(\frac{t_1}{c}\right)^{R-r-2}+\ldots\ldots+1\right)\right.$$

$$+(R-c+2)\left((N-r-1)_{R-r}\left(\frac{t_1}{c}\right)^{R-r}+(N-r-2)_{R-r-1}\left(\frac{t_1}{c}\right)^{R-r-1}+\ldots\ldots+1\right)$$

$$+\ldots\ldots+(N-c+1)\left((N-r-1)_{N-r-1}\left(\frac{t_1}{c}\right)^{N-r-1}+\right.$$

$$\left.\left.+(N-r-2)_{N-r-2}\left(\frac{t_1}{c}\right)^{N-r-2}+\ldots\ldots+1\right]\frac{A_1P_0(0)}{cA_2}\right.$$

$$-\left[(R+2-c)+(R+3-c)\left(1+(N-R)\left(\frac{t_1}{c}\right)\right)+(R+4-c)\left(1+(N-R)\left(\frac{t_1}{c}\right)+(N-R)_2\left(\frac{t_1}{c}\right)^2\right)+\right.$$

$$+\ldots\ldots+(N-c+1)\left(1+(N-R)\left(\frac{t_1}{c}\right)+(N-R)_2\left(\frac{t_1}{c}\right)^2+\ldots\ldots\right.$$

$$\left.\left.+(N-R)_{N-R-1}\left(\frac{t_1}{c}\right)^{N-R-1}\right)\right]\frac{A_3P_0(0)}{c}$$

$$+\left[(R+2-c)+(R+3-c)\left(1+(N-R-1)\left(\frac{t_1}{c}\right)\right)+\right.$$

$$(R+4-c)\left(1+(N-R-1)\left(\frac{t_1}{c}\right)+(N-R-1)_2\left(\frac{t_1}{c}\right)^2\right)$$

$$+\ldots\ldots+(N-c+1)\left(1+(N-R-1)\left(\frac{t_1}{c}\right)+(N-R-1)_2\left(\frac{t_1}{c}\right)^2+\ldots\ldots\right.$$

$$\left.+(N-R-1)_{N-R-1}\left(\frac{t_1}{c}\right)^{N-R-1}\right]\frac{A_1}{c}P_0(0) \tag{4.20}$$

where $P_0(0)$ is given by (4.14)

The expected waiting time $E[W_{B_1}]$ for the customer of type B when the system is in the slower rate of arrivals is given by [9].

$$E\left[W_{B_1}\right]=\frac{1}{\mu-\in}\left\{\left(\sum_{n=r+1}^{R-1}P_n(1)+\sum_{n=R}^{N}P_n(1)\right)\left(\frac{n-c+1}{c}+a_c\right)\right\} \tag{4.21}$$

where a_c is given by (4.15).

Using (4.7), (4.8), (4.9) , (4.10) and (4.11) in (4.21) ,

$$(\mu-\in)E\left[W_{B_1}\right]=\left(\sum_{j=1}^{c}\frac{1}{j}\right)\left[\left({}_2F_1\left(1,1+r-N;-;-\frac{t_1}{c}\right)+{}_2F_1\left(1,2+r-N;-;-\frac{t_1}{c}\right)+\ldots\ldots+1\right)\right.$$

$$\frac{A_1}{A_2}P_0(0)+\left((N-r-1)_{R-r-1}\left(\frac{t_1}{c}\right)^{R-r-1}+(N-r-2)_{R-r-2}\left(\frac{t_1}{c}\right)^{R-r-2}+\ldots\ldots+(N-R+1)\right)$$

$${}_2F_1\left(1, R-N;-;-\frac{t_1}{c}\right)\frac{A_1}{A_2}P_0(0) - \left({}_2F_1\left(1, R-N;-;-\frac{t_1}{c}\right) +\right.$$

$$\left. + {}_2F_1\left(1, R-N;-;-\frac{t_1}{c}\right) + \ldots\ldots + 1\right)A_3P_0(0) + \left({}_2F_1\left(1, 1+R-N;-;-\frac{t_1}{c}\right)\right.$$

$$\left. + {}_2F_1\left(1, 1+R-N;-;-\frac{t_1}{c}\right) + \ldots\ldots + 1\right)A_1P_0(0)\Bigg] + \Bigg[(r-c+2) +$$

$$+ (r-c+3)\left(1 + (N-r-1)\left(\frac{t_1}{c}\right)\right) + (r-c+4)\left(1 + (N-r-1)\left(\frac{t_1}{c}\right) + (N-r-1)_2\left(\frac{t_1}{c}\right)^2\right)$$

$$\left. + \ldots + (R-c)\left(1 + (N-r-1)\left(\frac{t_1}{c}\right) + (N-r-1)_2\left(\frac{t_1}{c}\right)^2 + \ldots + (N-r-1)_{R-r-2}\left(\frac{t_1}{c}\right)^{R-r-2}\right)\right]$$

$$\frac{A_1}{cA_2}P_0(0) + \Bigg[(R-c-1)\left((N-r-1)_{R-r-1}\left(\frac{t_1}{c}\right)^{R-r-1} + (N-r-2)_{R-r-2}\left(\frac{t_1}{c}\right)^{R-r-2} + \ldots + 1\right)$$

$$+ (R-c+2)\left((N-r-1)_{R-r}\left(\frac{t_1}{c}\right)^{R-r} + (N-r-2)_{R-r-1}\left(\frac{t_1}{c}\right)^{R-r-1} + \ldots\ldots + 1\right)$$

$$+ \ldots\ldots + (N-c+1)\left((N-r-1)_{N-r-1}\left(\frac{t_1}{c}\right)^{N-r-1}\right.$$

$$\left. + (N-r-2)_{N-r-2}\left(\frac{t_1}{c}\right)^{N-r-2} + \ldots\ldots + 1\right]\frac{A_1P_0(0)}{cA_2}$$

$$-\Bigg[(R+2-c) + (R+3-c)\left(1 + (N-R)\left(\frac{t_1}{c}\right)\right) + (R+4-c)\left(1 + (N-R)\left(\frac{t_1}{c}\right) + (N-R)_2\left(\frac{t_1}{c}\right)^2\right) +$$

$$+ \ldots\ldots + (N-c+1)\left(1 + (N-R)\left(\frac{t_1}{c}\right) + (N-R)_2\left(\frac{t_1}{c}\right)^2 + \ldots\right.$$

$$\left. + (N-R)_{N-R-1}\left(\frac{t_1}{c}\right)^{N-R-1}\right)\Bigg]\frac{A_3P_0(0)}{c}$$

$$+\left[(R+2-c)+(R+3-c)\left(1+(N-R-1)\left(\frac{t_1}{c}\right)\right)+\right.$$

$$(R+4-c)\left(1+(N-R-1)\left(\frac{t_1}{c}\right)+(N-R-1)_2\left(\frac{t_1}{c}\right)^2\right)$$

$$+\ldots\ldots+(N-c+1)\left(1+(N-R-1)\left(\frac{t_1}{c}\right)+(N-R-1)_2\left(\frac{t_1}{c}\right)^2+\ldots\ldots\right.$$

$$\left.\left.+(N-R-1)_{N-R-1}\left(\frac{t_1}{c}\right)^{N-R-1}\right)\right]\frac{A_1}{c}P_0(0) \quad (4.22)$$

where $P_0(0)$ is given by (4.14)

5. Conclusion

We have incorporated a new operating strategy , called controllable arrival rates (faster and slower) , for the loss and delay queueing model which is much useful in controlling congestion in buffers in the computer communication system. We have also derived the expected waiting time of a customer in the system on varying arrival rates(faster and slower).

Also this model includes the models studied earlier as particular cases. For example, when $\lambda_{02} = \in = 0$, this model reduces to finite population loss and delay queueing system with no passing [9]. When $\in = \lambda_{01} = \lambda_{11} = 0$, this model reduces to a limited multiserver queue with no passing [13].

The numerical results for the expected waiting times of the two classes of customers for varying arrival rates(faster and slower) and the effect of switching over to varying arrival rates have been provided in the tables.

Table 1.

c=1, r=6 , R=8, N=10, s_1=0.5 s_0 , t_1=0.4					
s_0	P_{00}	$(\mu-\in)E[W_{A0}]$	$(\mu-\in)E[W_{B0}]$	$(\mu-\in)E[W_{A1}]$	$(\mu-\in)E[W_{B1}]$
0.4	0.009629	3.814615	4.668849	1.342957	1.144900
0.6	0.001104	3.550843	4.253084	3.096742	2.392127
0.8	0.000265	4.096168	4.897846	3.345948	1.130330

Table 2.

c=1, r=6 , R=8, N=10, s_1=0.5 s_0 , s_0=0.4					
t_1	P_{00}	$(\mu-\in)E[W_{A_0}]$	$(\mu-\in)E[W_{B_0}]$	$(\mu-\in)E[W_{A_1}]$	$(\mu-\in)E[W_{B_1}]$
0.4	0.009629	3.814615	4.668849	1.342957	1.144900
0.6	0.006308	2.499043	3.058672	3.985145	4.083069
0.8	0.003450	1.366661	1.672707	6.318123	6.760707

Table 3.

c=2, r=6 , R=8, N=10, s_1=0.9 s_0 , t_1=0.8					
s_0	P_{00}	$(\mu-\in)E[W_{A_0}]$	$(\mu-\in)E[W_{B_0}]$	$(\mu-\in)E[W_{A_1}]$	$(\mu-\in)E[W_{B_1}]$
0.4	0.016343	2.935980	3.481986	0.441932	0.506144
0.6	0.001681	2.272402	2.626511	1.248901	1.420725
0.8	0.000357	2.236647	2.562523	1.077815	1.261254

Table 4.

c=2, r=6 , R=8, N=10, s_1=0.9 s_0 , s_0=0.8					
t_1	P_{00}	$(\mu-\in)E[W_{A_0}]$	$(\mu-\in)E[W_{B_0}]$	$(\mu-\in)E[W_{A_1}]$	$(\mu-\in)E[W_{B_1}]$
0.8	0.000357	2.236647	2.562523	1.077815	1.261254
0.9	0.000282	1.769531	2.027350	1.834841	2.084393
1.0	0.000222	1.389616	1.592081	2.466657	2.769980

References

[1] Aftab Begum, M.I., Maheswari, D., (2002) The M/M/c/∞ Interdependent queueing model with controllable arrival rates, *Opsearch* , Vol.39, No.2, 89-110.

[2] Akinpelu, J. and Shoog, R., (1985) Controlling transients and overloads in common channel signaling networks, *Tele traffic issue in advanced information Soc.*,**96**-1002.

[3] Borst,S.C. and Combe,M.B. (1992) Letters to the Editor: Busy period analysis of correlated queues, *Journal of Applied Probability,* Vol. 29, 482 – 483.

[4] Conolly, B.W. (1968) The waiting time process for a certain correlated queue, *Operational Research*, Vol.16, 1006-1015.

[5] Conolly, B.W. and Choo,Q.H. (1979) The waiting time process for a generalized correlated queue with exponential demand and service, *SIAM Journal of Applied Maths*, Vol. 37, 263-275.

[6] Conolly, B.W. and Hadidi, N.(1969) A correlated queue, *Journal of Applied probability*, Vol.6.

[7] Gross, D and Harris, C.M., (1974) *Fundamentals of queueing theory,* John Wiley, New York, p.118-125.

[8] Jain, M., Sharma, B. and Sharma, G.C.(1989) On no passing multiserver queueing model with two types of customers and discouragement, *J. Math. Phy. Sci.,*Vol. 23, No.4, 319-329.

[9] Jain, M., (1998) Finite population Loss and Delay Queueing System with no passing, *Opsearch*, Vol.35, No.3, 261-276.

[10] Langaris,C. (1987) Busy period analysis of a correlated queue, *Journal of Applied Probability*, Vol.24, 476 – 485.

[11] Mitchell,C. and Paulson,A. (1979) M/M/1 queues with interdependent arrival and service process, *Naval Research Logistics*, *Quart* .**26**.

[12] Norman, L., Johnson, Samuel Kotz (1969) *Discrete distributions*, John Wiley and Sons, New York, p.298.

[13] Sharma, M.B., Kumar, G.C., (1983) A limited multi server queue without passing, *Indian Journal of Pure & Applied Maths*. Vol.14,No.8, 935-939.

[14] Shawky, A.I (2000) The machine interference Model : M/M/c/K/N with balking, reneging and spares, *Opsearch*, Vol.37, No.1, 25-35.

[15] Shonick, W., (1970) A stochastic model for occupancy related random variable in general acute hospitals, *J. Am. stat. Asso.* **65**, 1975 – 1500.

[16] Shore, H., (1988) Sample approximation for GI/G/c queue II: The moments, the inverse distribution function and loss function of the number in the system and in the queue delay, *J. Appl. Res. Soc.*, **39**(4), 381 – 391.

[17] Srinivasa Rao, K., Shobha., Srinivasa Rao, P., (2000) The $M/M/1/\infty$ interdependent queueing model with controllable arrival rates, *Opsearch*, Vol.37, No.1, 14 – 24.

[18] Srinivasan, A., Thiagarajan., M., (2006) The M/M/1/K interdependent queueing model with controllable arrival rates, *International Journal of Management and systems*, Vol.22, No.1, 23-34.

[19] Srinivasan, A., Thiagarajan., M. (2007) The M/M/c/K interdependent queueing model with Controllable arrival rates, *Bulletin of Calcutta Mathematical Society*, Vol. 99, No.2, 173-182.

[20] Srinivasan, A., Thiagarajan., M. (2007) The M/M/c/K/N interdependent queueing model with Controllable arrival rates, balking, reneging and spares, *Journal of Statistics and Applications,* Vol.2, No.3, 151-166.

[21] Srinivasan, A., Thiagarajan., M., (2006) The $M^{[x]}/M/1/\infty$ interdependent queueing model with controllable arrival rates, *Journal of Decision And Mathematical Sciences*, Vol.11, No.1-3, 7-24.

[22] Washburn, A.R,(1974) A multi server queue with no passing, *Operational Research*, Vol.22, 428 – 434.

[23] Yuan Shin Chow, Henry Teicher (2004) *Probability Theory: Independence, Interchangeability, Martingales*, International Edition, 3rd Edition, pp.191.

In: New Developments in Applied Statistics
Editor: Mohammad Ahsanullah
ISBN 978-1-61324-648-1

Chapter 6

GENERALIZED LEAST SQUARES APPROACH FOR MOVING AVERAGE MODELS

Mohamed A. Ismail*
Department of Statistics, Cairo University,
and
Information and Decision Support Center (IDSC),
The Egyptian Cabinet

Abstract

A new Bayesian method for estimating moving average (MA) models is proposed. The proposed methodology is based on replacing lagged errors of the original MA model with appropriately lagged residuals from a long autoregression. Unlike Broemeling and Shaarawy (1988), the exact structure of the approximation error when replacing true errors with corresponding residuals is derived and used in deriving the posterior distribution of the model parameters. In addition, a modified version of Broemeling and Shaarawy's (1988) method is suggested. The proposed method and original Broemeling and Shaarawy's method and its modified version are compared using several simulation studies and a real data.

Key Words: GLS; posterior distribution; multivariate t; W6 series

1. Introduction

Bayesian analysis of moving average models is difficult since the likelihood function is highly nonlinear in the parameters, which complicates the prior specification and posterior calculations. Thus, the integrations involved in Bayesian analysis must be done numerically. This is not easy and time consuming especially in the case of the multiparameter models. Simplifying Bayesian analysis for MA models has been the objective of several authors, such as Newbold (1973) and Chip and Greenberg (1994) among others.

A successful solution to the problem is suggested by Broemeling and Shaarawy (1988) where the marginal posterior distribution of the model coefficients is approximated by T

*E-mail address: m.ismail@idsc.net.eg

distribution. Their approach is based on calculating the residuals recursively using nonlinear least squares estimates of the model coefficients. Then, the lagged errors of the model are replaced with their corresponding lagged residuals. A potential drawback of Broemeling and Shaarawy's (1988) approach is that it sets the true errors equal to the estimated errors or the residuals, i.e. it assumes there is no estimation error. In addition, the nonlinear least squares, which are required for calculating the residuals, are not easy to obtain and alternative accurate, fast and easy implemented estimation methods such as the Innovation Substitution method (IS) proposed by Koreisha and Pukkila (1989) can be used. The IS estimates for MA parameters are obtained by applying ordinary least squares (OLS) method to MA model after replacing lagged errors with correspondent lagged residuals from a long autoregression. The IS estimates for MA parameters are consistent see Koreisha and Pukkila (1989) and Frutos and Serrano (1997).

In this study a new Bayesian method for estimating MA models is proposed. The proposed methodology is based on replacing lagged errors of the original MA model with appropriately lagged residuals from a long autoregression. Unlike Broemeling and Shaarawy (1988), the exact structure of the approximation error when replacing true errors with corresponding residuals is used in approximating the conditional likelihood function by a normal gamma distribution. In addition, a modification to Broemeling and Shaarawy's method is suggested where IS estimates are used to calculate the residuals recursively instead of nonlinear least square estimates.

The rest of the paper is organized as follows. Section 2 presents the proposed approximate conditional likelihood function. Posterior distributions for the parameters of the MA model are derived in Section 3. In Section 4, the performance of the proposed technique is evaluated and compared to the original and modified methods of Broemeling and Shaarawy (1988) using several simulation studies. The Bayesian analysis results for the US. tobacco production (W6) series are discussed in Section 5. Finally, Section 6 contains some concluding remarks.

2. An Approximate Conditional Likelihood Function

A moving average model for a time series $\{y_t\}$ can be expressed as

$$y_t = \Theta_q(B)\varepsilon_t \tag{1}$$

where, $\{\varepsilon_t\}$ is a sequence of independent normal variates with zero mean and variance $\sigma^2 = \tau^{-1} > 0$. The backshift operator B is defined as $By_t = y_{t-1}$, and the moving average polynomial is $\Theta_q(B) = (1 + \theta_1 B + \theta_2 B^2 + \cdots + \theta_q B^q)$ and $\boldsymbol{\theta} = (\theta_1, \theta_1, \cdots, \theta_q)$ is the parameter vector. The model (1) is assumed invertible in this study, i.e. the equation $\Theta_q(B) = 0$ has its roots outside the unit circle. The model order q is assumed known and after observing the data $y_t, t = 1, 2, \cdots n$ the parameters $\boldsymbol{\theta}$ and τ are to be estimated. The inverible MA model (1) can be expressed as infinite autoregression

$$y_t = \sum_{i=1}^{\infty} \pi_i y_{t-i} + \varepsilon_t \tag{2}$$

Due to the invertibility, a long autoregression of order L might be a good approximation for model (1). That is

$$y_t = \sum_{i=1}^{L} \hat{\pi}_i y_{t-i} + \hat{\varepsilon}_t \tag{3}$$

where L is chosen as $\sqrt{n}$ and $\hat{\varepsilon}_t$ is an estimate of ε_t. The choice of L as $\sqrt{n}$ has been justified in Koreisha and Pukkila (1990a) and Koreisha and Pukkila (1990b) via extensive simulation studies. Define a_t as

$$\varepsilon_t = \hat{\varepsilon}_t + a_t \tag{4}$$

Using (3) and (4), the approximation error a_t can be expressed as

$$\begin{aligned} a_t &= \varepsilon_t - \hat{\pi}_L(B) y_t \\ &= \varepsilon_t - \hat{\pi}_L(B)\Theta(B)\varepsilon_t \\ &= [1 - \hat{\pi}_L(B)\Theta(B)]\,\varepsilon_t \end{aligned}$$

where $\hat{\pi}_L(B) = 1 - \hat{\pi}_1 B, -\hat{\pi}_2 B^2, \cdots - \hat{\pi}_L B^L$. Substituting for ε_t in (1) using (4)results in the model

$$y_t = \sum_{i=1}^{q} \theta_i \hat{\varepsilon}_{t-i} + \eta_t$$

where

$$\eta_t = \sum_{j=0}^{2q+L} \psi_j \varepsilon_{t-j} = \Psi_{2q+L}(B)\varepsilon_t \tag{5}$$

and $\psi_i' s$ are calculated by comparing the coefficients in

$$\Psi_{2q+L}(B) = \left[\Theta_q(B) + (1 - \Theta_q(B))\,\Theta_q(B)\hat{\pi}_L(B)\right].$$

Setting the initial errors zeros, i.e. $\varepsilon_0 = \varepsilon_{-1} = \cdots = \varepsilon_{1-2q-L} = 0$, the model (1) can be written in the following matrix form:

$$\mathbf{Y} = \mathbf{X}\boldsymbol{\theta} + \boldsymbol{\eta} \tag{6}$$

with $\mathbf{Y}^T = (y_1, y_2, \cdots, y_n)$, $\boldsymbol{\eta}^T = (\eta_1, \eta_2, \cdots, \eta_n)$, $\mathbf{X}$ is a n×q matrix with ti^{th} element $X_{ti} = \hat{\varepsilon}_{t-i}$. and the error vector $\boldsymbol{\eta}$ has a multivariate normal distribution with zero mean and covariance matrix $\tau^{-1}\boldsymbol{\Omega}$ where $\boldsymbol{\Omega}$ is a positive definite toeplitiz matrix. Using model (6), the conditional likelihood function can be written as:

$$L(\boldsymbol{\theta}, \tau \mid \mathbf{Y}) \propto \tau^{\frac{n}{2}} \exp\left[-\frac{\tau}{2} (\mathbf{Y} - \mathbf{X}\boldsymbol{\theta})^T \boldsymbol{\Omega}^{-1} (\mathbf{Y} - \mathbf{X}\boldsymbol{\theta})\right] \tag{7}$$

The conditional likelihood function (7) is a complicated function in $\boldsymbol{\theta}$ since the elements of $\boldsymbol{\Omega}$ are nonlinear functions in $\boldsymbol{\theta}$. Thus, numerical integration is a must to perform Bayesian inference for $\boldsymbol{\theta}$. As a solution to the problem, an estimated matrix $\hat{\boldsymbol{\Omega}}$ can be obtained by replacing the elements of $\boldsymbol{\theta}$ in the matrix $\boldsymbol{\Omega}$ by their estimates $\hat{\boldsymbol{\theta}}$ obtained by IS method.

Replacing the matrix $\mathbf{\Omega}$ by its estimate $\hat{\mathbf{\Omega}}$ in (7) results in an approximate conditional likelihood function which takes the form:

$$L^{\star}(\boldsymbol{\theta},\tau \mid \mathbf{Y}) \propto \tau^{\frac{n}{2}} \exp\left[-\frac{\tau}{2}(\mathbf{Y}-\mathbf{X}\boldsymbol{\theta})^{T}\hat{\mathbf{\Omega}}^{-1}(\mathbf{Y}-\mathbf{X}\boldsymbol{\theta})\right] \tag{8}$$

The direct use of (8) is difficult since it needs to invert the Toeplitz matrix $\hat{\mathbf{\Omega}}$ which can be computationally burdensome specially for large n. However, Galbraith and Zinde-Walsh (1992) derived the exact transformation matrix $\mathbf{R}$, such that $\mathbf{R}^{T}\mathbf{R} = \mathbf{\Omega}^{-1}$. Using the IS parameter estimates, it is possible to construct an estimated transformation matrix $\hat{\mathbf{R}}$, such that $\hat{\mathbf{\Omega}}^{-1} = \hat{\mathbf{R}}^{T}\hat{\mathbf{R}}$. Thus an approximate conditional likelihood function is given by

$$\begin{aligned} L^{\star}(\boldsymbol{\theta},\tau \mid \mathbf{Y}) &\propto \tau^{\frac{n}{2}} \exp\left[-\frac{\tau}{2}(\mathbf{Y}-\mathbf{X}\boldsymbol{\theta})^{T}\left(\hat{\mathbf{R}}^{T}\hat{\mathbf{R}}\right)(\mathbf{Y}-\mathbf{X}\boldsymbol{\theta})\right] \\ &= \tau^{\frac{n}{2}} \exp\left[-\frac{\tau}{2}(\mathbf{Y}^{\star}-\mathbf{X}^{\star}\boldsymbol{\theta})^{T}(\mathbf{Y}^{\star}-\mathbf{X}^{\star}\boldsymbol{\theta})\right] \end{aligned} \tag{9}$$

where $\mathbf{Y}^{\star} = \hat{\mathbf{R}}\mathbf{Y}$ and $\mathbf{X}^{\star} = \hat{\mathbf{R}}\mathbf{X}$.

The approximate conditional likelihood function used in Broemeling and Shaarawy (1988) is a simplification of (9) when $\hat{\mathbf{R}} = \hat{\mathbf{\Omega}} = I_n$ where I_n is the unit matrix of order n and the ti^{th} element of the matrix $\mathbf{X}$ is computed as follows

$$X_{ti} = \dot{\varepsilon}_{t-i} = y_{t-i} - \sum_{r=1}^{q} \dot{\theta}_{r}\ \dot{\varepsilon}_{t-i-r} \quad t = 1, 2, \cdots n,$$

where $\dot{\varepsilon}_{1-q} = \dot{\varepsilon}_{2-q} = \cdots \dot{\varepsilon}_{0} = 0$ and $\dot{\varepsilon}_{r}$ is a suitable estimate for θ_r such as nonlinear least squares estimates.

3. Posterior Analysis For Moving Average Models

Since the approximate conditional likelihood function (9) is a normal gamma density, a suitable choice for the proper prior distribution for the model parameters $\boldsymbol{\theta}$ and τ is a normal gamma distribution. That is $\zeta(\boldsymbol{\theta} \mid \tau) \sim N(\boldsymbol{\mu}, \tau^{-1}\mathbf{V}^{-1})$, $\zeta(\tau) \sim \Gamma(a, b)$ where $\mathbf{V}$ is a square positive definite matrix of order q, $a > 0$, $b > 0$. The joint prior of the model parameters, $\zeta(\boldsymbol{\theta}, \tau)$ may be written as :

$$\begin{aligned} \zeta(\boldsymbol{\theta},\tau) &= \zeta(\boldsymbol{\theta} \mid \tau)\,\zeta(\tau) \\ &\propto \tau^{a+q-1} \exp\left\{-\frac{\tau}{2}\left[2b + (\boldsymbol{\theta}-\boldsymbol{\mu})^{T}\mathbf{V}(\boldsymbol{\theta}-\boldsymbol{\mu})\right]\right\} \end{aligned} \tag{10}$$

Combining the approximate conditional likelihood function (9) with the normal gamma prior (10) via Bayes theorem results in an approximate joint posterior distribution for the parameters $\boldsymbol{\theta}$ and τ which can be written as:

$$\zeta^{\star}(\boldsymbol{\theta},\tau \mid \mathbf{Y}) \propto \tau^{\left(\frac{n+2a+2q}{2}\right)-1} \exp\left\{-\frac{\tau}{2}\left[2b + (\boldsymbol{\theta}-\boldsymbol{\mu})^{T}\mathbf{V}(\boldsymbol{\theta}-\boldsymbol{\mu}) + (\mathbf{Y}^{\star}-\mathbf{X}^{\star}\boldsymbol{\theta})^{T}(\mathbf{Y}^{\star}-\mathbf{X}^{\star}\boldsymbol{\theta})\right]\right\} \tag{11}$$

The approximate marginal posterior distribution of $\boldsymbol{\theta}$ can be found by integrating out the approximate joint posterior distribution (11) with respect to τ. The following theorem shows that the approximate marginal posterior distribution of $\boldsymbol{\theta}$ is a t density.

Theorem 1 *Given n observations* $\mathbf{Y} = (y_1, y_2, \cdots y_n)$ *from the MA(q) model given by (1), a conditional likelihood function approximated by (9) and a prior distribution given by (10), then the approximate marginal posterior distribution of* $\boldsymbol{\theta}$ *is a* T_q *distribution with* $n+2a$ *degrees of freedom, location vector* $E(\boldsymbol{\theta} \mid \mathbf{Y})$ *and precision matrix* $T(\boldsymbol{\theta} \mid \mathbf{Y})$*, where*

$$E(\boldsymbol{\theta}|\mathbf{Y}) = \left(\mathbf{X}^T\hat{\boldsymbol{\Omega}}^{-1}\mathbf{X}+\mathbf{V}\right)^{-1}\left(\mathbf{X}^T\hat{\boldsymbol{\Omega}}^{-1}\mathbf{Y}+\mathbf{V}\boldsymbol{\mu}\right) \quad ,$$

$$T(\boldsymbol{\theta} \mid \mathbf{Y}) = \frac{n+2a}{C}\left(\mathbf{X}^T\hat{\boldsymbol{\Omega}}^{-1}\mathbf{X}+\mathbf{V}\right) \quad ,$$

and

$$C = 2b + \mathbf{Y}^T\hat{\boldsymbol{\Omega}}^{-1}\mathbf{Y} + \boldsymbol{\mu}^T\mathbf{V}\boldsymbol{\mu} - \left(\mathbf{X}^T\hat{\boldsymbol{\Omega}}^{-1}\mathbf{Y}+\mathbf{V}\right)^T\left(\mathbf{X}^T\hat{\boldsymbol{\Omega}}^{-1}\mathbf{X}+\mathbf{V}\right)^{-1}\left(\mathbf{X}^T\hat{\boldsymbol{\Omega}}^{-1}\mathbf{Y}+\mathbf{V}\boldsymbol{\mu}\right).$$

Proof. The proof follows by integrating (11) over τ using the properties of gamma distribution and then completing the squares with respect to $\boldsymbol{\theta}$.

Similarly, integrating out the approximate joint posterior distribution with respect to $\boldsymbol{\theta}$ results in the approximate marginal distribution of τ. The next theorem shows that the marginal posterior of τ can be approximated by a gamma density.

Theorem 2 *Under the same conditions of theorem 1, the approximate posterior of* τ *is a gamma distribution with parameters* $(\frac{1}{2})(n+2a)$ *and* $(\frac{1}{2})C$*, where C is defined in theorem 1.*

Proof. The proof follows by completing the squares in the exponent of (11) with respect to $\boldsymbol{\theta}$ and then using the properties of multivatiate distribution to integrate over $\boldsymbol{\theta}$.

When there is little information about the hyperparameters, one may use Jeffreys' prior where $\boldsymbol{\theta}$ and τ are assumed independent and have vague priors. Thus, the Jeffreys' prior distribution is

$$\zeta(\boldsymbol{\theta}, \tau) \propto \frac{1}{\tau} \qquad \tau > 0. \tag{12}$$

The Jeffreys' prior (12), is a special case of the normal gamma class when $b = 0$, $\mathbf{V} = \mathbf{0}$, $a = -q$. The next two corollaries report the approximate posterior distributions of $\boldsymbol{\theta}$ and τ when little information about the parameters is known *a priori*.

Corollary 1 *If the parameters* $\boldsymbol{\theta}$, τ *are assumed independent a priori and have Jeffreys' prior distribution (12) and the other conditions of theorem 1 are not changed, the approximate posterior distribution of* $\boldsymbol{\theta}$ *is a T distribution with (n-q) degrees of freedom, location vector*

$$E(\boldsymbol{\theta}|\mathbf{Y}) = \left(\mathbf{X}^T\hat{\boldsymbol{\Omega}}^{-1}\mathbf{X}\right)^{-1}\left(\mathbf{X}^T\hat{\boldsymbol{\Omega}}^{-1}\mathbf{Y}\right)$$

and precision matrix

$$T(\boldsymbol{\theta} \mid \mathbf{Y}) = \frac{(n-q)\,\mathbf{X}^T\hat{\boldsymbol{\Omega}}^{-1}\mathbf{X}}{\mathbf{Y}^T\hat{\boldsymbol{\Omega}}^{-1}\mathbf{Y} - \left(\mathbf{X}^T\hat{\boldsymbol{\Omega}}^{-1}\mathbf{Y}\right)^T\left(\mathbf{X}^T\hat{\boldsymbol{\Omega}}^{-1}\mathbf{X}\right)^{-1}\left(\mathbf{X}^T\hat{\boldsymbol{\Omega}}^{-1}\mathbf{Y}\right)}.$$

Proof. The proof follows directly from Theorem 1 by setting $b = 0$, $\mathbf{V} = \mathbf{0}$, $a = -q$.

Corollary 2 *Under the same set-up as corollary 1, the approximate posterior distribution of* τ *is gamma with parameters* $(\frac{1}{2})(n - q)$ *and* $\frac{1}{2}\left(\mathbf{X}^T\hat{\Omega}^{-1}\mathbf{Y}\right)^T\left(\mathbf{X}^T\hat{\Omega}^{-1}\mathbf{X}\right)^{-1}\left(\mathbf{X}^T\hat{\Omega}^{-1}\mathbf{Y}\right)$.

Proof. The proof follows directly from Theorem 2 by setting $b = 0$, $\mathbf{V} = \mathbf{0}$, $a = -q$.

It worth mentioning that the posterior mean $E\left(\boldsymbol{\theta}|\mathbf{Y}\right)$ is equivalent to the feasible Generalized Least Squares (GLS) estimator derived in Frutos and Serrano (1997).

4. Simulation

In this section we evaluate and compare the quality of three Bayesian estimation methods. The first one is the proposed Bayesian generalized least squares (BGLS) method which is explained in section 3. The second method is a modified version of Broemeling and Shaarawy's (1988) method where IS estimates are used - instead of nonlinear linear least squares estimates - to estimate the errors and is denoted by BS-IS. The original method of Broemeling and Shaarawy's (1988) is the third one. The third method uses nonlinear least squares estimates of the errors and is denoted as BS-NL.

The comparison was based on several simulations experiments. Each experiment dealt with 1000 realizations for each of a variety of MA models with varying parameter values and different number of observations. Parameter values were selected to enable comparison with other published studies, e.g. Koreisha and Pukkila (1990a) and Koreisha Pukkila (1990b). In all experiments, the precision parameter τ was set to 1 and Jeffreys' prior was used. For nonlinear least squares method required for BS-NL method, all inital parameter values were set to their theoretical values. A Gauss macro was written to perform the necessary computations and executed on a Pantium PC.

Table 1 contains parameter estimates and Mean Square Error (MSE) of the three estimation methods for MA(1) model. The results were arranged according to the parameter value with the smallest first. For each model, the results were displayed into two panels. Each panel consists of two rows. The first row gives average of Bayesian estimates while the second row gives the sample standard errors of these estimates. Table 2 gives similar results for second order moving average MA(2) model.

In analyzing the results of Table 1 we observe that the proposed BGLS estimates on average match the true values for all parameter values and regardless the series length. Both of BS-LN and BS-IS estimates are close to true values for for small and moderate parameter values. However as the parameter approaches the invertibility boundary, both of BS-LN and BS-IS estimates become more baised. It is striking that the MSE of BGLS estimates is less than that for both of BS-LN and BS-IS estimates in all cases. Moreover, the BS-LN and BS-IS methods produce estimates that are relatively close to each other with greater precision (smaller MSE) for BS-IS in most cases.

Inspecting the results of Table 2 conveys similar conclusions to those drawn from Table 1 with two notable features. First, superiority of the BGLS method to BS-IS and BS-NL methods becomes more obvious when the model order increases. Second, the BS-IS

Table 1. Comparison of BGLS, BS-IS and BS-NL results for MA(1) Models

n	BGLS			BS-IS			BS-NL		
	θ	τ	MSE	θ	τ	MSE	θ	τ	MSE
				MA(1): θ=0.2					
50	0.198	1.023	0.021	0.2	1.026	0.023	0.203	1.026	0.023
	0.146	0.214		0.15	0.213		0.15	0.215	
100	0.202	1.012	0.01	0.203	1.013	0.011	0.204	1.013	0.011
	0.102	0.149		0.105	0.148		0.102	0.149	
				MA(1): θ=0.5					
50	0.489	1.016	0.021	0.475	0.997	0.021	0.49	1.009	0.022
	0.145	0.213		0.142	0.217		0.15	0.22	
100	0.497	1.008	0.01	0.491	0.998	0.01	0.499	1.004	0.01
	0.102	0.148		0.1	0.15		0.102	0.151	
				MA(1): θ=0.9					
50	0.846	0.983	0.04	0.647	0.836	0.136	0.641	0.845	0.148
	0.193	0.239		0.268	0.251		0.285	0.26	
100	0.864	0.977	0.017	0.717	0.873	0.089	0.746	0.894	0.073
	0.125	0.159		0.235	0.195		0.221	0.192	

Table 2. Comparison of BGLS, BS-IS and BS-NL results for selected MA(2) Models

n	BGLS				BS-IS				BS-NL			
	θ_1	θ_2	τ	MSE	θ_1	θ_2	τ	MSE	θ_1	θ_2	τ	MSE
					MA(2): θ_1=0 , θ_2=-0.73							
50	0	-0.664	0.978	0.049	0.001	-0.589	0.898	0.065	0	-0.546	0.894	0.093
	0.145	0.153	0.209		0.139	0.161	0.209		0.129	0.206	0.225	
100	0.003	-0.697	0.984	0.022	0.001	-0.652	0.935	0.028	0.001	-0.663	0.954	0.031
	0.102	0.101	0.146		0.097	0.111	0.144		0.098	0.128	0.152	
					MA(2): θ_1=1.42 , θ_2=-.73							
50	1.243	-0.595	0.908	0.151	0.912	-0.511	0.589	0.484	0.894	-0.619	0.598	0.52
	0.2	0.248	0.257		0.361	0.219	0.252		0.41	0.254	0.276	
100	1.301	-0.66	0.912	0.063	1.079	-0.628	0.688	0.244	1.108	-0.716	0.715	0.212
	0.127	0.166	0.17		0.295	0.174	0.223		0.302	0.154	0.24	

and BS-NL estimates become significantly less accurate (more biased) as the number of nonzero parameters increases.

Non of the three considered estimation methods guarantees the invertibility of the estimates. Table 3 shows the number of non invertible estimates for each of the three methods. Results of Table 3 show that the number of non invertible estimates is the smallest for BS-IS method and the highest for BS-NL method regardless the model and the series length. This results agrees with the findings of Koreisha and Pukkila (1990a) and Koreisha Pukkila (1990b). The proposed BGLS method has a slightly higher number of non invertible estimates than the BS-IS method. As expected, the number of non invertible estimates decreases as the series length increases.

Table 3. Number of noninveribile cases

Model	n	BGLS	BS-IS	BS-NL
MA(1)	50	0	0	0
θ_1=0.2	100	0	0	0
MA(1)	50	0	0	1
θ_1=0.5	100	0	0	0
MA(1)	50	50	48	79
θ_1=0.9	100	36	33	73
MA(1)	50	7	6	9
θ_1=0 ,θ_2=-0.73	100	0	0	1
MA(1)	50	36	24	62
θ_1=1.42 ,θ_2=-0.73	100	13	7	16

Table 4. Bayesian results for W6 series

Parameter	BGLS	BS-IS	BS-NL	NLSE
θ	0.646	0.633	0.645	0.61
	(0.097)	(0.095)	(0.094)	(0.076)
τ^{-1}	0.028	0.028	0.028	0.028

5. Real Data Example: W6

The W6 series is the yearly US. tobacco production from 1871 to 1984, i.e. 114 values and is given in the appendix to Wei's (1990) book. Using ACF and PACF, Wei (1990) has identified an ARIMA(0,1,1) for the natural logarithm of the series.

Assuming a non informative prior, the three Bayesian methods are applied to the changes of the log W6 series using a MA(1) model. The Bayesian and nonlinear least squares estimates (NLSE) of the coefficient and precision parameter are reported in Table 4. The numbers within parentheses are the standard errors. As can be seen from Table 4, the four methods give close estimates for the coefficient parameter and the same estimate for the precision parameter. This may be explained by the moderate value of the coefficient estimate.

6. Conclusion

In this study we have presented the Bayesian GLS approach to moving average models. Rather than ignoring the sampling errors that arise when replacing the errors with their estimates as in Broemeling and Shaarawy (1988), we proposed the use of IS estimation for estimating the errors and employ the stochastic structure of the sampling error in developing Bayesian estimation for MA models. In addition, a modified version of Broemeling and Shaarawy's (1988) method is suggested by using IS method to estimate the errors instead of using nonlinear least square estimates.

Simulation results indicate the superiority of the proposed method when compared with that of Broemeling and Shaarawy (1988) and its modified version. Future work will extend this approach to univariate ARMA and seasonal models.

References

[1] Broemeling, L.D. and Shaarawy, S. (1988). Time series: A Bayesian Analysis in Time Series Domain. In *Bayesian Analysis of Time Series and Dynamic Models,* (ed J. C. Spall), pp. 1-21. Marcel Dekker Inc., New York.

[2] Chib, S. and Greenberg, E. (1994). Bayes inference in regression models with ARMA(p,q) errors. *Journal of Econometrics*, **64**, 183-206.

[3] Galbraith J. and Zinde-Walsh, V. (1992). The GLS transformation matrix and a semi-recursive estimator for the linear regression model with ARMA errors. *Econometric Theory,* **8**, 95-111.

[4] Koreisha, S. and Pukkila, T. (1989). Fast linear estimation methods for vector autoregressive moving-average models. *Journal of Time Series Analysis*, **10**, 325-339.

[5] Koreisha, S. and Pukkila T. (1990a). Linear methods for estimating ARMA and regression models with serial correlation. *Communication in Statistics: Simulation,* **19**, 71-102.

[6] Koreisha, S. and Pukkila T. (1990b). A generalized least squares approach for estimating of autoregressive moving average models. *Journal of Time Series Analysis,* **11**, 139-151.

[7] Frutos, R. and Serrano, G. (1997). A generalized least squares estimation method for invertible vector moving average models. *Economics Letters,* **57**, 149-156.

[8] Newbold, P. (1973). Bayesian estimation of Box Jenkins transfer function noise models. *Journal of the Royal Statistical Society, Series B*, **35**, 323 336.

[9] Wei, W. (1990). *Time series analysis univariate and multivariate methods.* Addison Wesley Inc., U.K.

In: New Developments in Applied Statistics
Editor: Mohammad Ahsanullah

ISBN: 978-1-61324-648-1

Chapter 7

CONFIDENCE INTERVALS FOR QUANTILES USING RECORD VALUES FROM THE GENERALIZED EXPONENTIAL DISTRIBUTION

Ayman Baklizi *
Department of Mathematics and Physics, College of Arts and Sciences,
Qatar University, Doha - Qatar

Abstract

Likelihood-based and bootstrap confidence intervals for the quantiles of the generalized exponential distribution based on record values are developed. Intervals for the difference of quantiles based on two independent samples are also considered. Simulation studies are carried out to investigate and compare the performance of the intervals.

Key words: Confidence Interval, Generalized Exponential Distribution, Record Values, Quantiles, Bootstrap Intervals

1. Introduction

The generalized exponential distribution (GED) has been introduced by Gupta and Kundo (1999). It is found to be effective in model lifetime data of various types, Gupta and Kundo (2001a). Parameter estimation was considered by Gupta and Kundu (2001b) and by Raqab and Ahsanullah (2001). Bayesian inference in this model was considered by Raqab and Madi (2005.

Let $X_1, X_2, \ldots$ be a sequence of iid random variables. An observation X_j is called a lower record if its value is less than all previous observations. That is $X_j < X_i$ for every $i < j$. Raqab (2002) studied some inference problems in the GED based on record values

* E-mail address: a.baklizi@qu.edu.qa

while Jaheen considered empirical bayes inference for the GED based on record values. Baklizi (2008a) considered inference about the stress strength reliability in this model with record values.

The probability density function (pdf) and the cumulative distribution function (cdf) of the one parameter $GED(\theta)$ are given by

$$f(x) = \theta e^{-x}\left(1-e^{-x}\right)^{\theta-1}, x>0, \theta>0$$
$$F(x) = \left(1-e^{-x}\right)^{\theta} \tag{1}$$

In section 2 we discussed likelihood inference for the population quantiles based on records, while in section 3 we considered the two sample quantile problem. Bootstrap methods are considered in section 4. A simulation study is described in section 5. The conclusions are given in the final section.

2. Likelihood Inference: The One Sample Case

Let $X \sim GED(\theta)$ and let $x_p = -\ln\left(1-p^{1/\theta}\right)$ be the pth quantile of the distribution of X. We want to construct confidence intervals for x_p based on lower record data. Let $r_0,\ldots,r_n$ be a set of lower records from $GED(\theta)$. The likelihood function is given by (Arnold et al., 1998);

$$L(\theta \mid r_0,\ldots,r_n) = f(r_n)\prod_{i=1}^{n-1} f(r_i)/F(r_i)$$

Where f and F are the pdf and cdf of X. Substituting in the likelihood functions we obtain

$$L(\theta \mid r_0,\ldots,r_n) = \theta^n u_1(r_0,\ldots,r_n) e^{-\theta\tau(r_n)} \tag{2}$$

Where

$$u(r_0,\ldots,r_n) = \prod_{i=1}^{n} \frac{e^{-r_i}}{1-e^{-r_i}} \text{ and } \tau(r_n) = -\ln\left(1-e^{-r_n}\right).$$

The maximum likelihood estimator (MLE) of θ based on the record data is (Jaheen, 2004);

$$\hat{\theta} = \frac{n}{-\ln\left(1-e^{-r_n}\right)} \tag{3}$$

Hence the MLE of x_p is given by

$$\hat{x}_p = -\ln\left(1 - p^{1/\hat{\theta}}\right) \tag{4}$$

Consider $\hat{\theta} = \dfrac{n}{-\ln\left(1 - e^{-r_n}\right)}$, the pdf of R_n is given by (Ahsanullah, 1995)'

$$\begin{aligned} f_{R_n}(r_n) &= f(r_n)[-\ln F(r_n)]^{n-1}/(n-1)! \\ &= \frac{\theta^n e^{-r_n}}{\Gamma(n)}\left(1 - e^{-r_n}\right)^{\theta-1}\left(-\ln\left(1 - e^{-r_n}\right)\right)^{n-1}, r_n > 0 \end{aligned} \tag{5}$$

Consequently, the pdf of $Z = \hat{\theta} = \dfrac{n}{-\ln\left(1 - e^{-r_n}\right)}$ is given by;

$$f_Z(z) = \frac{(n\theta)^n}{\Gamma(n) z^{n+1}} e^{-n/z}, z > 0 \tag{6}$$

Which is the inverted gamma distribution. Note that, by the properties of the inverted gamma distribution and its relation with the gamma distribution we have $\dfrac{n\theta}{Z} \sim Gamma\,(n-1,1)$. Hence $\dfrac{2n\theta}{Z} \sim \chi^2_{2(n-1)}$. This fact can be used to construct the following $(1-\alpha)\%$ confidence interval for x_p;

$$\left\{ \ln\left(1 - p^{\left(\chi^2_{1-\alpha/2,2(n-1)}\frac{z}{2n}\right)^{-1}}\right), \ln\left(1 - p^{\left(\chi^2_{\alpha/2,2(n-1)}\frac{z}{2n}\right)^{-1}}\right) \right\} \tag{7}$$

We can use the asymptotic normality of MLEs to construct another (approximate) confidence interval. Note that

$$\sqrt{n}\left(\hat{\theta} - \theta\right) \xrightarrow{D} N\left(0, v^2\right) \text{ as } n \to \infty, \tag{8}$$

where v^2 is the asymptotic variance given by the inverse of the Fisher information given by

$$\left[-E\left(\frac{\partial^2 \ln L(\theta \mid r_0, \dots r_n)}{\partial \theta^2}\right)\right]^{-1} = \frac{\theta^2}{n}$$

Since $x_p = -\ln\left(1-p^{1/\theta}\right) = h(\theta)$ say, and $\hat{x}_p = -\ln\left(1-p^{1/\hat{\theta}}\right) = h\left(\hat{\theta}\right)$

$$\sqrt{n}\left(\hat{x}_p - x_p\right) = \sqrt{n}\left[h\left(\hat{\theta}\right) - h(\theta)\right] \xrightarrow{D} N\left(0, \eta^2\right) \tag{9}$$

Where $\eta^2 = \left(\frac{dh(\theta)}{d\theta}\right)^2 v^2$ and $\frac{dh(\theta)}{d\theta} = \frac{-p^{1/\theta}\ln p}{\theta^2\left(1-p^{1/\theta}\right)}$. A $(1-\alpha)\%$ confidence interval for x_p based on this asymptotic result is given by

$$\left\{\hat{x}_p - z_{1-\alpha/2}\hat{\eta}, \hat{x}_p + z_{1-\alpha/2}\hat{\eta}\right\} \tag{10}$$

Where $\hat{\eta}$ is obtained by substituting $\hat{\theta}$ in the asymptotic standard deviation η .

3. Likelihood Inference: The Two Sample Case

Let $r_0,\ldots,r_n$ be a set of lower records from $GED(\theta_1)$ and let $s_0,\ldots,s_m$ be an independent set of lower records from $GED(\theta_2)$. Consider estimating $\delta = x_p - y_p = -\ln\left(1-p^{1/\theta_1}\right) - \left(-\ln\left(1-p^{1/\theta_2}\right)\right) = h(\theta_1,\theta_2)$. The likelihood functions are given by (Arnold et al., 1998);

$$L_1(\theta_1 \mid r_0,\ldots,r_n) = f(r_n)\prod_{i=1}^{n-1} f(r_i)/F(r_i)$$

$$L_2(\theta_2 \mid s_0,\ldots,s_m) = g(s_m)\prod_{i=1}^{m-1} g(s_i)/G(s_i)$$

Where f and F are the pdf and cdf of $X \sim GED(\theta_1)$ and g and G F are the pdf and cdf of $Y \sim GED(\theta_2)$. Substituting in the likelihood functions and using the notation of Jaheen (2004) we obtain

$$L_1(\theta_1 \mid r_0,\ldots,r_n) = \theta_1^n u_1(r_0,\ldots,r_n) e^{-\theta_1\tau_1(r_n)}$$

$$L_2(\theta_2 \mid s_0,\ldots,s_m) = \theta_2^m u_2(s_0,\ldots,s_m) e^{-\theta_2\tau_2(s_m)} \tag{11}$$

Where

$$u_1(r_0,\ldots,r_n) = \prod_{i=1}^{n}\frac{e^{-r_i}}{1-e^{-r_i}},\ u_2(s_0,\ldots,s_m) = \prod_{j=1}^{m}\frac{e^{-s_j}}{1-e^{-s_j}},\ \tau_1(r_n) = -\ln\left(1-e^{-r_n}\right)$$

and $\tau_2(s_m) = -\ln\left(1-e^{-s_m}\right)$.

The maximum likelihood estimators (MLE) of θ_1 and θ_2 based on the record data are;

$$\hat{\theta}_1 = \frac{n}{-\ln\left(1-e^{-r_n}\right)},\ \hat{\theta}_2 = \frac{m}{-\ln\left(1-e^{-s_m}\right)} \tag{12}$$

Hence the MLE of $\delta = x_p - y_p$ is given by

$$\hat{\delta} = \hat{x}_p - \hat{y}_p = -\ln\left(1-p^{1/\hat{\theta}_1}\right) - \left(-\ln\left(1-p^{1/\hat{\theta}_2}\right)\right) \tag{13}$$

The exact distribution of $\hat{\delta}$ is quite complex so we use the asymptotic distribution. Note that

$$\sqrt{n}\left(\hat{\theta}_1 - \theta_1\right) \xrightarrow{D} N\left(0, v_1^2\right) \text{ as } n \to \infty$$

where $v_1^2 = \frac{\theta_1^2}{n}$. Similarly;

$$\sqrt{m}\left(\hat{\theta}_2 - \theta_2\right) \xrightarrow{D} N\left(0, v_2^2\right) \text{ as } m \to \infty$$

Where $v_2^2 = \frac{\theta_2^2}{m}$. Let $n \to \infty$, $m \to \infty$ such that $m/n \to p$ where $0 < p < 1$, $h(\theta_1,\theta_2) = -\ln\left(1-p^{1/\theta_1}\right) - \left(-\ln\left(1-p^{1/\theta_2}\right)\right)$, it follows that;

$$\sqrt{n}\left(\hat{\delta} - \delta\right) = \sqrt{n}\left[h\left(\hat{\theta}_1,\hat{\theta}_2\right) - h(\theta_1,\theta_2)\right] \xrightarrow{D} N\left(0,\eta^2\right) \tag{14}$$

Where $\eta^2 = \left(\frac{\partial h(\theta_1,\theta_2)}{\partial \theta_1}\right)^2 v_1^2 + \left(\frac{\partial h(\theta_1,\theta_2)}{\partial \theta_2}\right)^2 v_2^2 / p$, $\frac{dh(\theta_1,\theta_2)}{d\theta_1} = \frac{-p^{1/\theta_1}\ln p}{\theta_1^2\left(1-p^{1/\theta_1}\right)}$ and $\frac{dh(\theta_1,\theta_2)}{d\theta_2} = \frac{-p^{1/\theta_2}\ln p}{\theta_2^2\left(1-p^{1/\theta_2}\right)}$. A $(1-\alpha)\%$ confidence interval for R based on this asymptotic result is given by

$$\left\{\hat{R} - z_{1-\alpha/2}\hat{\eta},\ \hat{R} + z_{1-\alpha/2}\hat{\eta}\right\} \tag{15}$$

Where $\hat{\eta}$ is obtained by substituting m/n for p and the MLEs $\hat{\theta}_1$ and $\hat{\theta}_2$ in the asymptotic standard deviation η.

4. Bootstrap Intervals

Let $\hat{\delta}^*$ be the maximum likelihood estimator calculated from the bootstrap sample. Let z_α^* be the α quantile of the bootstrap distribution of $Z^* = (\hat{\delta}^* - \hat{\delta}) / \hat{\eta}^*$, where $\hat{\eta}^*$ is estimated variance of $\hat{\delta}$ calculated from the bootstrap sample. The bootstrap-*t* interval for δ is given by

$$\left(\hat{\delta} - z_{1-\alpha/2}^* \hat{\eta}\, , \hat{\delta} - z_{\alpha/2}^* \hat{\eta}\right) \tag{16}$$

where z_α^* is determined by simulation. Alternatively we can use the quantiles of the bootstrap distribution of $\varepsilon^* = (\hat{\delta}^* - \hat{\delta}) / sd^*\left(\hat{\delta}^*\right)$ where $sd^*\left(\hat{\delta}^*\right)$ is the estimated standard deviation of $\hat{\delta}^*$ obtained from a further bootstrap sample. The resulting bootstrap-t interval is given by;

$$\left(\hat{\delta} - \varepsilon_{1-\alpha/2}^* \hat{\eta}\, , \hat{\delta} - \varepsilon_{\alpha/2}^* \hat{\eta}\right) \tag{17}$$

We will also consider the percentile intervals obtained from the quantiles of the bootstrap distribution of $\hat{\delta}$.

5. A Simulation Study

To investigate and compare the performance of the likelihood and bootstrap intervals presented in this paper, we used a simulation design similar to the one used by Baklizi (2008b) with n = 5, 10, 15 in the one sample case and $n = m$ = 5, 10, 15 in the two sample case. The records are based on $GED(1)$ in the one sample case and on $GED(1)$ and $GED(2)$ in the two sample case. The confidence level taken is $(1-\alpha) = 0.95$. For each combination of the simulation indices we used 2000 replications. For each replication we calculated the following intervals for x_p, p = 0.1, 0.3, … , 0.9;

1) AN: The interval based on the asymptotic normality of the MLE.
2) PR: The percentile interval
3) T1: The bootstrap-t interval using bootstrap variance estimate based on the asymptotic distribution.
4) T2: The bootstrap-t interval using bootstrap variance estimate based on the nested bootstrap.

Table 1. Coverage probabilities and widths of the intervals for x_p

p	CI	$n=5$				$n=10$				$n=15$			
		L	U	T	W	L	U	T	W	L	U	T	W
0.1	AN	090	014	103	0.552	069	013	082	0.356	068	012	079	0.275
0.1	PR	000	172	172	0.981	-	118	118	0.492	004	077	081	0.346
0.1	T1	209	001	210	0.981	188	-	188	0.492	181	001	182	0.346
0.1	T2	221	001	222	0.433	140	006	146	0.303	104	007	111	0.247
0.3	AN	051	039	090	0.946	033	032	065	0.660	039	026	064	0.532
0.3	PR	000	168	168	1.353	001	111	111	0.788	003	092	094	0.599
0.3	T1	173	002	175	1.353	114	002	116	0.788	109	004	113	0.599
0.3	T2	174	009	183	0.845	075	008	083	0.625	058	011	068	0.519
0.5	AN	025	050	075	1.240	026	041	066	0.863	027	028	055	0.705
0.5	PR	000	180	180	1.617	001	120	121	0.980	003	088	091	0.765
0.5	T1	122	002	123	1.617	097	008	105	0.980	081	003	084	0.765
0.5	T2	126	008	134	1.190	069	011	080	0.854	053	005	057	0.710
0.7	AN	011	063	074	1.472	015	041	056	1.033	016	042	057	0.845
0.7	PR	000	178	178	1.821	001	115	116	1.140	001	100	101	0.899
0.7	T1	087	003	090	1.821	063	006	069	1.140	050	009	059	0.899
0.7	T2	098	010	108	1.481	054	009	062	1.054	041	014	054	0.868
0.9	AN	008	064	072	1.665	015	046	060	1.176	015	048	063	0.960
0.9	PR	000	169	169	1.991	001	107	107	1.274	001	092	093	1.009
0.9	T1	060	005	065	1.991	059	010	068	1.274	046	011	057	1.009
0.9	T2	079	008	086	1.732	057	010	067	1.225	043	013	056	1.001

Table 2. Coverage probabilities and widths of the intervals for $\rho_p = x_p - y_p$

p	CI	$n=5$				$n=10$				$n=15$			
		L	U	T	W	L	U	T	W	L	U	T	W
0.1	AN	028	033	060	1.155	028	033	061	0.777	022	029	051	0.626
0.1	PR	000	000	000	1.835	000	000	000	0.995	000	003	003	0.740
0.1	T1	001	001	001	1.835	003	001	004	0.995	003	009	011	0.740
0.1	T2	007	050	057	1.063	012	053	064	0.727	011	041	052	0.599
0.3	AN	034	024	058	1.626	034	026	060	1.129	027	027	054	0.917
0.3	PR	000	000	000	2.265	000	000	000	1.331	000	005	005	1.024
0.3	T1	006	005	010	2.265	011	010	020	1.331	013	016	028	1.024
0.3	T2	016	021	036	1.615	024	027	050	1.131	018	032	050	0.928
0.5	AN	036	031	067	1.932	029	027	056	1.356	029	026	055	1.106
0.5	PR	000	000	000	2.520	000	000	000	1.538	000	006	006	1.200
0.5	T1	009	008	017	2.520	012	015	027	1.538	018	017	035	1.200
0.5	T2	021	024	045	1.982	016	024	040	1.393	021	026	047	1.142

Table 2. Continued

p	CI	$n=5$				$n=10$				$n=15$			
		L	U	T	W	L	U	T	W	L	U	T	W
0.7	AN	038	032	070	2.175	027	034	060	1.536	027	023	050	1.252
0.7	PR	000	000	000	2.715	000	000	000	1.703	000	005	005	1.338
0.7	T1	015	012	027	2.715	014	025	039	1.703	018	021	039	1.338
0.7	T2	024	027	050	2.270	019	030	049	1.602	019	022	040	1.308
0.9	AN	034	027	061	2.386	026	033	059	1.686	028	029	056	1.376
0.9	PR	000	000	000	2.891	000	000	000	1.841	000	016	016	1.455
0.9	T1	016	011	027	2.891	016	021	037	1.841	021	025	046	1.455
0.9	T2	026	022	048	2.523	021	024	045	1.776	021	026	046	1.445

In the bootstrap calculations we used 1000 bootstrap sequence of records. The nested bootstrap variance estimate is based on a further 25 resamples from the bootstrap sample. The coverage rates and the expected lengths of the intervals are approximated using the results of the 2000 replications. The results of our simulations are given in Tables 1 and 2.

6. Results and Conclusions

For tables 1 and 2, it is clear that the length of the intervals increase as p increases, while it decreases as the sample size increases. This is true for all intervals under study. It also appears that the error rates of the intervals tend to be larger than the nominal ones when the sample size is small, while this violation is small for (AN) intervals, it is large for bootstrap intervals, especially the percentile interval (PR) and the bootstrap-t interval with variance estimates from the Fisher information matrix (T1). The other bootstrap-t interval (T2) has a very similar performance to (AN) for moderate to large sample sizes, however it tends to be longer. All bootstrap intervals are asymmetric, having unequal lower and upper error rates. However, (AN) intervals are reasonably symmetric in most situations under study. A very similar pattern holds in the two sample case. The performance of (AN) interval continues to be the best with (T2) has about a similar performance for moderate to large sample sizes.

References

Ahsanullah, M. (2004). Record values: *Theory and applications*. University Press of America Inc., Lanham, Maryland, USA.

Arnold, B.C., Balakrishnan, N. and Nagaraja, H.N. (1998). *Records*. Wiley.

Baklizi, A. (2008a). Confidence Intervals for the Quantiles of the Weibull Distribution Based on Record Values, *Journal of Applied Statistical Science,* **16**(2), *119 – 130.*

Baklizi, A. (2008b). Likelihood and Bayesian Estimation of Pr(X <Y) Using Lower Record Values From the Generalized Exponential Distribution. *Computational Statistics and Data Analysis,* **52**(7), 3468 – 3473.

Gupta, R.D. and Kundu, D. (2001a). Exponentiated exponential distribution, an alternative to Gamma and Weibull distributions. *Biometrical Journal*, **43**(1), 117-130.

Gupta, R.D. and Kundu, D. (2001b). Exponentiated exponential distribution. Different methods of estimation. *Journal of Statistical Computation and Simulation*, **69**(4), 315-338.

Jaheen, Z.F. (2004). Empirical Bayes inference for generalized exponential distribution based on records. *Communications in Statistics, Theory and Methods*, **33**(8), 1851 – 1861.

Raqab, M.Z. (2002). Inferences for generalized exponential distribution based on record statistics. *Journal of Statistical Planning and Inference*, **104**(2), 339 – 350.

Raqab, M.Z. and Ahsanullah, M. (2001). Estimation of the location and scale parameters of generalized exponential distribution based on order statistics. *Journal of Statistical Computation and Simulation,* **69**(2), 109 – 123.

Raqab, M.Z. and Madi, M.T. (2005). Bayesian inference for the generalized exponential distribution. *Journal of Statistical Computation and Simulation*, **75**(10), 841– 852.

In: New Developments in Applied Statistics
Editor: Mohammad Ahsanullah

ISBN 978-1-61324-648-1

Chapter 8

MARSHALL-OLKIN EXPONENTIAL DISTRIBUTION: MOMENTS OF ORDER STATISTICS

***Mukhtar M. Salah*[a,*], *Mohammad Z. Raqab*[a] *and M. Ahsanullah*[b]**
[a]Department of Mathematics, University of Jordan,
Amman 11942, Jordan;
[b]Department of Management, Rider University,
Lawrenceville, NJ 08648-3099, USA

Abstract

Order statistics from Marshall-Olkin exponential distribution is studied. We derive exact forms for means, variances and covariances of order statistics from Marshall-Olkin exponential distribution. We also establish several recurrence relations satisfied by the single moments and the product moments for order statistics from the Marshall-Olkin exponential distribution. These relations are used in a simple recursive manner in order to evaluate the single and the product moments of all order statistics for all sample sizes.

Keywords: *Order statistics; exponential distribution; Marshall-Olkin exponential distribution; moments of order statistics,* polygamma functions; hypergeometric functions.

AMS Subject Classification: 62G30, 62E15, 60E05.

1. Introduction

Consider the exponential distribution with cumulative distribution function (cdf)

$$H(x) = 1 - e^{-x}, \quad 0 \leq x < \infty. \tag{1}$$

Marshall and Olkin (1997) introduced a powerful method of adding new parameters to an existing distribution. Their method can be summarized as follow: if we have a family of distributions with cdf say $H(x)$, the Marshall-Olkin extended distribution is defined by

$$F(x) = \frac{H(x)}{\alpha + (1-\alpha)H(x)}, \qquad -\infty < x < \infty,\ \alpha > 0, \tag{2}$$

or equivalently if we have a survival function $\bar{H}(x)$, then the Marshall-Olkin extended survival distribution is defined by

$$\bar{F}(x) = \frac{\alpha \bar{H}(x)}{1-(1-\alpha)\bar{H}(x)}, \qquad -\infty < x < \infty,\ \alpha > 0.$$

If we substitute Eq.(1) into Eq.(2) we get the so called Marshall-Olkin exponential distribution. For simplicity, let us denote it by MOE distribution. The cdf of MOE distribution is given as follow:

$$F(x) = 1 - \frac{\alpha e^{-x}}{(1-(1-\alpha)e^{-x})}, \qquad 0 \le x < \infty,\ \alpha > 0. \tag{3}$$

Its probability density function (pdf) is given as follow:

$$f(x) = \frac{\alpha e^{-x}}{(1-(1-\alpha)e^{-x})^2},\ 0 \le x < \infty,\ \alpha > 0. \tag{4}$$

When $\alpha = 1$, the MOE distribution reduces to the standard exponential distribution. The MOE distribution is used recently (Ghitany et al. , 2005) for analyzing censored samples.

Let $X_1, X_2, ..., X_n$ be a random sample of the MOE distribution with cdf and pdf as given in Eqs.(3) and (4) respectively, and let $X_{1:n} \le X_{2:n} \le ... \le X_{n:n}$ be the order statistics obtained from this sample. Denote the single moments $E(X_{r:n}^i)$ by $\mu_{r:n}^{(i)}$ for $1 \le r \le n$ and $i = 0, 1, 2, ...$ and the product moments $E(X_{r:n}^i X_{s:n}^j)$ by $\mu_{r,s}^{(i,j)}$ for $1 \le r < s \le n$, $i = 0, 1, 2, ..., j = 0, 1, 2, ...$ Exact moments of order statistics from other distributions can be found in the literature. For example, one may refer to Raqab (2004) and Raqab and Ahsanullah (2001).

From Eqs.(3) and (4) it can be readily seen the following three relations

$$f(x) = [1-F(x)] + \frac{1-\alpha}{\alpha}[1-F(x)]^2,\ \alpha > 0, \tag{5}$$

$$f(x) = F(x)[1-F(x)] + \frac{1}{\alpha}[1-F(x)]^2,\ \alpha > 0, \tag{6}$$

$$f(x) = \frac{1}{\alpha}\left[1 + (\alpha-2)F(x) + (1-\alpha)F^2(x)\right],\ \alpha > 0. \tag{7}$$

We use these three relations in the following sections in order to establish several recurrence relations satisfied by the single and product moments of all order statistics for all sample sizes via the recursive computations method.

The paper is organized as follow. In the next section, we derive exact expressions for the means, and variances of order statistics from MOE distribution. In Section 3, we establish two recurrence relations for evaluating the single moments of order statistics from the MOE distribution. In Section 4, we establish four recurrence relations for evaluating the product moments of order statistics from the MOE distribution. Finally, in Section 5, we show how the recursive computational algorithm used for computing the single and product moments.

2. Exact Moments for MOE Distribution

Here in, this section, we derive exact expressions for the moments of order statistics from the MOE distribution in terms of polygamma and hypergeometric functions.

The pdf of $X_{i:n}$ $(i = 1, 2, ..., n)$ can be written (see, Arnold *et al.*, 1992, p.10) as

$$f_{i:n}(x) = \frac{n!}{(i-1)!(n-i)!}[F(x)]^{i-1}[1-F(x)]^{n-i}f(x), \quad -\infty < x < \infty, \tag{8}$$

where $F(x)$ and $f(x)$ are given in Eqs.(3) and (4) respectively. The first moment of the MOE distribution is given by

$$\mu_{1:1} = E(X_{1:1}) = E(X) = \frac{\alpha ln\ \alpha}{\alpha - 1}, \quad \alpha > 0,\ \alpha \neq 1 \tag{9}$$

From Eq. (8), the MGF of $X_{i:n}$ can be written as

$$M_{i:n}(t) = E(e^{tX_{i:n}}) = \frac{n!}{(i-1)!(n-i)!}\int_0^\infty e^{tx}[F(x)]^{i-1}[1-F(x)]^{n-i}f(x)dx. \tag{10}$$

Substituting the cdf and pdf in Eqs.(3) and (4) respectively into Eq.(10) and simplifying the result, we obtain

$$M_{i:n}(t) = \frac{n!}{(i-1)!(n-i)!}\alpha^{n-i+1}\sum_{r=0}^{i-1}(-1)^r\binom{i-1}{r}\int_0^\infty \frac{e^{-(n-t-r)x}}{(1-(1-\alpha)\,e^{-x})}dx.$$

Taking the transformation $w = (\alpha - 1)e^{-x}$, we can get

$$M_{i:n}(t) = \frac{n!}{(i-1)!(n-i)!}\alpha^{n-i+1}\sum_{r=0}^{i-1}\frac{(-1)^r\binom{i-1}{r}}{(\alpha-1)^{(n-t-r)}}\int_0^{(\alpha-1)}\frac{w^{(n-t-r)-1}}{(1+w)^{n+1}}dw.$$

Noteworthy here is that the above MGF can be expressed in terms of the hypergeometric function. With ${}_pF_q(\gamma_1, ..., \gamma_p; \beta_1, ..., \beta_q; u)$ denoting the generalized hypergeometric function

$${}_pF_q(\gamma_1, ..., \gamma_p; \beta_1, ..., \beta_q; u) = \sum_{i=0}^{\infty}\frac{(\gamma_1)_i \cdots (\gamma_p)_i\, u^i}{(\beta_1)_i \cdots (\beta_q)_i\, i!}, \tag{11}$$

for the nonnegative integers p and q, with

$$(\gamma)_i = \gamma(\gamma+1)\ldots(\gamma+i-1) = \frac{\Gamma(\gamma+i)}{\Gamma(\gamma)}, \ \gamma \neq 0, i = 1, 2, ...$$

The hypergeometric functions ${}_2F_1$ and ${}_3F_2$ are special cases of the generalized hypergeometric functions. Using the fact that (see, Gradshteyn *el at.*, 1994)

$$\int_0^u \frac{x^{\kappa-1}}{(1+\beta x)^\theta}dx = \frac{u^\kappa}{\kappa}\,{}_2F_1(\theta, \kappa; 1+\kappa; -\beta u), \ \kappa > 0,$$

we immediately obtain the moment generating function of $X_{i:n}$ from MOE distribution as

$$M_{i:n}(t) = \frac{n!}{(i-1)!(n-i)!}\alpha^{n-i+1}\sum_{r=0}^{i-1}\frac{(-1)^r\binom{i-1}{r}}{(n-t-r)}\,{}_2F_1(n+1, n-t-r\,; n-t-r+1\,; 1-\alpha). \tag{12}$$

The derivative of $(t)_k$ with respect to t is derived to be

$$\frac{d\,(t)_k}{dt} = [\Psi\,(t+k) - \Psi\,(t)]\,(t)_k\,,$$

where $\Psi\,(t)$ is diagamma function defined by $\Psi\,(t) = \frac{d}{dt}(\ln\,\Gamma(t)) = \acute{\Gamma}(t)/\,\Gamma(t),\ t \neq 0, -1, -2, ...$ As special cases, $\Psi\,(1) = -\gamma;\ \gamma = 0.577215...$ is the Euler's constant and $\acute{\Psi}\,(1) = \pi^2/6$. Upon using Eq.(11) we can write Eq.(12) as

$$\begin{aligned} M_{i:n}(t) &= \frac{n!}{(i-1)!(n-i)!}\alpha^{n-i+1}\sum_{j=0}^{\infty}\sum_{r=0}^{i-1}\frac{(-1)^r\binom{i-1}{r}(n+1)_j\,(1-\alpha)^j}{j!} \\ &\quad \frac{(n-t-r)_j}{(n-t-r)\,(n-t-r+1)_j}. \end{aligned} \tag{13}$$

On differentiating Eq.(13) with respect to t and simplifying the result, we get

$$\mu_{i:n} = \frac{\partial M_{i:n}(t)}{\partial t}\Big|_{t=0} = \frac{n!\alpha^{n-i+1}}{(i-1)!(n-i)!}\sum_{j=0}^{\infty}\sum_{r=0}^{i-1}\frac{(-1)^r\binom{i-1}{r}(n+1)_j}{j!(n-r)(n-r+j)}\Phi\,(n,r,j)\,, \tag{14}$$

where

$$\Phi\,(n,r,j) = 1 + (n-r)[\Psi(n-r+1+j) - \Psi\,(n-r+j) + \Psi\,(n-r) - \Psi\,(n-r+1)].$$

To derive the variance of $X_{i:n}$, we differentiate Eq.(13) with respect to t twice to get

$$\mu_{i:n}^{(2)} = \frac{\partial^2 M_{i:n}(t)}{\partial t^2}\Big|_{t=0} = \frac{n!\alpha^{n-i+1}}{(i-1)!(n-i)!}\sum_{j=0}^{\infty}\sum_{r=0}^{i-1}\frac{(-1)^r\binom{i-1}{r}(n+1)_j}{j!}G\,(n,r,j)\,, \tag{15}$$

where

$$\begin{aligned} G\,(n,r,j) \;=\; & \{\{(n-r)_j[\Psi(n-r) - \Psi(n-r+j)][(n-r)[\Psi(n-r+1+j) - \Psi(n-r+j) \\ & -\Psi(n-r+1) + \Psi(n-r)] + 1] + (n-r)_j[(n-r)[\Psi'(n-r) - \Psi'(n-r+1) \\ & +\Psi'(n-r+1+j) - \Psi'(n-r+j)] + \Psi(n-r+1) - \Psi(n-r) \\ & -\Psi(n-r+1+j) + \Psi(n-r+j)]\}(n-r)^2(n-r+1)_j - (n-r)_j[(n-r) \\ & \times[\Psi(n-r+1+j) - \Psi(n-r+j) - \Psi(n-r+1) + \Psi(n-r)] + 1] \\ & \times[(n-r)^2(n-r+1)_j[\Psi(n-r+1) - \Psi(n-r+1+j)] - 2(n-r)(n-r+1)_j]\} \\ & \times\frac{1}{(n-r)^4[(n-r+1)_j]^2}. \end{aligned}$$

Therefore, the variance of $X_{i:n}$ is given by

$$Var(X_{i:n}) = \beta_{i,i:n} = \mu_{i:n}^{(2)} - (\mu_{i:n})^2, \tag{16}$$

where $\mu_{i:n}$ and $\mu_{i:n}^{(2)}$ are defined in Eqs.(14) and (15) respectively.

3. Relations for Single Moments

In this section, we introduce some simple recurrence relations satisfied by the single moments of order statistics based on Eqs.(5) - (7).

Theorem 1 *For the MOE distribution with cdf and pdf as given in Eqs.(3) and (4) respectively, we have*

$$\mu_{1:n+1}^{(k+1)} = \frac{\alpha}{1-\alpha}\left[\frac{k+1}{n}\mu_{1:n}^{(k)} - \mu_{1:n}^{(k+1)}\right], \; n \geq 1, k \geq 0, \; \alpha > 0, \; \alpha \neq 1 \qquad (17)$$

with $\mu_{1:n}^{(0)} = 1$.

Proof. For $n \geq 1$, we have from Eq.(8) that

$$\mu_{1:n}^{(k)} = n\int_0^\infty x^k \left[1 - F(x)\right]^{n-1} f(x)dx.$$

Upon using the relation in Eq.(5), we have

$$\begin{aligned} \mu_{1:n}^{(k)} &= n\int_0^\infty x^k \left[1 - F(x)\right]^n \left[1 + \frac{1-\alpha}{\alpha}\left[1 - F(x)\right]\right] dx \qquad (18) \\ &= n\int_0^\infty x^k \left[1 - F(x)\right]^n dx + n\frac{(1-\alpha)}{\alpha}\int_0^\infty x^k \left[1 - F(x)\right]^{n+1} dx. \end{aligned}$$

Integrating the RHS of Eq.(18) by parts, treating x^k for integration and the rest of the integrand for differentiation, we get

$$\begin{aligned} \mu_{1:n}^{(k)} &= \frac{n}{k+1}[n\int_0^\infty x^{k+1}\left[1 - F(x)\right]^{n-1} f(x)dx \\ &\quad + (n+1)\frac{(1-\alpha)}{\alpha}\int_0^\infty x^{k+1}\left[1 - F(x)\right]^n f(x)dx]. \qquad (19) \end{aligned}$$

Simplifying Eq.(19) yields the result. ■

Theorem 2 *For the MOE distribution with cdf and pdf as given in Eqs.(3) and (4) respectively, we have*

$$\begin{aligned} \mu_{i+1:n+1}^{(k+1)} &= \frac{1}{i}[\frac{(n+1)(k+1)}{n-i+1}\mu_{i:n}^{(k)} - \frac{n-\alpha i+1}{\alpha}\mu_{i:n+1}^{(k+1)} \\ &\quad + \frac{n+1}{\alpha}\mu_{i-1:n}^{(k+1)}], \; 1 \leq i \leq n, \; n \geq 1, \; k \geq 0, \; \alpha > 0, \qquad (20) \end{aligned}$$

with $\mu_{0:r}^{(k)} \equiv 0$ *for* $r \geq 1, k \geq 0$ *and* $\mu_{i:r}^{(0)} \equiv 1$ *for* $1 \leq i \leq r$.

Proof. From Eq.(8), we write the *kth* moment of the $X_{i:n}$ $(1 \leq i \leq n,\, k \geq 0)$, as

$$\mu_{i:n}^{(k)} = \frac{n!}{(i-1)!(n-i)!} \int_0^\infty x^k \left[F(x)\right]^{i-1} \left[1 - F(x)\right]^{n-i} f(x) dx. \tag{21}$$

By substituting the relation in Eq.(6) into Eq.(21), we obtain

$$\begin{aligned} \mu_{i:n}^{(k)} &= \frac{n!}{(i-1)!(n-i)!} \left[\int_0^\infty x^k \left[F(x)\right]^{i} \left[1 - F(x)\right]^{n-i+1} dx \right. \\ &\quad \left. + \frac{1}{\alpha} \int_0^\infty x^k \left[F(x)\right]^{i-1} \left[1 - F(x)\right]^{n-i+2} dx \right]. \end{aligned} \tag{22}$$

Let

$$\begin{aligned} I_1 &= \frac{n!}{(i-1)!(n-i)!} \int_0^\infty x^k \left[F(x)\right]^{i} \left[1 - F(x)\right]^{n-i+1} dx, \\ I_2 &= \frac{1}{\alpha} \frac{n!}{(i-1)!(n-i)!} \int_0^\infty x^k \left[F(x)\right]^{i-1} \left[1 - F(x)\right]^{n-i+2} dx. \end{aligned}$$

Integrating I_1 by parts, x^k for integration and the rest of the integrand for differentiation with $F(\infty) = 1, F(0) = 0$, we get

$$I_1 = \frac{i\,(n-i+1)}{(n+1)\,(k+1)} \left[\mu_{i+1:n+1}^{(k+1)} - \mu_{i:n+1}^{(k+1)}\right].$$

Similarly, I_2 can be written as

$$I_2 = \frac{1}{\alpha} \frac{(n-i+1)\,(n-i+2)}{(n+1)\,(k+1)} \left[\mu_{i:n+1}^{(k+1)} - \mu_{i-1:n+1}^{(k+1)}\right].$$

Substituting I_1 and I_2 into Eq.(22) to yield

$$\begin{aligned} \mu_{i:n}^{(k)} &= \frac{i\,(n-i+1)}{(n+1)\,(k+1)} \left[\mu_{i+1:n+1}^{(k+1)} - \mu_{i:n+1}^{(k+1)}\right] \\ &\quad + \frac{1}{\alpha} \frac{(n-i+1)\,(n-i+2)}{(n+1)\,(k+1)} \left[\mu_{i:n+1}^{(k+1)} - \mu_{i-1:n+1}^{(k+1)}\right]. \end{aligned} \tag{23}$$

From Relation 3.3.1 (see, Balakrishnan *et al.*,1991, p.24) for $i = 1, 2, ..., n-1$ and for $k \geq 1$ that

$$i\mu_{i+1:n}^{(k)} + (n-i)\mu_{i:n}^{(k)} = n\mu_{i:n-1}^{(k)},$$

we have

$$(n-i+2) \left[\mu_{i:n+1}^{(k+1)} - \mu_{i-1:n+1}^{(k+1)}\right] = (n+1) \left(\mu_{i:n+1}^{(k+1)} - \mu_{i-1:n}^{(k+1)}\right). \tag{24}$$

Substituting Eq.(24) into Eq.(23), we immediately obtain

$$\frac{(n+1)\,(k+1)}{i\,(n-i+1)} \mu_{i:n}^{(k)} = \mu_{i+1:n+1}^{(k+1)} - \mu_{i:n+1}^{(k+1)} + \frac{(n+1)}{\alpha i} \mu_{i:n+1}^{(k+1)} - \frac{(n+1)}{\alpha i} \mu_{i-1:n}^{(k+1)}. \tag{25}$$

From Eq.(25), the result follows. ∎

Remark 1: Marshall *et al.* (1997) have proved that

$$\mu_{1:1}^{(k)} = E(X^k) = \Gamma(k+1)\,\alpha \sum_{j=0}^{\infty} \frac{(1-\alpha)^j}{(j+1)^k},\ 0 < \alpha < 2. \tag{26}$$

Applying Eq. (17) recursively, we immediately obtain

$$\begin{aligned}\mu_{1:n+1}^{(k+1)} &= \frac{\alpha(k+1)}{1-\alpha} \sum_{i=0}^{n-1} (-1)^i \left(\frac{\alpha}{1-\alpha}\right)^i \frac{\mu_{i:n-i}^{(k)}}{n-i} + (-1)^n \left(\frac{\alpha}{1-\alpha}\right)^n \mu_{1:1}^{(k+1)}, \\ &\alpha > 0,\ \alpha \neq 1,\ n \geq 1.\end{aligned}$$

Using the moments in Eq.(9) and Eq.(26) and Theorem 1 and 2, one can start computing the *k* raw moments of all order statistics for all sample sizes in a very simple recursive way.

4. Relations for Product Moments

In this section, we present four simple recurrence relations for the product moments of order statistics from MOE distribution. The joint density function (see, Arnold *et al.*, 1992, p. 16) of $X_{i:n}, X_{j:n}$ $(1 \leq i < j \leq n)$, is

$$\begin{aligned} f_{i,j:n}(x,y) &= c_{i,j:n} \left[F(x)\right]^{i-1} \left[F(y) - F(x)\right]^{j-i-1} \\ &\quad \times \left[1 - F(y)\right]^{n-j} f(x) f(y), \\ &\qquad -\infty < x < y < \infty, \end{aligned} \tag{27}$$

where

$$c_{i,j:n} = \frac{n!}{(i-1)!\,(j-i-1)!\,(n-j)!},$$

and $F(x)$ and $f(x)$ are defined in Eqs.(3), (4) respectively.

Theorem 3 *For the MOE distribution, we have*

$$\begin{aligned}\mu_{i,i+1:n+1} &= \mu_{i:n+1}^{(2)} + \frac{\alpha\,(n+1)}{(1-\alpha)\,(n-i+1)} \left[\mu_{i:n}^{(2)} - \mu_{i,i+1:n} + \frac{1}{n-i}\mu_{i:n}\right], \\ &1 \leq i \leq n-1,\ n \geq 2,\ \alpha > 0,\ \alpha \neq 1,\end{aligned} \tag{28}$$

Proof. From Eq.(27), for $1 \leq i \leq n-1$, we may write

$$\begin{aligned}\mu_{i:n} &= E(X_{i:n} X_{i+1:n}^0) \\ &= c_{i,i+1:n} \int_0^{\infty} x \left[F(x)\right]^{i-1} f(x) D(x) dx,\end{aligned} \tag{29}$$

where

$$D(x) = \int_x^\infty [1 - F(y)]^{n-i-1} f(y)dy. \tag{30}$$

Now substituting Eq.(5) into Eq.(30) and integrating by parts, we obtain

$$\begin{aligned} D(x) &= (n-i)\int_x^\infty y\,[1-F(y)]^{n-i-1} f(y)dy - x[1-F(x)]^{n-i} \\ &+\frac{(1-\alpha)}{\alpha}(n-i+1)\int_x^\infty y\,[1-F(y)]^{n-i} f(y)dy - \frac{(1-\alpha)}{\alpha}x[1-F(x)]^{n-i+1}. \end{aligned} \tag{31}$$

By substituting Eq.(31) into Eq.(29), we obtain

$$\begin{aligned} \mu_{i:n} &= c_{i,i+1:n}\Bigg[(n-i)\int_0^\infty\int_x^\infty xy\,[F(x)]^{i-1}\,[1-F(y)]^{n-i-1}\,f(x)\,f(y)\,dy\,dx \\ &- \int_0^\infty x^2\,[F(x)]^{i-1}\,[1-F(x)]^{n-i}\,f(x)\,dx \\ &+\frac{(1-\alpha)}{\alpha}(n-i+1)\int_0^\infty\int_x^\infty xy\,[F(x)]^{i-1}\,[1-F(y)]^{n-i}\,f(x)\,f(y)\,dy\,dx \\ &-\frac{(1-\alpha)}{\alpha}\int_0^\infty x^2\,[F(x)]^{i-1}\,[1-F(x)]^{n-i+1}\,f(x)\,dx\Bigg]. \end{aligned} \tag{32}$$

On simplifying Eq.(32), the result in Eq.(28) follows. ■

Theorem 4 *For the MOE distribution with the cdf and pdf given in Eqs.(3) and (4), respectively, we have*

$$\begin{aligned} \mu_{2,3:n+1} &= \mu^{(2)}_{3:n+1} + \frac{n+1}{2(1-\alpha)}\left[n\mu^{(2)}_{1:n-1} - \alpha\mu_{2:n}\right] + \frac{(\alpha-2)(n+1)}{2(1-\alpha)} \\ &\cdot\left[\mu^{(2)}_{2:n} - \mu_{1,2:n}\right],\ n\geq 2,\ \alpha>0, \alpha\neq 1, \end{aligned} \tag{33}$$

and

$$\begin{aligned} \mu_{i+1,i+2:n+1} &= \mu^{(2)}_{i+2:n+1} + \frac{n+1}{i(i+1)(1-\alpha)}\left[-\alpha\mu_{i+1:n} + n\left(\mu^{(2)}_{i:n-1} - \mu_{i-1,i:n-1}\right)\right] \\ &+\frac{(\alpha-2)(n+1)}{(i+1)(1-\alpha)}\left[\mu^{(2)}_{i+1:n} - \mu_{i,i+1:n}\right], \\ &2\leq i\leq n-1,\ n\geq 3,\ \alpha>0,\ \alpha\neq 1. \end{aligned} \tag{34}$$

Proof. From Eq.(27), we can write for $1 \leq i \leq n-1$,

$$\begin{aligned} \mu_{i+1:n} &= E(X^0_{i:n}X_{i+1:n}) \\ &= c_{i,i+1:n}\int_0^\infty y\,[1-F(y)]^{n-i-1} f(y)L(y)dy, \end{aligned} \tag{35}$$

where

$$L(y) = \int_0^y [F(x)]^{i-1} f(x)dx. \tag{36}$$

Substituting Eq.(7) into Eq.(36) and integrating by parts, we obtain the following

$$\begin{aligned} L(y) &= \frac{1}{\alpha}\Big[y\,[F(y)]^{i-1} - (i-1)\int_0^y x\,[F(x)]^{i-2} f(x)dx + (\alpha-2)\; y\; [F(y)]^{i} \\ &\quad -i\,(\alpha-2)\int_0^y x\,[F(x)]^{i-1}\; f(x)\; dx \\ &\quad +(1-\alpha)\; y\; [F(y)]^{i+1} - (1-\alpha)\,(i+1)\int_0^y x\,[F(x)]^{i}\; f(x)\; dx\Big]. \end{aligned} \tag{37}$$

For $i = 1$ in Eqs.(35) and (37) with simplifying the result we obtain Eq.(33). For $2 \le i \le n-1$, substituting Eq.(37) into Eq.(35) we have

$$\begin{aligned} \alpha\mu_{i+1:n} &= c_{i,i+1:n}\Big[\int_0^\infty y^2\,[F(y)]^{i-1}\,[1-F(y)]^{n-i-1}\, f(y)dy \\ &\quad -(i-1)\int_0^\infty\int_0^y xy\,[F(x)]^{i-2}\,[1-F(y)]^{n-i-1}\, f(x)\; f(y)\; dy\; dx \\ &\quad +(\alpha-2)\int_0^\infty y^2\,[F(y)]^{i}\,[1-F(y)]^{n-i-1}\; f(y)\; dy \\ &\quad -i\,(\alpha-2)\int_0^\infty\int_0^y xy\,[F(x)]^{i-1}\,[1-F(y)]^{n-i-1}\, f(x)\; f(y)\; dy\; dx \\ &\quad +(1-\alpha)\int_0^\infty y^2\,[F(y)]^{i+1}\,[1-F(y)]^{n-i-1}\; f(y)\; dy \\ &\quad -(1-\alpha)\,(i+1)\int_0^\infty\int_0^y xy\,[F(x)]^{i}\,[1-F(y)]^{n-i-1}\; f(x)\; f(y)\; dy\; dx\Big]. \end{aligned} \tag{38}$$

One can easily simplify Eq.(38) to get the result in Eq.(34). ■

Theorem 5 *For the MOE distribution with pdf given in Eq.(4), we have*

$$\mu_{i,j:n+1} = \mu_{i,j-1:n+1} + \left(\frac{\alpha}{1-\alpha}\right)\left(\frac{n+1}{n-j+2}\right)\left[\mu_{i,j-1:n} - \mu_{i,j:n} + \frac{1}{n-j+1}\mu_{i:n}\right],$$
$$1 \le i < j \le n,\, j-i \ge 2,\ \alpha > 0,\ \alpha \ne 1. \tag{39}$$

Proof. From the joint density of $X_{i:n}$, and $X_{j:n}$ in Eq.(27), we may write, for $1 \le i < j \le n$

$$\begin{aligned} \mu_{i:n} &= E(X_{i:n}X^0_{j:n}) \\ &= c_{i,j:n}\int_0^\infty x\,[F(x)]^{i-1}\, I(x) f(x)dx, \end{aligned} \tag{40}$$

where

$$I(x)=\int_{x}^{\infty}\left[F(y)-F(x)\right]^{j-i-1}\left[1-F(y)\right]^{n-i}f(y)dy. \tag{41}$$

Upon using Eq.(6) and the fact that $F(y)=1-[1-F(y)]$ and integrating by parts we can simply write for $j-i\geq 2$ that

$$\begin{aligned}I(x) &= (n-j+1)\int_{x}^{\infty}y\left[F(y)-F(x)\right]^{j-i-1}\left[1-F(y)\right]^{n-i}f(y)\,dy\\ &-(j-i-1)\int_{x}^{\infty}y\left[F(y)-F(x)\right]^{j-i-2}\left[1-F(y)\right]^{n-i+1}f(y)\,dy\\ &+\left(\frac{1-\alpha}{\alpha}\right)(n-j+2)\int_{x}^{\infty}y\left[F(y)-F(x)\right]^{j-i-1}\left[1-F(y)\right]^{n-i+1}f(y)\,dy\\ &-\left(\frac{1-\alpha}{\alpha}\right)(j-i-1)\int_{x}^{\infty}y\left[F(y)-F(x)\right]^{j-i-2}\left[1-F(y)\right]^{n-i+2}f(y)\,dy.\end{aligned} \tag{42}$$

The result in (39) follows by substituting Eq.(42) into Eq.(40). ■

Theorem 6 *For the MOE distribution, we have*

$$\begin{aligned}\mu_{2,j+1:n+1} &= \mu_{3,j+1:n+1}+\frac{n+1}{2(1-\alpha)}\left[n\mu_{1,j-1:n-1}-\alpha\mu_{j:n}\right]\\ &+\frac{(n+1)(\alpha-2)}{2(1-\alpha)}\left[\mu_{2,j:n}-\mu_{1,j:n}\right],\\ &3\leq j\leq n, \alpha>0,\ \alpha\neq 1,\end{aligned} \tag{43}$$

and

$$\begin{aligned}\mu_{i+1,j+1:n+1} &= \mu_{i+2,j+1:n+1}+\frac{n+1}{(1-\alpha)\,i\,(i+1)}\left[n(\mu_{i,j-1:n-1}-\mu_{i-1,j-1:n-1})-\alpha\mu_{j:n}\right]\\ &+\frac{(n+1)(\alpha-2)}{(1-\alpha)(i+1)}\left[\mu_{i+1,j:n}-\mu_{i,j:n}\right],\\ &2\leq i\leq n-2, j-i\geq 2,\ \alpha>0,\ \alpha\neq 1.\end{aligned} \tag{44}$$

Proof. From the joint density function of $X_{i:n}$ and $X_{j:n}$ in Eq.(27), we may write, for $1\leq i<j\leq n$,

$$\begin{aligned}\mu_{j:n} &= E(X_{i:n}^{0}X_{j:n})\\ &= c_{i,j:n}\int_{0}^{\infty}y[1-F(y)]^{n-j}f(y)L(y)dy,\end{aligned} \tag{45}$$

where

$$L(y)=\int_{0}^{y}[F(x)]^{i-1}[F(y)-F(x)]^{j-i-1}f(x)dx.$$

Upon using the relation in Eq.(7) and integrating by part, we obtain for $i = 1, 3 \le j \le n$

$$\begin{aligned}
L(y) &= \frac{1}{\alpha}\Bigg[(j-2)\int_0^y x[F(y)-F(x)]^{j-3} f(x)dx \\
&\quad +(\alpha-2)(j-2)\int_0^y xF(x)[F(y)-F(x)]^{j-3} f(x)dx \\
&\quad -(\alpha-2)\int_0^y x[F(y)-F(x)]^{j-2} f(x)dx \\
&\quad +(1-\alpha)(j-2)\int_0^y x[F(x)]^2[F(y)-F(x)]^{j-3} f(x)dx \\
&\quad -2(1-\alpha)\int_0^y xF(x)[F(y)-F(x)]^{j-2} f(x)dx\Bigg]. \qquad (46)
\end{aligned}$$

For $j - i \ge 2$, we have

$$L(y) = \frac{1}{\alpha}\left[L_1 + (\alpha-2)L_2 + (1-\alpha)L_3\right], \qquad (47)$$

where

$$\begin{aligned}
L_1 &= (j-i-1)\int_0^y x\,[F(x)]^{i-1}\,[F(y)-F(x)]^{j-i-2} f(x)dx \\
&\quad -(i-1)\int_0^y x\,[F(x)]^{i-2}\,[F(y)-F(x)]^{j-i-1} f(x)dx, \\
L_2 &= (j-i-1)\int_0^y x\,[F(x)]^{i}\,[F(y)-F(x)]^{j-i-2} f(x)dx \\
&\quad -i\int_0^y x\,[F(x)]^{i-1}\,[F(y)-F(x)]^{j-i-1} f(x)dx, \\
L_3 &= (j-i-1)\int_0^y x\,[F(x)]^{i+1}\,[F(y)-F(x)]^{j-i-2} f(x)dx \\
&\quad -(i+1)\int_0^y x\,[F(x)]^{i}\,[F(y)-F(x)]^{j-i-1} f(x)dx.
\end{aligned}$$

The recurrence relations in Eqs.(43) and (44) follow upon substituting the expressions of $L(y)$ in Eqs.(46) and (47), respectively, into Eq.(45) and then simplifying the resulting equations. ■

Remark 2: With the help of the fact that (see, Arnold *et al.*, 1992, p. 111)

$$\mu_{1,2:2} = E(X_{1:2}X_{2:2}) = (\mu_{1:1})^2 = \left(\frac{\alpha\ Ln\ \alpha}{1-\alpha}\right)^2, \ \alpha > 0,\ \alpha \ne 1, \qquad (48)$$

the relationships proved in Theorems 3-6, enable one to evaluate all the product moments of all order statistics for all sample sizes in a computational recursive manner. Further, one can also note that the recurrence relations for the half logistic distribution can be obtained as special cases when $\alpha = 2$ (see, for example, Balakrishnan (1985)).

5. Recursive Computational Algorithm

By starting with the values of $\mu_{1:1} = E(X)$ and $\mu_{1:1}^{(2)} = E(X^2)$ which are given in Eqs.(9) and (26) respectively, $\mu_{1:2}$ and $\mu_{1:2}^{(2)}$ can be determined from Eq.(17), while $\mu_{2:2}$ and $\mu_{2:2}^{(2)}$ can be evaluated from Eq.(20). For the sample of size 2, $\mu_{1:3}$ and $\mu_{1:3}^{(2)}$ can be determined from Eq.(17), $\mu_{2:3}$ and $\mu_{2:3}^{(2)}$ are obtained from Eq.(20). This process may be followed similarly to determine $\mu_{i:n}$ and $\mu_{i:n}^{(2)}$ and for $n = 4, 5, \ldots$. For these values, variances of order statistics can be readily computed.

From (48), $\mu_{1,2:2} = (\mu_{1:1})^2$, $\mu_{1,2:3}$ can be determined from Eq.(28), $\mu_{2,3:3}$ can be obtained from Eq.(33), $\mu_{1,3:3}$ from Theorem 5.3.5 (see, Arnold *et al.*, 1992, p. 115). This process may be followed similarly to determined $\mu_{i,j:n}$ for $1 \le i < j \le n$ and for $n = 4, 5, \ldots$. For these values, covariances of order statistics can be readily computed. The numerical computations for the means, variances, and covariances are not presented here but they will be presented in another future work in the context of estimation problems.

Acknowledgement

This research work is a part of the Ph. D thesis of Mukhtar Salah worked under the supervision of Professor Mohammad Raqab at the University of Jordan. The first and second authors would like to thank the University of Jordan for supporting this research work.

References

[1] Arnold, B.C., Balakrishnan, N. and Nagaraja, H.N. (1992). *A First Course in Order Statistics,* John Wiley, New York.

[2] Balakrishnan, N. (1985). Order Statistics from the Half Logistic Distribution. *Journal of Statistical Computation and Simulation* **20**, 287-309.

[3] Ghitany, M. E., Al-Hussaini, E. K. and A-Jarallah (2005). Marshall-Olkin Extended Weibull Distribution and its Application to Censored Data, *Journal of Applied Statistics* **32**(10), 1025-1034.

[4] Gradshteyn, I. S. and Ryzhik, I. M. (1994). *Tables of Integrals, Series, and Products*, 5th Edition, Alan Jeffrey, Ed., Academic Press, San Diego, USA.

[5] Marshall, A. W. and Olkin, I. (1997). A new Method for Adding a Parameter to a Family of Distributions with Application to the Exponential and Weibull Families, *Biometrika* **84**, 641-652.

[6] Raqab, M. Z. (2004). Generalized Exponential Distribution: Moments of Order Statistics, *Statistics* **38**(1), 29-41.

[7] Raqab, M. Z. and Ahsanullah, M. (2001). Estimation of Location and Scale Parameters of Generalized Exponential Distribution Based on Order Statistics, *Journal of Statistical Computation and Simulation* **69**(2), 109-124.

In: New Developments in Applied Statistics
Editor: Mohammad Ahsanullah

ISBN 978-1-61324-648-1

Chapter 9

SHRINKAGE TEST - ESTIMATION FOR MEASURE OF DISPERSION IN THE INVERSE GAUSSIAN MODEL

Gyan Prakash*[*] *and D.C. Singh
Department of Statistics, Harish Chandra P. G. College,
Varanasi, U. P., India

Abstract

This paper deals with the problem of the shrinkage test - estimation for the Measure of Dispersion of the inverse Gaussian distribution when a prior point information of the parameter is available. The performances of the shrinkage test - estimators based on a single stage and double stage samples are studied.

Keywords: Shrinkage factor; Shrinkage test - estimator; Cost function; Level of significance; Relative bias; Relative efficiency and Effective interval.

1. Introduction

The inverse Gaussian distribution plays an important role in reliability theory and life testing problems. It has useful applications in a wide variety of fields such as Biology, Economics and Medicine. For details see Folks and Chhikara (1978, 1989) and Seshadri (1998). The probability density function of x, the inverse Gaussian variate with the parameter μ (Measure of Location) and Measure of Dispersion $\frac{1}{\lambda} = \sigma$ (say) is given by

$$f\left(x;\mu,\lambda\right) = \sqrt{\frac{\lambda}{2\pi x^3}}\, exp\left(-\frac{\lambda\left(x-\mu\right)^2}{2\mu^2 x}\right) \;\; ; x>0,\ \mu>0,\ \lambda>0. \tag{1}$$

Let $x_1, x_2, x_3, ---, x_n$ be a random sample of size n drawn from inverse Gaussian distribution (1.1). The maximum likelihood estimates of μ and λ are

$$\hat{\mu} = \frac{1}{n}\sum_{i=1}^{n} x_i = \bar{x} \quad and \quad \hat{\lambda} = \frac{n}{v} \;; \tag{2}$$

[*]E-mail address: ggyanji@yahoo.com. (Corresponding author)

where $v = \sum_{i=1}^{n} \left(\frac{1}{x_i} - \frac{1}{\bar{x}} \right)$.

$\bar{x}$ and $\tilde{\sigma} = \frac{v}{n-1}$ are unbiased estimate of μ and σ respectively. It is well known that the distribution of $\bar{x}$ is the inverse Gaussian with parameter μ and $n\lambda$ and λv is distributed as a chi - square with $(n-1)$ degrees of freedom. Also, $\bar{x}$ and v are the stochastically independent and are jointly sufficient for μ and λ if both are unknown.

Varian (1975) and Zellner (1986) proposed an asymmetric loss function called as the LINEX loss function and its invariant form for any parameter θ is given as

$$L(\Delta) = \left\{ e^{(a\Delta)} - a\Delta - 1 \right\} \; ; \; a \neq 0 \; and \; \Delta = \left(\frac{\hat{\theta} - \theta}{\theta} \right). \tag{3}$$

Here $\hat{\theta}$ is any estimate of the parameter θ. The sign and magnitude of $'a'$ represents the direction and degree of asymmetry respectively. The positive (negative) value of $'a'$ is used when overestimation is more (less) serious than underestimation. For $'a'$ close to zero, the LINEX loss is approximately square error loss and therefore almost symmetric. Sadooghi - Alvandi and Parsian (1992), Pandey et al. (1996), Pandey (1997), Parsian and Farsipour (1999), Singh et al. (2002), Misra and Meulen (2003), Ahmadi et al. (2005), Xiao et al. (2005), Singh et al. (2007) and others have used the LINEX loss function in various estimation and prediction problems.

Thompson (1968) suggested a procedure which makes the use of prior information of the parameter in form of a guess value by shrinking the usual estimator towards the guess value of the parameter with the help of a shrinkage factor $k \, (0 \leq k \leq 1)$. The value of shrinkage factor is specified by the experimenter according to his belief in the guess value. The shrinkage estimator for the measure of dispersion σ of the inverse Gaussian distribution when a guess value σ_0 (say) of σ is available, is given by

$$T = k\tilde{\sigma} + (1-k)\,\sigma_0. \tag{4}$$

Some estimators for σ have been obtained by Pandey and Malik (1988) and they have studied their properties under MSE criterion. Recently, some shrinkage testimators for the inverse measure of dispersion have been proposed by Prakash and Singh (2006). Palmer (1973), Banerjee and Bhattacharya (1979) have discussed the Bayesian inference about the parameters of the inverse Gaussian distribution.

In the present article, we have studied the performances of the shrinkage testimators for σ when a guess value σ_0 is available under different choices of the shrinkage factor. A double stage shrinkage estimator is also proposed and its properties studied with the cost function. All the comparisons have been made in terms of the relative efficiencies under the invariant version of the LINEX loss function.

2. A Class of Estimators and Its Properties

The proposed class of estimators for the measure of dispersion σ of the inverse Gaussian distribution (1.1) is

$$Y = c\tilde{\sigma}\ ;\ c \in R^{+}. \tag{5}$$

The risk of Y under the invariant form of the LINEX loss (1.2) is given by

$$R(Y) = e^{-a}\left(1 - \frac{2ac}{n-1}\right)^{-\frac{n-1}{2}} - a\,c + (a-1)\,. \tag{6}$$

The value of c for which (2.2) is minimum,

$$\frac{(n-1)}{2a}\left\{1 - exp\left(-\frac{2a}{n+1}\right)\right\} = c_1\ (say). \tag{7}$$

Hence, the minimum risk estimator of σ in the class $Y = c\tilde{\sigma}$ is

$$Y_1 = c_1\tilde{\sigma} \tag{8}$$

with the risk

$$R(Y_1) = n' + a\ ;\ n' = \frac{n+1}{2}\left(exp\left(-\frac{2a}{n+1}\right) - 1\right). \tag{9}$$

The value of c which minimizes the mean square error of the class $Y = c\,\tilde{\sigma}$ is $c' = \frac{n-1}{n+1}$ and the minimum mean square error estimator among the class (2.1) is $Y_2 = c'\,\tilde{\sigma}$ with $MSE(Y_2) = \frac{2\sigma^2}{n+1}$ and is inadmissible under the LINEX loss. Also, as $a \rightarrow 0$ $c_1 \rightarrow c'$.

3. The Shrinkage Estimator and Its Properties

The risk of the shrinkage estimator T (1.3) under the LINEX loss (1.2) is obtained as

$$R(T) = e^{a((1-k)\beta-1)}\left(1 - \frac{2ak}{n-1}\right)^{-\frac{n-1}{2}} - a(k-1)(1-\beta) - 1\ ;\ \beta = \frac{\sigma_0}{\sigma}. \tag{10}$$

The relative bias of T is obtained as

$$RB(T) = \frac{E(T)}{\sigma} - 1 = (1-k)(\beta-1)\,. \tag{11}$$

This expression clearly shows that the relative bias is zero at $\beta = 1$ and has a tendency of being negative for $0 < \beta < 1$ and positive for $\beta > 1$.

The relative efficiency of T with respect to Y_1 is defined as

$$RE(T, Y_1) = \frac{R(Y_1)}{R(T)}. \tag{12}$$

The value of $RE(T, Y_1)$ is a function of n, a, k and β. For the selected values of $n = 05, 10, 15, 25$; $a = \pm 0.50, \pm 0.25$; $k = 0.20(0.20)0.80$ and $\beta = 0.40(0.20)1.60$, we have calculated the relative efficiency (not presented here). The estimator T has smaller risk than the estimator Y_1 when σ_0 is in the vicinity of the true value σ. More specially, the proposed shrinkage estimator is preferable over the improved estimator under the LINEX loss for the small sample size n, small $k \leq 0.60$ and $0.40 \leq \beta \leq 1.40$. The effective interval of β decreases as the sample size n increases and the efficiency attains maximum at $\beta = 1.00$. The efficiency decreases with the n increase (except $\beta = 1.00$) for small k. The nominal gain in efficiency has been recorded when the shrinkage factor k increases. In addition, as $'a'$ increases the efficiency decreases for the all considered values of parametric space when $\beta \geq 1.00$.

The value of shrinkage factor k, k_1 (say) which minimizes the risk of the shrinkage estimator T, is obtained numerically by solving the equality :

$$(1-\beta) = e^{a((1-k)\beta-1)}\left(1-\frac{2ak}{n-1}\right)^{-\frac{n-1}{2}}\left(\left(1-\frac{2ak}{n-1}\right)^{-1}-\beta\right) \tag{13}$$

for a given set of values.

Hence, the improved shrinkage estimator among the class (1.3) is

$$T^* = k_1\tilde{\sigma} + (1-k_1)\,\sigma_0. \tag{14}$$

The expressions of the relative bias and the risk are obtained from equations (3.2) and (3.1) respectively by replacing k_1 in place of k. The relative efficiency for the estimator T^* with respect to Y_1 is defined as

$$RE\,(T^*, Y_1) = \frac{R\,(Y_1)}{R\,(T^*)}. \tag{15}$$

The performances of the relative bias for the estimator T^* are similar to T. The value of $RE\,(T^*, Y_1)$ is a function of n, a and β. For the similar set of values as considered earlier we have calculated the relative efficiency but not presented here. The shrinkage estimator T^* is more efficient than Y_1 in the interval $0.40 \leq \beta \leq 1.60$ for the all considered values of parametric space. Other properties are similar to the shrinkage estimator T.

4. The Shrinkage Test - Estimators and Their Properties

We have seen that the shrinkage estimators T and T^* have performed well than the improved estimator Y_1 when the guessed value σ_0 is in the vicinity of the true value σ. This suggested that when the guess value σ_0 for σ is given, the null hypothesis $H_0 : \sigma = \sigma_0$ against $H_1 : \sigma \neq \sigma_0$ is carried out first and on the acceptance of the H_0, the shrinkage estimator is used as an estimator for the parameter σ; otherwise Y_1 is taken as an estimator. Therefore, we proposed shrinkage test - estimator for σ as

$$\hat{\sigma}_{SH} = \begin{cases} k\tilde{\sigma} + (1-k)\,\sigma_0 & if\ v_1 \leq v \leq v_2 \\ Y_1 = c_1\tilde{\sigma} & if\ (v \leq v_1) \cup (v \geq v_2), \end{cases}$$

where $v_1 = r_1\sigma_0$, $v_2 = r_2\sigma_0$ and r_1 and r_2 being the values of the lower and upper $100\alpha/2\%$ points of the chi - square distribution with $(n-1)$ degrees of freedom.

The relative bias of the shrinkage test - estimator $\hat{\sigma}_{SH}$ is given by

$$RB\left(\hat{\sigma}_{SH}\right) = I\left(f\right) + \beta\, I\left(1\right) - I\left(f_0\right) + c_1 - 1, \tag{16}$$

where $z_1 = \frac{r_1\beta}{2}$, $z_2 = \frac{r_2\beta}{2}$, $f = k\left(\frac{2z}{n-1} - \beta\right)$, $f_0 = c_1\frac{2z}{n-1}$, $I\left(q\right) = \left(\Gamma\left(\frac{n-1}{2}\right)\right)^{-1}$ $\int_{z_1}^{z_2}\left(q\right)\, e^{-z}\, z^{\frac{n-3}{2}}dz$ and q may be the function of z.

The risk of $\hat{\sigma}_{SH}$ under the LINEX loss is given by

$$R\left(\hat{\sigma}_{SH}\right) = e^{a(\beta-1)}\, I\left(e^{af}\right) - aI\left(f\right) - e^{-a}\, I\left(e^{af_0}\right) \tag{17}$$

$$-a\beta\, I\left(1\right) + aI\left(f_0\right) + n' + a. \tag{18}$$

The relative efficiency of the shrinkage test - estimator $\hat{\sigma}_{SH}$ with respect to Y_1 is

$$RE\left(\hat{\sigma}_{SH}, Y_1\right) = \frac{R\left(Y_1\right)}{R\left(\hat{\sigma}_{SH}\right)}. \tag{19}$$

The expressions of the relative bias and the relative efficiency are the functions of n, a, k, β and α (level of significance). For the similar set of values as considered earlier with $\alpha = 0.01$ and $\alpha = 0.05$, the relative biases and the relative efficiencies have been calculated but not presented here.

The relative bias $RB\left(\hat{\sigma}_{SH}\right)$ lies between -0.203 and 0.351 and decreases with increase in the shrinkage factor k. The absolute relative biases (ARB) decreases as the sample size n increase (except $\beta = 1.20$). The ARB increase when $\beta \leq 1.20$ and decreases otherwise when $'a'$ increases. Similar trend has been seen when level of significance α increases for the all considered values of β (except $\beta = 1.00$).

The proposed test - estimator $\hat{\sigma}_{SH}$ is more efficient than Y_1 in the interval $0.60 \leq \beta \leq 1.40$ for small $n\,(n \leq 15)$ and the effective interval of β decreases as n or k increases. For the moderate value of k, efficiency is maximum at $\beta = 1.00$. The efficiency decreases for $\beta \geq 1.00$ when $'a'$ as well as α increases. The efficiency also decreases when n increases for $k < 0.60$ (except $\beta = 1.00$). The gain in efficiency decreases as k increases near $\beta = 1.00$, for small n. Hence the proposed test - estimator performs well for the moderate values of n and k.

Based on k_1 (3.3), the proposed shrinkage test - estimator for σ is defined as

$$\hat{\sigma}_{SH\,1} = \begin{cases} k_1\tilde{\sigma} + \left(1-k_1\right)\sigma_0 & if\; v_1 \leq v \leq v_2 \\ Y_1 = c_1\tilde{\sigma} & if\; \left(v \leq v_1\right) \cup \left(v \geq v_2\right). \end{cases}$$

One may interested to obtain the value of k that minimizes the risk of the shrinkage test - estimator $\hat{\sigma}_{SH}$. For this the value of $k = k_2$ (say), obtained numerically by solving the equality

$$e^{a(\beta-1)}\, I\left(e^{af}\frac{f}{k}\right) = I\left(\frac{f}{k}\right). \tag{20}$$

Based on k_2 (4.3), the proposed shrinkage test - estimator for σ is given as

$$\hat{\sigma}_{SH\,2} = \begin{cases} k_2\tilde{\sigma} + (1-k_2)\,\sigma_0 & if\ v_1 \le v \le v_2 \\ Y_1 = c_1\tilde{\sigma} & if\,(v \le v_1) \cup (v \ge v_2)\,. \end{cases}$$

On the other hand, the null hypothesis $H_0 : \sigma = \sigma_0$ is accepted if

$$r_1 \le \frac{v}{\sigma_0} \le r_2 \Longleftrightarrow 0 \le \frac{1}{r_2 - r_1}\left(\frac{v}{\sigma_0} - r_1\right) = k_3\,(say)\ \le 1 \tag{21}$$

and

$$r_1 \le (n-1) \le r_2 \Longleftrightarrow \frac{r_1}{n-1} \le 1. \tag{22}$$

For the smaller values of the shrinkage factor k_3 given in (4.4) can be rewritten after making it positive as

$$k_3 = \frac{n-1}{r_2 - r_1}\left|\frac{v}{\sigma_0\,(n-1)} - 1\right|. \tag{23}$$

Adke et al. (1987) considered this type of shrinkage factor. Based on this shrinkage factor the proposed test - estimator is

$$\hat{\sigma}_{SH\,3} = \begin{cases} k_3\tilde{\sigma} + (1-k_3)\,\sigma_0 & if\ v_1 \le v \le v_2 \\ Y_1 = c_1\tilde{\sigma} & if\,(v \le v_1) \cup (v \ge v_2)\,. \end{cases}$$

The value of constant $c_1 = k_4$ (say) (2.3) also lies between 0 and 1. Substituting this choice for k, we define a new shrinkage test - estimator as

$$\hat{\sigma}_{SH\,4} = \begin{cases} k_4\tilde{\sigma} + (1-k_4)\,\sigma_0 & if\ v_1 \le v \le v_2 \\ Y_1 = c_1\tilde{\sigma} & if\,(v \le v_1) \cup (v \ge v_2)\,. \end{cases}$$

The expressions of relative biases and the risks of these shrinkage test - estimator $\hat{\sigma}_{SH\,i}$; $i = 1, 2, 3, 4$ are obtained as

$$RB\,(\hat{\sigma}_{SH\,i}) = I\,(f_i) + \beta\, I\,(1) - I\,(f_0) + c_1 - 1 \tag{24}$$

and

$$R\,(\hat{\sigma}_{SH\,i}) = e^{a(\beta-1)}\, I\left(e^{af_i}\right) - aI\,(f_i) - e^{-a}\, I\left(e^{af_0}\right) \tag{25}$$

$$-a\beta\, I\,(1) + aI\,(f_0) + n' + a. \tag{26}$$

where $f_i = k_i\left(\frac{2z}{n-1} - \beta\right)$ and $i = 1, 2, 3, 4$ and $I(.)$ is defined in (4.1).

The relative efficiency for $\hat{\sigma}_{SH\,i}$; $i = 1, 2, 3, 4$ with respect to Y_1 is given as

$$RE\,(\hat{\sigma}_{SH\,i}, Y_1) = \frac{R\,(Y_1)}{R\,(\hat{\sigma}_{SH\,i})}. \tag{27}$$

Table 01. << **RE** $(\hat{\sigma}_{SH\,1}, Y_1)$

		β						
a	α	0.40	0.60	0.80	1.00	1.20	1.40	1.60
$n = 05$								
- 0.50	0.01	1.5283	2.3703	5.6195	15.643	6.5220	3.4408	2.2376
	0.05	1.5126	2.0074	3.2906	5.8120	3.3838	2.8379	2.0953
- 0.25	0.01	1.5752	2.3918	5.5387	15.563	6.4413	3.3423	2.1628
	0.05	1.5623	2.0592	3.3228	5.7802	3.3729	2.7967	2.0464
0.25	0.01	1.6503	2.4352	5.3872	15.429	6.2135	3.1421	2.0191
	0.05	1.6409	2.1608	3.4062	5.1200	3.3668	2.7028	1.9460
0.50	0.01	1.6771	2.4548	5.3239	15.213	6.0719	3.0421	1.9509
	0.05	1.6689	2.2070	3.4597	5.0828	3.3539	2.6511	1.8953
$n = 10$								
- 0.50	0.01	1.3154	1.6840	3.6039	15.614	4.2353	2.3796	1.6810
	0.05	1.3388	1.5528	2.5019	4.9265	2.5942	2.1780	1.6678
- 0.25	0.01	1.3609	1.7232	3.5808	15.463	4.1848	2.3316	1.6454
	0.05	1.3767	1.6045	2.5212	4.8663	2.5883	2.1487	1.6384
0.25	0.01	1.4169	1.7797	3.5374	15.367	4.0632	2.2353	1.5764
	0.05	1.4244	1.6851	2.5682	4.8369	2.5659	2.0874	1.5800
0.50	0.01	1.4263	1.7939	3.5186	15.164	3.9932	2.1874	1.5433
	0.05	1.4346	1.7108	2.5965	4.5705	2.5497	2.0557	1.5512
$n = 15$								
- 0.50	0.01	1.2737	1.4492	2.6673	15.445	3.1072	1.9642	1.4772
	0.05	1.2958	1.3923	2.0443	4.7589	2.1190	1.8868	1.5065
- 0.25	0.01	1.3149	1.4955	2.6640	15.432	3.0698	1.9327	1.4537
	0.05	1.3237	1.4450	2.0612	4.6867	2.1081	1.8633	1.4841
0.25	0.01	1.3520	1.5491	2.6586	15.248	2.9862	1.8697	1.4081
	0.05	1.3506	1.5119	2.0996	4.6341	2.0812	1.8158	1.4401
0.50	0.01	1.3508	1.5524	2.6569	15.085	2.9404	1.8385	1.3861
	0.05	1.3523	1.5228	2.1214	4.5552	2.0653	1.7919	1.4186
$n = 25$								
- 0.50	0.01	1.2629	1.2886	1.7779	15.231	2.0319	1.6321	1.3136
	0.05	1.2690	1.2876	1.5329	4.6934	1.5728	1.6239	1.3820
- 0.25	0.01	1.2900	1.3429	1.7858	15.229	2.0063	1.6129	1.2988
	0.05	1.2822	1.3418	1.5481	4.6507	1.5596	1.6063	1.3649
0.25	0.01	1.3056	1.3865	1.8018	15.098	1.9525	1.5855	1.2702
	0.05	1.2908	1.3858	1.5806	4.6192	1.5317	1.5715	1.3316
0.50	0.01	1.2976	1.3871	1.8102	14.988	1.9244	1.5663	1.2563
	0.05	1.2883	1.3813	1.5980	4.5111	1.5170	1.5543	1.3156

Table 02. << **RE** $(\hat{\sigma}_{SH\,2}, Y_1)$

		β						
a	α	0.40	0.60	0.80	1.00	1.20	1.40	1.60
$n = 05$								
- 0.50	0.01	1.1092	2.4747	5.8103	15.762	6.7383	2.6094	1.3363
	0.05	1.2084	2.8256	3.3461	4.7663	3.3992	1.9061	1.1518
- 0.25	0.01	1.1408	2.2852	5.7264	15.766	6.6609	2.5094	1.2590
	0.05	1.2464	2.3623	3.3801	4.8317	3.4098	1.8669	1.1034
0.25	0.01	1.7578	2.3903	5.5689	15.514	6.4343	2.6099	1.4052
	0.05	1.8563	2.1220	3.4677	4.9577	3.4061	1.7483	1.2592
0.50	0.01	1.2412	1.7279	4.0311	15.296	5.6530	1.7376	1.2382
	0.05	1.3578	1.6805	2.9619	5.0253	2.7127	1.7250	1.1519
$n = 10$								
- 0.50	0.01	0.7731	2.2232	3.7831	15.106	4.4682	1.4875	0.7375
	0.05	0.9506	2.2330	2.5676	4.6890	2.6574	1.2360	0.7069
- 0.25	0.01	0.8013	2.2431	3.7566	15.084	4.4207	1.4387	0.7009
	0.05	0.9814	2.1625	2.5876	4.7276	2.6535	1.2073	0.6774
0.25	0.01	1.7410	1.7710	3.7067	14.852	4.3023	1.7578	1.2527
	0.05	1.0410	1.7441	2.6362	4.7958	2.6347	1.5964	1.2325
0.50	0.01	0.8874	1.6585	3.4226	14.649	4.0181	1.5410	1.1416
	0.05	1.0696	1.6750	2.4922	4.8281	2.3520	1.1445	1.1286
$n = 15$								
- 0.50	0.01	0.7031	1.9091	2.8178	14.973	3.3045	1.0416	0.5467
	0.05	0.9179	1.9951	2.1095	4.6244	2.1868	0.9342	0.5368
- 0.25	0.01	0.7291	1.9285	2.8123	14.956	3.2689	1.0094	0.5224
	0.05	0.9440	1.9205	2.1266	4.6514	2.1774	0.9109	0.5141
0.25	0.01	1.6183	1.4541	2.8025	14.772	3.1877	1.2992	1.0745
	0.05	0.9936	1.5458	2.1658	4.6973	2.1534	1.1104	1.0605
0.50	0.01	0.8065	1.6001	2.9570	14.615	3.4238	1.4174	1.0975
	0.05	1.0170	1.2100	2.2332	4.7177	2.1738	1.0613	1.0118
$n = 25$								
- 0.50	0.01	0.7609	1.6598	1.8890	14.791	2.1759	0.6614	0.4387
	0.05	1.0069	1.8090	1.5914	4.5613	1.6367	0.6591	0.4046
- 0.25	0.01	0.7841	1.6766	1.8959	14.779	2.1510	0.6418	0.3935
	0.05	1.0248	1.8302	1.6067	4.5780	1.6243	0.6414	0.3872
0.25	0.01	0.8292	1.1622	1.9100	14.657	2.0982	0.8584	0.7951
	0.05	1.0577	1.3678	1.6396	4.6057	1.5980	0.8200	0.7837
0.50	0.01	0.8508	1.5617	2.3411	14.552	2.5295	1.2856	1.0334
	0.05	1.0727	1.1948	1.9040	4.6172	1.7897	1.0495	0.9521

Table 03. << **RE** $(\hat{\sigma}_{SH\,3}, Y_1)$

		β						
a	α	0.40	0.60	0.80	1.00	1.20	1.40	1.60
$n = 05$								
- 0.50	0.01	1.5741	1.9556	2.7510	3.4495	2.9190	1.8770	1.1712
	0.05	1.6981	1.9737	2.2344	3.2606	1.8950	1.3914	0.9889
- 0.25	0.01	1.6115	1.9019	2.5367	3.3049	2.6699	1.7429	1.0866
	0.05	1.7530	1.9906	2.1613	3.2304	1.7799	1.3095	0.9274
0.25	0.01	1.6886	1.7841	2.3021	2.9084	2.2333	1.4568	0.9173
	0.05	1.8610	2.0228	2.0997	2.8396	1.6144	1.1225	0.7947
0.50	0.01	1.7270	1.7186	2.1843	2.7661	2.1495	1.3023	0.8312
	0.05	1.9128	2.0361	2.0905	2.6822	1.5667	1.0173	0.7229
$n = 10$								
- 0.50	0.01	1.3498	1.6199	2.4822	3.5242	2.5975	1.3517	0.7493
	0.05	1.4268	1.6928	2.1379	2.3402	1.7580	1.0895	0.6955
- 0.25	0.01	1.3934	1.6282	2.3892	3.3108	2.4751	1.2933	0.7105
	0.05	1.4639	1.7319	2.1184	2.2648	1.6997	1.0505	0.6637
0.25	0.01	1.4817	1.6456	2.1983	2.8756	2.2116	1.1730	0.6342
	0.05	1.5336	1.8152	2.0767	2.1998	1.5678	0.9668	0.5996
0.50	0.01	1.5260	1.6544	2.1010	2.6572	2.0716	1.1111	0.5968
	0.05	1.5658	1.8591	2.0544	2.1112	1.4947	0.9222	0.5673
$n = 15$								
- 0.50	0.01	1.2736	1.4369	2.2491	3.5013	2.2928	1.0428	0.5494
	0.05	1.3266	1.5233	2.0108	2.3502	1.6068	0.8938	0.5512
- 0.25	0.01	1.3083	1.4584	2.1986	3.3412	2.2165	1.0067	0.5241
	0.05	1.3502	1.5612	2.0079	2.2942	1.5679	0.8668	0.5277
0.25	0.01	1.3758	1.5036	2.0941	3.0133	2.0531	0.9334	0.4745
	0.05	1.3928	1.6405	2.0026	2.1731	1.4818	0.8111	0.4813
0.50	0.01	1.4084	1.5273	2.0405	2.8474	1.9663	0.8962	0.4503
	0.05	1.4118	1.6816	2.0001	2.1086	1.4347	0.7824	0.4584
$n = 25$								
- 0.50	0.01	1.2398	1.2644	1.9354	3.4521	1.8428	0.7126	0.3669
	0.05	1.2658	1.3558	1.8114	2.3499	1.3675	0.6746	0.4244
- 0.25	0.01	1.2583	1.2908	1.9177	3.3397	1.8022	0.6918	0.3511
	0.05	1.2751	1.3865	1.8206	2.3105	1.3435	0.6560	0.4067
0.25	0.01	1.2923	1.3455	1.8805	3.1092	1.7163	0.6501	0.3204
	0.05	1.2912	1.4487	1.8406	2.2263	1.2923	0.6188	0.3720
0.50	0.01	1.3078	1.3738	1.8612	2.9919	1.6712	0.6293	0.3054
	0.05	1.2981	1.4799	1.8513	2.1818	1.2652	0.6002	0.3552

Table 04. << **RE** $(\hat{\sigma}_{SH\,4}, Y_1)$

		β						
a	α	0.40	0.60	0.80	1.00	1.20	1.40	1.60
$n = 05$								
- 0.50	0.01	1.6412	1.7318	1.7716	1.7839	1.7570	1.6869	1.5819
	0.05	1.6452	1.7664	1.7478	1.6673	1.5573	1.4304	1.2975
- 0.25	0.01	1.6933	1.7950	1.8204	1.8083	1.7506	1.6467	1.5099
	0.05	1.6933	1.8442	1.8124	1.6991	1.5551	1.3995	1.2449
0.25	0.01	1.7942	1.9274	1.9110	1.8287	1.6933	1.5171	1.3238
	0.05	1.7799	2.0077	1.9492	1.7525	1.5267	1.3100	1.1158
0.50	0.01	1.8421	1.9979	1.9554	1.8276	1.6469	1.4351	1.2195
	0.05	1.8177	2.0934	2.0234	1.7763	1.5032	1.2549	1.0735
$n = 10$								
- 0.50	0.01	1.4839	1.5268	1.5072	1.4929	1.4670	1.4158	1.3400
	0.05	1.4627	1.5751	1.5337	1.4490	1.3539	1.2488	1.1404
- 0.25	0.01	1.5173	1.5699	1.5355	1.5059	1.4646	1.3967	1.3039
	0.05	1.4889	1.6284	1.5755	1.4659	1.3486	1.2256	1.1132
0.25	0.01	1.5808	1.6596	1.5871	1.5147	1.4344	1.3291	1.2036
	0.05	1.5352	1.7387	1.6617	1.4913	1.3232	1.1639	1.1071
0.50	0.01	1.6105	1.7068	1.6111	1.5116	1.4078	1.2830	1.1427
	0.05	1.5553	1.7954	1.7068	1.5007	1.3039	1.1265	1.0692
$n = 15$								
- 0.50	0.01	1.4125	1.4682	1.4300	1.4068	1.3787	1.3256	1.2472
	0.05	1.3779	1.5053	1.4666	1.3801	1.2879	1.1860	1.0843
- 0.25	0.01	1.4347	1.5041	1.4516	1.4156	1.3764	1.3108	1.2193
	0.05	1.3929	1.5474	1.5002	1.3915	1.2807	1.1638	1.0502
0.25	0.01	1.4758	1.5787	1.4915	1.4213	1.3544	1.2620	1.1460
	0.05	1.4187	1.6330	1.5692	1.4082	1.2564	1.1106	1.0368
0.50	0.01	1.4945	1.6176	1.5102	1.4185	1.3353	1.2290	1.1019
	0.05	1.4296	1.6763	1.6050	1.4137	1.2397	1.0799	1.0178
$n = 25$								
- 0.50	0.01	1.3284	1.4222	1.3735	1.3417	1.3099	1.2474	1.1585
	0.05	1.2976	1.4333	1.4148	1.3268	1.2354	1.1349	1.0473
- 0.25	0.01	1.3381	1.4513	1.3901	1.3470	1.3070	1.2340	1.1335
	0.05	1.3027	1.4626	1.4417	1.3337	1.2263	1.1124	1.0146
0.25	0.01	1.3553	1.5109	1.4216	1.3503	1.2908	1.1973	1.0764
	0.05	1.3113	1.5205	1.4969	1.3435	1.2026	1.0641	1.0139
0.50	0.01	1.3628	1.5414	1.4367	1.3484	1.2778	1.1742	1.0446
	0.05	1.3148	1.5490	1.5254	1.3465	1.1881	1.0384	1.0119

The relative biases and the relative efficiencies are the functions of n, β, a and α. For the similar set of values as considered earlier, we have computed the relative biases (not presented here) and the relative efficiencies (Table 01 - 04).

The relative biase lies between 0.165 and -0.175 for $\hat{\sigma}_{SH\,1}$, between 0.416 and -0.218 for $\hat{\sigma}_{SH\,2}$, between 0.368 and -0.174 for $\hat{\sigma}_{SH\,3}$ and for test - estimator $\hat{\sigma}_{SH\,4}$ it lies between 0.168 and -0.170. The ARB decrease with the sample size n increases for the all considered values of parametric space. The ARB also decrease when level of significance α increases for all the considered values of β (except $\beta = 1.00$) for the test - estimator ($\hat{\sigma}_{SH\,1}, \hat{\sigma}_{SH\,2}$ and $\hat{\sigma}_{SH\,3}$) and in the interval $\beta \geq 1.00$ for the test - estimator $\hat{\sigma}_{SH\,4}$. Opposite trend has been seen when $'a'$ increases in $\beta \leq 1.00$ for the test - estimator ($\hat{\sigma}_{SH\,1} and \hat{\sigma}_{SH\,3}$) and for the all considered values of β for the test estimator $\hat{\sigma}_{SH\,4}$.

The tables of the relative efficiencies show that the testimators $\hat{\sigma}_{SH\,1}$ and $\hat{\sigma}_{SH\,4}$ perform uniformly better than improved estimator Y_1 for the all considered values of β, whereas $\hat{\sigma}_{SH\,2}$ performs well in the interval $0.60 \leq \beta \leq 1.20$ and $\hat{\sigma}_{SH\,3}$ in the interval $0.40 \leq \beta \leq 1.20$ when they are compared with Y_1. The efficiency reaches maximum at $\beta = 1.00$ for all the testimators (except $\hat{\sigma}_{SH\,4}$) and the efficiency decreases when n increases in the interval $0.40 \leq \beta \leq 1.60$ (except $\beta = 1.00$ for $\hat{\sigma}_{SH\,3}$). Further, the efficiency decreases when the level of significance α increases in the interval $0.80 \leq \beta \leq 1.40$ for $\hat{\sigma}_{SH\,1}$ and $\hat{\sigma}_{SH\,2}$. Similar trend is also seen in the interval $1.00 \leq \beta \leq 1.40$ for the testimators $\hat{\sigma}_{SH\,3}$ and $\hat{\sigma}_{SH\,4}$. In addition, the efficiency decreases as $'a'$ increases when $\beta \geq 1.00$ for the all testimators (except $\hat{\sigma}_{SH\,2}$).

5. Double Stage Shrinkage Test - Estimator and Its Properties

Let $x_{1i}\ (i = 1, 2, ---, n_1)$ and $x_{2j}\ (j = 1, 2, ---, n_2)$ be the two independent random samples of sizes n_1 and n_2 respectively, drawn from the inverse Gaussian distribution (1.1). The unbiased pooled estimator for σ, based on the two samples of sizes n_1 and n_2 is

$$\bar{\sigma} = \frac{v_1 + v_2}{n_1 + n_2 - 2}. \tag{28}$$

The proposed class of estimators for the pooled estimate of σ is given by

$$P = m\bar{\sigma}\ ;\ m \in R^+. \tag{29}$$

The value of m for which $R\,(P)$, the risk of P is minimum, given by

$$m_1 = \frac{n_1 + n_2 - 2}{2a}\left(1 - exp\left(-\frac{2a}{n_1 + n_2}\right)\right). \tag{30}$$

Hence the improved estimator among the class (5.2) is

$$P_1 = m_1\bar{\sigma} \tag{31}$$

with

$$R\,(P_1) = n'' + a\ ;\ n'' = \frac{n_1 + n_2}{2}\left(exp\left(-\frac{2a}{n_1 + n_2}\right) - 1\right). \tag{32}$$

In the previous section, we observed that the performances of the shrinkage test - estimator $\hat{\sigma}_{SH\,1}$ are better than the other shrinkage test - estimators in terms of the effective interval and the magnitude of efficiency when they are compared with Y_1. Hence, we have considered $\hat{\sigma}_{SH\,1}$ in double stage setup. The double stage technique is to obtain first a sample of size n_1 and compute $\tilde{\sigma}_1$. If $\tilde{\sigma}_1$ implies that our prior estimate σ_0 is reasonable, we stop the sampling and estimate the parameter with the shrinkage estimator. Otherwise, we obtain n_2 additional observations and then use an improved estimate based on all $(n_1 + n_2)$ observations. The proposed double stage shrinkage test - estimator is given as

$$\hat{\sigma}_{DSH} = \begin{cases} k_1 \tilde{\sigma} + (1 - k_1)\,\sigma_0 & if\ v_1 \leq v \leq v_2 \\ P_1 = m_1 \tilde{\sigma} & if\ (v \leq v_1) \cup (v \geq v_2)\,. \end{cases}$$

where $\tilde{\sigma}_1 = \frac{v''}{n_1 - 1}$, $v'' = \sum_{i=1}^{n_1} \left(\frac{1}{x_{1i}} - \frac{1}{\bar{x}_1} \right)$ and $\bar{x}_1 = \frac{1}{n_1} \sum_{i=1}^{n_1} x_{1i}$.

The risk of $\hat{\sigma}_{DSH}$ under the LINEX loss is given as

$$R\left(\hat{\sigma}_{DSH}\right) = e^{a(\beta - 1)}\, I\,\left(e^{a f_D}\right) - a\, I\,\left(f_D\right) - a\beta\, I\,\left(1\right) + a\, I\,\left(f_{D0}\right) \tag{33}$$

$$-e^{a\frac{(n_1+1)}{(n_1+n_2)}}\, I\,\left(e^{a f_{D0}}\right) - \frac{n_2 - 1}{n_1 + n_2}\, n''\, I\,\left(1\right) + n'' + a, \tag{34}$$

where $f_D = k_1 \left(\frac{2z}{n_1 - 1} - \beta \right)$, $f_{D0} = \frac{2m_1}{n_1 + n_2 - 2} z$ and $I(.)$ is defined in (4.1).

The problem is considered as a sequential estimation problem with stopping random variable N defined as

$$N = \begin{cases} n_1 & if\ v_1 \leq v \leq v_2 \\ n_1 + n_2 & else. \end{cases}$$

Let us introduce a cost $d\,(> 0)$ for each observation. Then the risk of $\hat{\sigma}_{DSH}$ and P_1 are modified as

$$\tilde{R}\left(\hat{\sigma}_{DSH}\right) = R\left(\hat{\sigma}_{DSH}\right) + d\,E\,\left(N\right) \tag{35}$$

and

$$\tilde{R}\left(P_1\right) = R\left(P_1\right) + d\,\left(n_1 + n_2\right). \tag{36}$$

The relative efficiency of $\hat{\sigma}_{DSH}$ with respect to P_1 is given by

$$RE\,\left(\hat{\sigma}_{DSH}, P_1\right) = \frac{\tilde{R}\left(P_1\right)}{\tilde{R}\left(\hat{\sigma}_{DSH}\right)}. \tag{37}$$

The expression of relative efficiency involves $n_1, n_2,\ \beta, a,\ d$ and α. For the similar set of values as considered earlier with $n_1 = 05, 10, 15$, $n_2 = 04, 06, 10$ and the cost $d = 0.50, 1.00, 5.00, 50.00$, the relative efficiency has been calculated and presented in Table 05 only for $n_1 = 05$ and $d = 0.50$.

Table 05. << **RE** $(\hat{\sigma}_{DSH}, P_1)$

		$n_1 = 05,\ d = 0.50$						
		β						
a	α	0.40	0.60	0.80	1.00	1.20	1.40	1.60
$n_2 = 04$								
- 0.50	0.01	1.5103	2.3934	3.2639	3.5281	3.2371	2.4447	1.7967
	0.05	1.2640	2.0793	2.9262	3.1684	2.8626	2.1170	1.5241
- 0.25	0.01	1.5214	2.3993	3.2461	3.4947	3.2173	2.4525	1.8274
	0.05	1.2700	2.0835	2.9150	3.1461	2.8506	2.1238	1.5447
0.25	0.01	1.5219	2.3997	3.2458	3.4943	3.2168	2.4515	1.8255
	0.05	1.2703	2.0839	2.9149	3.1459	2.8503	2.1232	1.5435
0.50	0.01	1.5140	2.3960	3.2619	3.5249	3.2334	2.4372	1.7819
	0.05	1.2665	2.0819	2.9258	3.1667	2.8605	2.1122	1.5149
$n_2 = 06$								
- 0.50	0.01	2.0185	3.4264	4.7689	5.1580	4.7089	3.5334	2.5794
	0.05	1.5997	2.8173	4.0856	4.4305	3.9613	2.8897	2.0523
- 0.25	0.01	2.0382	3.4476	4.7634	5.1325	4.7010	3.5591	2.6324
	0.05	1.6105	2.8310	4.0840	4.4157	3.9584	2.9069	2.0831
0.25	0.01	2.0388	3.4483	4.7633	5.1322	4.7006	3.5580	2.6298
	0.05	1.6109	2.8314	4.0840	4.4156	3.9582	2.9062	2.0817
0.50	0.01	2.0234	3.4313	4.7683	5.1560	4.7059	3.5243	2.5598
	0.05	1.6027	2.8209	4.086	4.4297	3.9597	2.8844	2.0415
$n_2 = 10$								
- 0.50	0.01	2.7748	5.2664	7.6107	8.2427	7.4546	5.5271	3.9855
	0.05	2.0387	3.9525	6.0126	6.5373	5.7455	4.1006	2.8517
- 0.25	0.01	2.8018	5.3084	7.6223	8.2259	7.4627	5.5793	4.0715
	0.05	2.0521	3.9750	6.0210	6.5295	5.7517	4.1279	2.8915
0.25	0.01	2.8025	5.3095	7.6225	8.2258	7.4624	5.5778	4.0679
	0.05	2.0525	3.9755	6.0211	6.5295	5.7515	4.1271	2.8899
0.50	0.01	2.7804	5.2744	7.6119	8.2417	7.4518	5.5147	3.9580
	0.05	2.0417	3.9569	6.0137	6.5370	5.7441	4.0945	2.8393

The double stage shrinkage test - estimator $\hat{\sigma}_{DSH}$ is more efficient than improved pooled estimator P_1 for the all considered values of parametric space and attains maximum efficiency at $\beta = 1.00$. The efficiency decreases as the sample size n_1 increases whereas the reverse trend has been seen when n_2 increases. It is also noted that the efficiency decreases when α increases. In addition, the efficiency increases as the cost d increases in the interval $0.80 \leq \beta \leq 1.20$ whereas the gain in efficiency is nominal.

References

[1] Adke, S. R., Waikar, V. B. and Schuurmann, F. J. (1987). A two stage shrinkage testimator for the mean of an Exponential distribution. *Communication in Statistics - Theory and Methods*, **16**, 1821 - 1834.

[2] Ahmadi, J., Doostparast, M. and Parsian, A. (2005). Estimation and prediction in a two - parameter Exponential distribution based on k - record values under LINEX loss function. *Communications in Statistics - Theory and Methods*, **34**, 795 - 805.

[3] Banerjee, A. K. and Bhattacharya, G. K. (1979). Bayesian results for the inverse Gaussian distribution with an application. *Technometrices*, **21**, 247 - 251.

[4] Folks, J. L. and Chhikara, R. S. (1978). The inverse Gaussian distribution and its statistical application - A Review. *Journal of the Royal Statistical Society, Series - B*, **40**, 263 - 289.

[5] Folks, J. L. and Chhikara, R. S. (1989). *The inverse Gaussian distribution.* Marcel Dekker, New York.

[6] Misra, N. and Meulen, E. Van Der (2003). On estimating the mean of the selected Normal population under the LINEX loss function. *Metrika*, **173 - 184.**

[7] Palmer, T. (1973). *Certain non - classical inference procedures applied to the inverse Gaussian distribution.* Ph. D. Dissertation, Oklahoma State University, Stillwater.

[8] Pandey, B. N. (1997). Testimator of the scale parameter of the exponential distribution using LINEX loss function. *Communications in Statistics - Theory and Methods*, **26**, 2191 - 2200.

[9] Pandey, B. N. and Malik, H. J. (1988). Some improved estimators for a Measure of Dispersion of an inverse Gaussian Distribution. *Communications in Statistics - Theory and Methods*, **17**, 3935 - 3949.

[10] Pandey, B. N., Singh, B. P. and Mishra, C. S. (1996). Bayes estimation of shape parameter of classical Pareto distribution under LINEX loss function. *Communication in Statistics - Theory and Methods*, **25 (12)**, 3125 - 3145.

[11] Parsian, A. and Farsipour, N. S. (1999). Estimation of the mean of the selected population under asymmetric loss function. *Metrika*, **89 - 107.**

[12] Prakash, G. and Singh, D. C. (2006). Shrinkage testimators for the inverse dispersion of the inverse Gaussian distribution under the LINEX loss function. *Austrian Journal of Statistics*, **35(4)**, 463 - 470.

[13] Sadooghi - Alvandi, S. M. and Parsian, A. (1992). Estimation of the Binomial parameter in using a LINEX loss function. *Communication in Statistics - Theory and Methods*, **21**, 1427 - 1439.

[14] Seshadri, V. (1998). The inverse Gaussian distribution. *Statistical Theory and Applications*, Springer - Verlag, New York.

[15] Singh, D. C., Prakash, G. and Singh, P. (2007). Shrinkage testimators for the shape parameter of Pareto distribution using the LINEX loss function. *Communication in Statistics - Theory and Methods*, **36 (4)**, 741 - 753.

[16] Singh, U., Gupta, P. K. and Upadhyay, S. K. (2002). Estimation of exponentiated Weibull shape parameters under LINEX loss function. *Communication in statistics - Simulation*, **31(4)**, 523 - 537.

[17] Thompson, J. R. (1968). Some shrinkage techniques for estimating the mean.*Journal of the American Statistical Association*, **63**, 113 - 122.

[18] Varian, H. R (1975). *A Bayesian approach to real estate assessment.* In studies in Bayesian econometrics and statistics in honor of L. J. Savage, Eds S. E. Feinberg and A, Zellner, Amsterdam North Holland, **195 - 208**.

[19] Xiao, Y., Takada, Y. and Shi, N. (2005). Minimax confidence bound of the Normal mean under an asymmetric loss function. *Annals of Statistical Mathematics*, **57(1)**, 167 - 182.

[20] Zellner, A. (1986). Bayesian estimation and prediction using asymmetric loss function. *Journal of the American Statistical Association*, **81**, 446 - 451.

In: New Developments in Applied Statistics
Editor: Mohammad Ahsanullah

ISBN 978-1-61324-648-1

Chapter 10

INFERENCE AND PREDICTION FOR A GENERALIZED LOGISTIC DISTRIBUTION BASED ON THE kTH LOWER RECORDS

***Iwona Malinowska*[1,*] *and Dominik Szynal*[2,†]**
[1]Department of Mathematics, Lublin University of Technology
Nadbystrzycka 38a 20-618 Lublin, Poland
[2]Department of Mathematics, Maria Curie-Skłodowska University
Pl. M Curie-Sklodowska 1, 20-031 Lublin, Poland

Abstract

The minimum variance unbiased estimator, the maximum likelihood estimator and the Bayesian estimator for the parameter of the generalized logistic distribution are obtained based on kth lower record values. The Bayes estimators are obtain using both the symmetric squared error, squared log error loss function and the asymmetric LINEX and General Entropy loss functions and two new loss functions which we call Modified General Entropy (MGE) loss function and Kullback-Leibler divergence type loss function (KLD). Interval prediction for future kth upper record values is also presented from a Bayesian point of view. Numerical computations are given to illustrate these procedures.

Key Words and Phrases: Type I generalized logistic distribution, Minimum variance estimators, Maximum likelihood estimators, Bayes estimators, kth record values, LINEX loss function, General Entropy loss function, Modified General Entropy, Kullback-Leibler divergence, Prediction.

AMS 2000 Subject Classifications: 62F10, 62F15, 62G32.

1. Introduction

Let $\{X_n, n \geq 1\}$ be a sequence of independent identically distributed (iid) random variables with cumulative distribution function (cdf) $F(x)$ and probability density function

*E-mail address: i.malinowska@pollub.pl
†E-mail address: szynal@poczta.umcs.lublin.pl

(pdf) $f(x)$. The jth order statistic of a sample $(X_1, ..., X_n)$ is denoted by $X_{j:n}$. For a fixed $k \geq 1$ we define the sequence $L_k(n)$, $n \geq 1$, of kth lower record times of $\{X_n, n \geq 1\}$ as follows

$$L_k(1) = 1, \quad L_k(n+1) = \min\left\{j > L_k(n) : X_{k:L_k(n)+k-1} > X_{k:j+k-1}\right\}, \; n \geq 1.$$

The sequence $\{Z_n^{(k)}, n \geq 1\}$ with $Z_n^{(k)} = X_{k:L_k(n)+k-1}$, $n \geq 1$, is called the sequence of kth lower record values of $\{X_n, n \geq 1\}$. Note that $Z_1^{(k)} = \max\{X_1, ..., X_k\}$ and $Z_n^{(1)} = X_{L(n)}$, $n \geq 1$, are lower record values. It is known that

$$f_{Z_n^{(k)}}(z) = \frac{k^n}{(n-1)!}\left[-\ln F(z)\right]^{n-1}(F(z))^{k-1} f(z), \quad z \in \mathbb{R}, \tag{1}$$

and

$$\begin{aligned} & f_{Z_1^{(k)},...,Z_n^{(k)}}(z_1, z_2, ...z_n) \\ & = k^n (F(z_n))^{k-1} f(z_n) \prod_{i=1}^{n-1} \frac{f(z_i)}{F(z_i)}, \; z_1 > ... > z_n, \text{ (cf. [13]).} \end{aligned} \tag{2}$$

The probability density function and the corresponding cumulative distribution function of the one parameter type I generalized logistic distribution is

$$f(x|\theta) = \frac{\theta e^{-x}}{(1+e^{-x})^{\theta+1}}, \quad -\infty < x < +\infty, \; (\theta > 0), \tag{3}$$

and

$$F(x|\theta) = \frac{1}{(1+e^{-x})^{\theta}}, \quad -\infty < x < +\infty, \; (\theta > 0). \tag{4}$$

Estimators of parameters for various classes of distributions for example Gumbel, power, Weibull, Rayleigh, logistic, Pareto, based on record values were given in [1]. Moreover, two types of predictors of the sth record values based on the first m $(m < s)$ record values were presented there. The Bayesian estimators of parameters for the Gumbel distribution and Burr X distribution based on record values and Bayesian prediction of the sth lower record were discussed in [3], [12] and [11]. In [4] are obtained point and interval estimations for a generalized logistic distribution under progressive type II censoring.

This note contains the minimum variance unbiased estimator, the maximum likelihood estimator and the Bayesian estimator for the parameter of the type I generalized logistic distribution expressed in terms of kth lower record values. The Bayes estimators which assume that the prior distribution is a gamma distribution, are obtained using both symmetric and asymmetric error loss functions. The classical squared error, squared log error loss functions and a new Kullback-Leibler divergence type loss function are applied in the symmetric case. As asymmetric functions we use LINEX, General Entropy loss function and a new Modified General Entropy function. Interval prediction for future kth upper record values is also presented from a Bayesian point of view. A simulation study is included.

2. Conclusion

2.1. Non-Bayesian Inference

2.1.1. Point Estimation

Suppose we observe the first m kth lower record values $Z_1^{(k)} = x_1^{(k)}$, $Z_2^{(k)} = x_2^{(k)}, \ldots,$ $Z_m^{(k)} = x_m^{(k)}$ from the generalized logistic distribution type I, with cdf and pdf given by (3) and (4), respectively. By (2) the likelihood function is as follows

$$L(\theta|\underline{x}^{(k)}) = k^m \left(\prod_{i=1}^{m-1} \frac{f(x_i^{(k)})}{F(x_i^{(k)})}\right) [F(x_m^{(k)})]^{k-1} f(x_m^{(k)})$$

$$= k^m \left[1 + e^{-x_m^{(k)}}\right]^{-k\theta} \prod_{i=1}^{m} \frac{\theta e^{-x_i^{(k)}}}{1 + e^{-x_i^{(k)}}} \tag{5}$$

$$\propto k^m \theta^m e^{-k\theta T_m^{(k)}}, \quad x_1^{(k)} > x_2^{(k)} > \ldots > x_m^{(k)}$$

where

$$\underline{x}^{(k)} = (x_1^{(k)}, x_2^{(k)}, \ldots, x_m^{(k)}),$$

$$T_m^{(k)} = \ln\left(1 + e^{-x_m^{(k)}}\right). \tag{6}$$

Then using (5) the maximum likelihood (ML) estimator of θ is given by

$$\hat{\theta}_{ML}^{(k,m)} = \frac{m}{kT_m^{(k)}} \tag{7}$$

where $T_m^{(k)}$ is given by (6).

From (5) one can see that the statistic $T_m^{(k)}$ is sufficient and complete for the parameter θ and from (1) that it is distributed as gamma (m, kθ) with pdf

$$f(T_m^{(k)}|\theta) = \frac{(k\theta)^m}{\Gamma(m)} \left(T_m^{(k)}\right)^{m-1} e^{-k\theta T_m^{(k)}}, \quad T_m^{(k)} > 0. \tag{8}$$

So, the random variable $Z = \left(1/T_m^{(k)}\right)$ has the inverted gamma (m, $k\theta$) distribution with

$$E(Z|\theta) = \frac{k\theta}{m-1}, \quad Var(Z|\theta) = \frac{k^2\theta^2}{(m-1)^2(m-2)}, \qquad \text{(cf. [9])}, \tag{9}$$

and therefore

$$E(\hat{\theta}_{ML}^{(k,m)}|\theta) = \frac{m\theta}{(m-1)}, \quad Var(\hat{\theta}_{ML}^{(k,m)}|\theta) = \frac{m^2\theta^2}{(m-1)^2(m-2)}. \tag{10}$$

As mentioned earlier, the statistic $T_m^{(k)}$ is sufficient and complete for θ. It then follows from the Rao–Blackwell and Lehmann–Scheffe theorems that the minimum variance unbiased estimate (MVU) of θ is

$$\hat{\theta}_{MU}^{(k,m)} = \frac{m-1}{kT_m^{(k)}}, \tag{11}$$

with

$$E(\hat{\theta}_{MU}^{(k,m)}|\theta) = \theta, \quad Var(\hat{\theta}_{MU}^{(k,m)}|\theta) = \frac{\theta^2}{(m-2)}. \tag{12}$$

2.1.2. Interval Estimation

Writing $Q = 2k\theta T_m^{(k)}$, then from (8), Q is distributed as a Chi-squared rv with 2m degrees of freedom. Q is therefore a pivotal quantity for the parameter θ, and a $(1-\tau)100\%$ classical confidence interval for θ is given by

$$\left(\frac{q_1}{2kT_m^{(k)}}, \frac{q_2}{2kT_m^{(k)}}\right)$$

where $q_1 = \chi^2\left((\tau/2), 2m\right)$ and $q_2 = \chi^2\left(1-(\tau/2), 2m\right)$.

2.2. Bayesian Inference

Assume that the parameter θ is a realization of a rv Θ which has the gamma conjugate prior distribution of the form

$$\pi(\theta) = \frac{\beta^\alpha}{\Gamma(\alpha)}\theta^{\alpha-1}e^{-\beta\theta}, \quad \theta > 0, \, (\alpha, \, \beta > 0). \tag{13}$$

Combining (5) and (13), the posterior density of θ is a gamma distribution with parameters $(m+\alpha, kT_m^{(k)} + \beta)$, given by

$$\pi^*(\theta|\underline{x}^{(k)}) = \frac{(\beta + kT_m^{(k)})^{m+\alpha}}{\Gamma(m+\alpha)}\theta^{m+\alpha-1}e^{-(\beta+kT_m^{(k)})\theta}, \quad \theta > 0. \tag{14}$$

2.2.1. Symmetric Bayes Estimation

Assuming a squared error loss function

$$L(\theta, \widehat{\theta}) = (\theta - \widehat{\theta})^2$$

the Bayes estimate of θ is its posterior mean, or

$$\hat{\theta}_B^{(k,m)} = \frac{m+\alpha}{\beta + kT_m^{(k)}}. \tag{15}$$

Note that

$$\hat{\theta}_B^{(1,m)} = \frac{m+\alpha}{\beta + T_m^{(1)}}.$$

For $m = 1$

$$\hat{\theta}_B^{(1,1)} = \frac{1+\alpha}{\beta + T_1^{(1)}}$$

is the estimator based on a sample of size 1. Our approach allows us to give the Bayesian estimator of θ using a sample of size k. Namely, for $m = 1$ we have

$$\hat{\theta}_B^{(k,1)} = \frac{1+\alpha}{\beta + kT_1^{(k)}}.$$

Note that when α and β tend to zero then the estimator (15) tends to the estimator

$$\hat{\theta}_B^{(k,m)} = \frac{m}{kT_m^{(k)}}$$

which is the maximum likelihood estimator $\hat{\theta}_{ML}^{(k,m)}$ in (7). Combining (8) and (13) the marginal density of $T_m^{(k)}$ is

$$\begin{aligned} f(T_m^{(k)}) &= \int_0^\infty \frac{\beta^\alpha k^m}{\Gamma(m)\Gamma(\alpha)} \left(T_m^{(k)}\right)^{m-1} \theta^{\alpha+m-1} e^{-\theta(\beta+kT_m^{(k)})} d\theta \\ &= \frac{\beta^\alpha k^m}{B(m,\alpha)} \frac{\left(T_m^{(k)}\right)^{m-1}}{(\beta+kT_m^{(k)})^{m+\alpha}}, \quad T_m^{(k)} > 0 \end{aligned} \tag{16}$$

where $B(a,b)$ is the beta function, i.e.

$$B(a,b) = \int_0^1 t^{a-1}(1-t)^{b-1}dt, \quad a>0,\ b>0.$$

It follows from (16) that for $\hat{\theta}_B^{(k,m)}$ in (15) we have

$$\begin{aligned} E(\hat{\theta}_B^{(k,m)}) &= E\left(\frac{m+\alpha}{\beta+kT_m^{(k,m)}}\right) = \frac{\alpha}{\beta}, \\ Var(\hat{\theta}_B^{(k,m)}) &= Var\left(\frac{m+\alpha}{\beta+kT_m^{(k)}}\right) = \frac{\alpha m}{(m+\alpha+1)\beta^2}. \end{aligned} \tag{17}$$

Now we study the estimation of θ under the squared log error loss (cf. [10]). The squared log error has the form

$$L(\widehat{\theta},\theta) = (\ln\widehat{\theta} - \ln\theta)^2, \tag{18}$$

whose minimum occurs at $\widehat{\theta} = \theta$. The Bayes estimate $\widehat{\theta}_{BLG}$ of θ under the squared log error loss (18) is

$$\widehat{\theta}_{BLG} = \exp\left[\mathrm{E}_\theta\left(\ln\theta\right)\right],$$

provided that $\mathrm{E}_\theta\left(\ln\theta\right)$ exists and is finite. From (14) we have

$$\widehat{\theta}_{BLG}^{(k,m)} = \frac{\exp(\psi(m+\alpha))}{\beta+kT_m^{(k)}}$$

where $\psi(x) = \frac{d}{dx}\ln\Gamma(x)$.

We consider a new symmetric error loss function, called the Kullback-Leibler divergence type loss function for θ, given by

$$L\left(\widehat{\theta},\theta\right) = \theta^r \ln\left(\theta/\widehat{\theta}\right) + \widehat{\theta}^r \ln\left(\widehat{\theta}/\theta\right) \tag{19}$$

where $\widehat{\theta}$ is an estimate of θ and r is a positive number. The posteriori expectation of this loss function in (19) is

$$\mathrm{E}_\theta\left(L\left(\widehat{\theta},\theta\right)\right)=\mathrm{E}_\theta\left(\theta^r\ln\theta\right)-\ln\widehat{\theta}\,\mathrm{E}_\theta\,\theta^r+\left(\widehat{\theta}\right)^r\ln\widehat{\theta}-\left(\widehat{\theta}\right)^r\mathrm{E}_\theta\ln\theta$$

where $\mathrm{E}_\theta\left(.\right)$ denotes the posteriori expectation with respect to the posterior density of θ. The Bayes estimator of θ, denoted by $\widehat{\theta}:=\widehat{\theta}_{BKL}$ under the error loss function (19) is the solution of the equation

$$r\left(\widehat{\theta}_{BKL}\right)^r\ln\widehat{\theta}_{BKL}+\left(\widehat{\theta}_{BKL}\right)^r-r\left(\widehat{\theta}_{BKL}\right)^r\mathrm{E}_\theta\ln\theta=\mathrm{E}_\theta\left(\theta^r\right),$$

provided that the above expectations exist and are finite. They are obtained from (14). So far from the type I generalized logistic distribution the Bayes estimate $\widehat{\theta}_{BKL}$ for θ is the solution of the equation

$$\begin{aligned}&r\left(\widehat{\theta}^{(k,m)}_{BKL}\right)^r\ln\widehat{\theta}^{(k,m)}_{BKL}+\left(\widehat{\theta}^{(k,m)}_{BKL}\right)^r-r\left(\widehat{\theta}^{(k,m)}_{BKL}\right)^r\left(\psi(m+\alpha)-\ln\left(\beta+kT^{(k)}_m\right)\right)\\&=\frac{\Gamma(m+\alpha+p)}{\Gamma(m+\alpha)}\left(\beta+kT^{(k)}_m\right)^{-r},\quad a.s.\text{ (almost surely).}\end{aligned}$$

2.2.2. Bayes Estimation Based on the LINEX Loss Function

An useful asymmetric loss function, known as the LINEX loss function (linear-exponential), was introduced in [15] and was widely used by several authors. This function rises approximately exponentially on one side of zero and approximately linearly on other side. Under the assumption that the minimal loss occurs at $\widehat{\theta}=\theta$, the LINEX loss function for θ can be expressed as

$$L(\triangle)=e^{a\triangle}-a\triangle-1,\quad a\neq 0 \tag{20}$$

where $\triangle=\widehat{\theta}-\theta$, and $\widehat{\theta}$ is an estimate of θ ([cf. [15]]). The sign of a represents the direction and its magnitude represents the degree of symmetry. First, for $a=1$ the LINEX loss function is quite asymmetric about zero with overestimation being more costly than underestimation. Second, if $a<0$, $L(\triangle)$ rise exponentially when $\triangle<0$ (underestimation) and almost linearly when $\triangle>0$ (overestimation). For a near zero $L(\triangle)$ approximates the squared error loss function, and so is almost symmetric.

The posteriori expectation of $L(\triangle)$ is

$$\mathrm{E}_\theta\left(L(\triangle)\right)=e^{a\widehat{\theta}}\,\mathrm{E}_\theta\left(e^{-a\theta}\right)-a\left(\widehat{\theta}-\mathrm{E}_\theta(\theta)\right)-1$$

where $\mathrm{E}_\theta\left(.\right)$ denoting posteriori expectation with respect to the posterior density of θ. The Bayes estimator of θ, denoted by $\widehat{\theta}_{BL}$ under the LINEX loss function is the value $\widehat{\theta}$, that minimizes (20). It is given by

$$\widehat{\theta}_{BL}=-\frac{1}{a}\ln\left(\mathrm{E}_\theta\,e^{-a\theta}\right),$$

provided that $\mathrm{E}_\theta\left(e^{-a\theta}\right)$ exists and is finite (see [6]).
From (14) we have

$$\widehat{\theta}_{BL}^{(k,m)} = -\frac{1}{a}\ln\left[\left(\frac{\beta + kT_m^{(k)}}{\beta + kT_m^{(k)} + a}\right)^{\alpha+m}\right].$$

2.2.3. Bayes Estimators Using General Entropy Loss

Another useful asymmetric loss function is the General Entropy loss, for which

$$L(\widehat{\theta},\theta) \propto (\widehat{\theta}/\theta)^q - q\ln(\widehat{\theta}/\theta) - 1 \tag{21}$$

and its minimum occurs at $\widehat{\theta} = \theta$. This function is a generalization of the entropy loss used by several authors, where the shape parameter q is equal to 1 (see for example, [7], [14]). The Bayes estimate $\widehat{\theta}_{BGE}$ of θ under the General Entropy loss is

$$\widehat{\theta}_{BGE} = \left[\mathrm{E}_\theta\left(\theta^{-q}\right)\right]^{-\frac{1}{q}},$$

provided that $\mathrm{E}_\theta(\theta)^{-q}$ exists and is finite.
From (14) we have

$$\widehat{\theta}_{BGE}^{(k,m)} = \left[\frac{\Gamma(\alpha+m)}{\Gamma(\alpha+m-q)}\right]^{\frac{1}{q}} \frac{1}{\left(\beta + kT_m^{(k)}\right)}.$$

Now we consider a negative loss function for which the condition $L(\widehat{\theta},\theta) \geq -c > -\infty$ (cf. [10]) is satisfied, and we modify (21) as follows

$$L(\widehat{\theta},\theta) \propto (\widehat{\theta}/\theta)^q - q\theta\ln(\widehat{\theta}/\theta) - 1. \tag{22}$$

We call it the Modified General Entropy loss function. The Bayes estimate $\widehat{\theta}_{BMGE}$ of θ under the Modified General Entropy loss is

$$\widehat{\theta}_{BMGE} = \left[\frac{\mathrm{E}_\theta\left(\theta\right)}{\mathrm{E}_\theta\left(\theta^{-q}\right)}\right]^{\frac{1}{q}},$$

provided that $\mathrm{E}_\theta\left(\cdot\right)$ exists and is finite.
From (14) we have

$$\widehat{\theta}_{BMGE}^{(k,m)} = \left[\frac{\Gamma(\alpha+m)}{\Gamma(\alpha+m-q)} \frac{m+\alpha}{\left(\beta + kT_m^{(k)}\right)^{q+1}}\right]^{\frac{1}{q}}.$$

2.2.4. Interval Estimation

Writing $R = 2(\beta + kT_m^{(k)})\theta$, then from (14), R is distributed as a Chi-squared rv with $2(m+\alpha)$ degrees of freedom. R is therefore a pivotal quantity for the parameter θ and a $(1-\tau)100\%$ Bayesian confidence interval for θ is given by

$$\left(\frac{r_1}{2(\beta + kT_m^{(k)})}, \frac{r_2}{2(\beta + kT_m^{(k)})}\right)$$

where $r_1 = \chi^2\left((\tau/2), 2(m+\alpha)\right)$ and $r_2 = \chi^2\left(1-(\tau/2), 2(m+\alpha)\right)$.

2.3. Bayesian Prediction

Assume that we have the first m kth lower records $Z_1^{(k)} = x_1^{(k)}$, $Z_2^{(k)} = x_2^{(k)}, \ldots, Z_m^{(k)} = x_m^{(k)}$ from the type I generalized logistic distribution. Based on that sample, we would like to predict the sth one of the kth lower record values, $1 < m < s$. Now let $Y^{(k)} = Z_s^{(k)}$ be the sth lower record value, $1 < m < s$. The conditional pdf of $Y^{(k)}$ given the parameters θ and the observed value $x_m^{(k)}$ of $Z_m^{(k)}$ is

$$f^*(y^{(k)}|\theta) = \frac{k^{s-m}}{\Gamma(s-m)}[H(y^{(k)}) - H(x_m^{(k)})]^{s-m-1}\left(\frac{F(y^{(k)})}{F(x_m^{(k)})}\right)^{k-1}\frac{f(y_s^{(k)})}{F(x_m^{(k)})} \tag{23}$$

where $f(\cdot)$, $F(\cdot)$ are the pdf and cdf, respectively, and

$$H(y_s^{(k)}) = -\ln F(y_s^{(k)}) = \theta T_s^{(k)}, \quad H(x_m^{(k)}) = -\ln F(x_m^{(k)}) = \theta T_m^{(k)}$$

where $T_s^{(k)}$ is obtained from (6). Applying (3) and (4) in (23) we get

$$f^*(y_s^{(k)}|\theta) = \frac{k^{s-m}}{\Gamma(s-m)}[T_s^{(k)} - T_m^{(k)}]^{s-m-1}\theta^{s-m}\left(\frac{y_s^{(k)}e^{-y_s^{(k)}}}{1+e^{-y_s^{(k)}}}\right)e^{-(k\theta(T_s^{(k)}-T_m^{(k)}))},$$
$$-\infty < y_s^{(k)} < x_m^{(k)} < \infty.$$

Combining the posterior density, given by (14), and integrating out the parameters θ, we obtain the Bayesian predictive density function of $Y_s^{(k)} = Z_s^{(k)}$ given the past m kth lower record values, in the form

$$f(y_s^{(k)}|\underline{x}^{(k)}) = \int_0^\infty f^*(y_s^{(k)}|\theta)\pi^*(\theta|\underline{x}^{(k)})d\theta$$
$$= Ak^{s-m}\left(\frac{y_s^{(k)}e^{-y_s^{(k)}}}{1+e^{-y_s^{(k)}}}\right)\left[\frac{T_s^{(k)} - T_m^{(k)}}{\beta + kT_s^{(k)}}\right]^{s-m-1}\left[\frac{\beta + kT_m^{(k)}}{\beta + kT_s^{(k)}}\right]^{\alpha+m+1}$$
$$-\infty < y_s^{(k)} < x_m^{(k)} < \infty$$

where

$$A^{-1} = (\beta + kT_m^{(k)})B(s-m, \alpha+m).$$

The Bayesian prediction bounds for $Y_s^{(k)} = Z_s^{(k)}$ are obtained by evaluating

$$Pr(Y_s^{(k)} \geq \gamma | \underline{x}^{(k)}) = \int_{\gamma}^{x_m^{(k)}} f(y_s^{(k)} | \underline{x}^{(k)}) dy_s^{(k)}$$

$$= Ak \int_{\gamma}^{x_m^{(k)}} \left(\frac{y_s^{(k)} e^{-y_s^{(k)}}}{1 + e^{-y_s^{(k)}}} \right) \left[\frac{kT_s^{(k)} - kT_m^{(k)}}{\beta + kT_s^{(k)}} \right]^{s-m-1} \left[\frac{\beta + kT_m^{(k)}}{\beta + kT_s^{(k)}} \right]^{\alpha+m+1} dy_s^{(k)}.$$

Using the transformation

$$w = \frac{kT_s^{(k)} - kT_m^{(k)}}{\beta + kT_s^{(k)}},$$

the above integral simplifies to

$$Pr(Y_s^{(k)} \geq \gamma | \underline{x}^{(k)}) = \frac{1}{B(s-m, \alpha+m)} \int_0^{w_\gamma} w^{s-m-1} (1-w)^{m+\alpha-1} dw = F_B(w_\gamma) \tag{24}$$

where

$$\underline{x}^{(k)} = (x_1^{(k)}, x_2^{(k)}, ..., x_m^{(k)}), \quad w_\gamma = \frac{kT_\gamma^{(k)} - kT_m^{(k)}}{\beta + kT_\gamma^{(k)}}, \quad T_\gamma^{(k)} = \ln\left(1 + e^{-\gamma}\right), \tag{25}$$

and $F_B(\cdot)$ is the Beta cdf with parameters $(s-m, \alpha+m)$.

The $(1-\tau)100\%$ predictive interval for $Y_s^{(k)} = Z_s^{(k)}$ is obtained by evaluating both the lower, $L(\underline{x}^{(k)})$, and upper, $U(\underline{x}^{(k)})$, limits which satisfy

$$Pr(Y_s^{(k)} > L(\underline{x}^{(k)}) | \underline{x}^{(k)}) = 1 - \frac{\tau}{2}, \quad Pr(Y_s^{(k)} > U(\underline{x}^{(k)}) | \underline{x}^{(k)}) = \frac{\tau}{2}.$$

Thus, applying (24), one can obtain the lower and upper prediction bounds of $Y_s^{(k)} = Z_s^{(k)}$, analytically, in the forms

$$L(\underline{x}^{(k)}) = -\ln\left\{-1 + \exp\left(\frac{kT_m^{(k)} + \beta\delta_1(s)}{k(1-\delta_1(s))}\right)\right\}$$

$$U(\underline{x}^{(k)}) = -\ln\left\{-1 + \exp\left(\frac{kT_m^{(k)} + \beta\delta_2(s)}{k(1-\delta_2(s))}\right)\right\}$$

where

$$\delta_1(s) = F_B^{-1}\left(1 - \frac{\tau}{2}\right), \quad \delta_2(s) = F_B^{-1}\left(\frac{\tau}{2}\right).$$

For the special case $s = m+1$, when predicting the next kth lower record $Y_{m+1}^{(k)} = Z_{m+1}^{(k)}$, we see that (24) reduces to

$$\begin{aligned} Pr(Y_{m+1}^{(k)} \geq \gamma | \underline{x}^{(k)}) &= (\alpha+m) \int_0^{w_\gamma} (1-w)^{m+\alpha-1} dw \\ &= 1 - (1-w_\gamma)^{m+\alpha} \end{aligned} \tag{26}$$

where w_γ is given by (25). By using (26), the lower and upper prediction bounds for $Y_{m+1}^{(k)}$ are given by

$$L_1(\underline{x}^{(k)}) = -\ln\left\{-1+\exp\left(\frac{kT_m^{(k)}+\beta(1-\zeta_1)}{k\zeta_1}\right)\right\}$$

$$U_1(\underline{x}^{(k)}) = -\ln\left\{-1+\exp\left(\frac{kT_m^{(k)}+\beta((1-\zeta_2))}{k\zeta_2}\right)\right\}$$

where

$$\zeta_1 = \left(\frac{\tau}{2}\right)^{1/m+\alpha}, \quad \zeta_2 = \left(1-\frac{\tau}{2}\right)^{1/m+\alpha}.$$

Table 1. The estimators $\widehat{\theta}_{ML}^{(k,m)}$, $\widehat{\theta}_{MU}^{(k,m)}$, $\widehat{\theta}_{B}^{(k,m)}$, $\widehat{\theta}_{BLG}^{(k,m)}$, $\widehat{\theta}_{BKL}^{(k,m)}$, $\widehat{\theta}_{BL}^{(k,m)}$, $\widehat{\theta}_{BGE}^{(k,m)}$ and $\widehat{\theta}_{BMGE}^{(k,m)}$ of $\theta = 1.163$

m	k	$\widehat{\theta}_{ML}$	$\widehat{\theta}_{MU}$	$\widehat{\theta}_{B}$	$\widehat{\theta}_{BLG}$	$\widehat{\theta}_{2,BKL}$	$\widehat{\theta}_{3,BKL}$	$\widehat{\theta}_{4,BKL}$	$\widehat{\theta}_{BL}$	$\widehat{\theta}_{BGE}$	$\widehat{\theta}_{BMGE}$
3	1	1.803	**1.202**	1.364	**1.230**	1.362	1.432	1.506	0.996	1.870	1.757
	2	1.525	**1.017**	1.260	**1.137**	1.259	1.322	1.391	0.939	1.728	1.649
	3	**1.199**	0.799	1.110	1.001	1.109	**1.165**	1.226	0.851	1.522	1.490
	4	**1.124**	0.749	1.071	0.965	1.069	1.123	**1.182**	0.827	1.468	1.448
	5	0.899	0.599	0.937	0.845	0.935	**0.983**	**1.034**	0.743	1.284	1.301
	6	0.769	0.513	0.848	0.764	0.846	0.889	0.936	0.685	**1.162**	**1.201**
4	1	2.330	1.747	1.614	**1.482**	1.612	1.681	1.754	**1.183**	2.118	1.924
	2	2.019	1.514	1.507	**1.383**	1.505	1.569	1.637	**1.123**	1.977	1.821
	3	1.356	1.017	1.212	**1.113**	**1.211**	1.262	1.317	0.947	1.590	1.530
	4	1.199	0.899	**1.124**	1.032	1.123	**1.171**	1.222	0.892	1.475	1.441
	5	0.959	0.719	0.972	0.892	0.971	1.012	**1.056**	0.792	**1.276**	1.283
	6	0.999	0.449	0.999	0.917	0.998	**1.041**	**1.086**	0.810	1.311	1.311
5	1	2.256	1.805	1.660	**1.543**	1.659	1.719	1.783	**1.254**	2.107	1.904
	2	2.183	1.746	1.631	**1.516**	1.630	1.689	1.752	**1.237**	2.071	1.878
	3	1.682	1.346	1.408	**1.308**	1.407	1.459	1.512	**1.101**	1.787	1.669
	4	1.444	1.155	1.281	**1.191**	1.280	1.327	1.376	**1.021**	1.627	1.548
	5	1.155	0.924	1.106	1.028	1.105	**1.145**	**1.188**	0.905	1.404	1.376
	6	0.999	0.799	0.999	0.928	0.998	1.034	**1.073**	0.831	**1.268**	1.269
6	1	1.628	1.357	1.407	**1.320**	1.406	1.451	1.498	**1.130**	1.741	1.626
	2	1.921	1.601	1.561	**1.465**	1.560	1.610	1.662	**1.229**	1.932	1.767
	3	1.746	1.455	1.472	**1.381**	1.471	1.517	1.567	**1.172**	1.821	1.685
	4	1.525	1.271	1.348	**1.265**	1.347	1.390	1.435	**1.091**	1.668	1.571
	5	1.220	1.017	**1.156**	1.085	**1.156**	**1.192**	1.231	0.961	1.431	1.390
	6	**1.155**	0.962	1.112	1.043	1.111	**1.146**	1.184	0.929	1.376	1.347

$\widehat{\theta}_{2,BKL}$ estimators $\widehat{\theta}_{BKL}^{(k,m)}$ with $r=2$,
$\widehat{\theta}_{3,BKL}$ estimators $\widehat{\theta}_{BKL}^{(k,m)}$ with $r=3$,
$\widehat{\theta}_{4,BKL}$ estimators $\widehat{\theta}_{BKL}^{(k,m)}$ with $r=4$.

2.4. Numerical Illustration

Here we discuss some numerical results of a simulation study testing the performance of non-Bayesian and Bayesian methods in estimating the parameter of the type I generalized logistic distribution. We generate the first m ($m = 6$) upper kth record values ($k = 1, 2, 3, 4, 5, 6$) from the generalized logistic distribution with the parameter $\theta = 1.163$ which was generated via the gamma conjugate priori distribution with the parameters $\alpha = 2.0$ and $\beta = 2.0$. In the prediction we take $m = 7$. The values of the shape parameter a in the LINEX loss function, and value of q in GE and MGE loss functions, respectively, are $a = 3$ and $q = -5$. In the Kullback-Leibler divergence type loss function we take with $r = 2$, 3 and 4. The point estimators $\widehat{\theta}_{ML}^{(k,m)}$, $\widehat{\theta}_{MU}^{(k,m)}$, $\widehat{\theta}_{B}^{(k,m)}$, $\widehat{\theta}_{BKL}^{(k,m)}$, $\widehat{\theta}_{BL}^{(k,m)}$, $\widehat{\theta}_{BGE}^{(k,m)}$, $\widehat{\theta}_{BMGE}^{(k,m)}$ and $\widehat{\theta}_{BLG}^{(k,m)}$ of θ are given in Table 1, where we omit (k, m) (bold numbers in rows are for the two closest estimates to the estimated values). Table 2 contains the MSE of those estimators. The 95% classical and Bayesian confidence intervals for θ and also the 95% Bayesian prediction bounds for the next kth record $x_{m+1}^{(k)}$ are presented in the Table 3. The MSE based on 1000 replications.

Table 2. MSE of the estimators $\widehat{\theta}_{ML}^{(k,m)}$, $\widehat{\theta}_{MU}^{(k,m)}$, $\widehat{\theta}_{B}^{(k,m)}$, $\widehat{\theta}_{BLG}^{(k,m)}$, $\widehat{\theta}_{BKL}^{(k,m)}$, $\widehat{\theta}_{BL}^{(k,m)}$, $\widehat{\theta}_{BGE}^{(k,m)}$ and $\widehat{\theta}_{BMGE}^{(k,m)}$ of $\theta = 1.163$

m	k	$\widehat{\theta}_{ML}$	$\widehat{\theta}_{MU}$	$\widehat{\theta}_{B}$	$\widehat{\theta}_{BLG}$	$\widehat{\theta}_{2,BKL}$	$\widehat{\theta}_{3,BKL}$	$\widehat{\theta}_{4,BKL}$	$\widehat{\theta}_{BL}$	$\widehat{\theta}_{BGE}$	$\widehat{\theta}_{BMGE}$
3	1	2.128	0.678	0.104	**0.063**	0.103	0.137	0.186	**0.059**	0.578	0.386
	2	1.987	0.815	0.097	**0.089**	**0.096**	0.111	0.136	0.111	0.384	0.259
	3	1.524	0.681	**0.105**	0.108	**0.105**	0.113	0.130	0.137	0.328	0.221
	4	1.117	0.548	**0.108**	0.118	**0.108**	0.113	0.125	0.151	0.295	0.199
	5	1.038	0.543	**0.116**	0.129	**0.116**	0.118	0.127	0.164	0.277	0.187
	6	0.997	0.546	**0.118**	0.135	**0.118**	0.119	0.125	0.172	0.259	0.175
4	1	1.052	0.415	0.109	**0.068**	0.108	0.138	0.179	**0.044**	0.495	0.319
	2	0.781	0.389	0.102	**0.089**	0.102	0.117	0.139	**0.091**	0.343	0.221
	3	0.632	0.358	**0.106**	**0.103**	**0.106**	0.115	0.129	0.115	0.291	0.188
	4	0.511	0.332	**0.103**	0.108	**0.103**	0.107	0.117	0.130	0.243	0.158
	5	0.765	0.476	**0.120**	0.126	**0.120**	0.124	0.133	0.143	0.254	0.163
	6	0.587	0.402	**0.118**	0.129	**0.118**	0.119	0.125	0.151	0.227	0.146
5	1	0.595	0.260	0.103	**0.068**	0.102	0.128	0.160	**0.038**	0.409	0.257
	2	0.512	0.293	0.100	**0.087**	0.100	0.113	0.131	**0.080**	0.294	0.184
	3	0.409	0.270	**0.100**	**0.097**	**0.100**	0.107	0.118	0.102	0.239	0.150
	4	0.332	0.246	**0.096**	0.099	**0.096**	0.100	0.107	0.113	0.204	0.129
	5	0.340	0.263	**0.105**	0.111	**0.105**	0.107	0.113	0.125	0.201	0.126
	6	0.302	0.255	**0.106**	0.115	**0.106**	**0.106**	0.109	0.134	0.182	0.115
6	1	0.486	0.256	0.111	**0.082**	0.110	0.130	0.155	**0.048**	0.348	0.213
	2	0.356	0.225	0.096	**0.084**	0.095	0.106	0.120	**0.075**	0.248	0.152
	3	0.284	0.208	**0.093**	**0.091**	**0.093**	0.098	0.107	**0.093**	0.200	0.123
	4	0.262	0.212	**0.096**	0.099	**0.096**	0.0.98	0.104	0.107	0.177	0.109
	5	0.261	0.221	**0.102**	0.107	**0.102**	0.103	0.107	0.117	0.172	0.106
	6	0.221	0.206	**0.098**	0.106	**0.098**	**0.097**	0.099	0.122	0.151	0.094

$\widehat{\theta}_{2,BKL}$ estimators $\widehat{\theta}_{BKL}^{(k,m)}$ with $r = 2$.

$\widehat{\theta}_{3,BKL}$ estimators $\widehat{\theta}_{BKL}^{(k,m)}$ with $r = 3$.

$\widehat{\theta}_{4,BKL}$ estimators $\widehat{\theta}_{BKL}^{(k,m)}$ with $r = 4$.

Table 3. The 95% classical and Bayesian confidence intervals for θ, the 95% Bayesian prediction bounds for the next kth record $x_{m+1}^{(k)}$.

m	k	θ	Conf.int.	Bayes pred. int.
3	1	1.163	$(0.283, 3.313)^a$	-1.789^c
			$(0.388, 2.449)^b$	$(-4.621, -0.822)^d$
	2		$(0.530, 6.075)^a$	-1.550^c
			$(0.509, 3.211)^b$	$(-3.510, 0.815)^d$
	3		$(0.378, 4.414)^a$	-0.799^c
			$(0.446, 2.816)^b$	$(-1.869, 0.149)^d$
	4		$(0.283, 3.311)^a$	-0.722^c
			$(0.388, 2.448)^b$	$(-1.656, 0.151)^d$
	5		$(0.228, 2.648)^a$	0.541^c
			$(0.343, 2.166)^b$	$(-1.103, 0.537)^d$
	6		$(0.289, 2.207)^a$	0.548^c
			$(0.307, 1.942)^b$	$(-0.287, 1.333)^d$
4	1		$(0.380, 3.057)^a$	-2.057^c
			$(0.452, 2.397)^b$	$(-5.288, -1.809)^d$
	2		$(0.801, 6.445)^a$	-1.789^c
			$(0.655, 3.472)^b$	$(-4.055, -1.564)^d$
	3		$(0.611, 4.915)^a$	-1.550^c
			$(0.582, 3.083)^b$	$(-2.663, -0.810)^d$
	4		$(0.458, 3.866)^a$	-0.799^c
			$(0.503, 2.665)^b$	$(-2.406, -0.732)^d$
	5		$(0.366, 2.949)^a$	0.163^c
			$(0.443, 2.346)^b$	$(-0.824, 0.532)^d$
	6		$(0.305, 2.457)^a$	0.541^c
			$(0.395, 2.096)^b$	$(-0.734, 0.540)^d$
5	1	1.163	$(0.456, 2.879)^a$	-2.091^c
			$(0.506, 2.350)^b$	$(-5.070, -2.074)^d$
	2		$(1.133, 6.7.147)^a$	-2.057^c
			$(0.820, 3.804)^b$	$(-3.968, -1.802)^d$
	3		$(0.795, 5.018)^a$	-1.789^c
			$(0.695, 3.232)^b$	$(-3.381, -1.561)^d$
	4		$(0.597, 3.764)^a$	-1.158^c
			$(0.596, 2.766)^b$	$(-2.227, -0.808)^d$
	5		$(0.477, 3.011)^a$	-0.559^c
			$(0.521, 2.418)^b$	$(-1.007, -0.447)^d$
	6		$(0.398, 2.509)^a$	0.164^c
			$(0.463, 2.147)^b$	$(-0.554, 0.534)^d$
6	1		$(0.521, 2.762)^a$	-2.166^c
			$(0.505, 2.317)^b$	$(-4.662, -2.106)^d$
	2		$(1.087, 5.762)^a$	-2.091^c
			$(0.858, 3.583)^b$	$(-4.021, -2.068)^d$
	3		$(1.024, 5.428)^a$	-2.057^c
			$(0.832, 3.475)^b$	$(-3.442, -1.799)^d$
	4		$(0.768, 4.071)^a$	-1.239^c
			$(0.709, 2.964)^b$	$(-2.483, -1.166)^d$
	5		$(0.614, 3.257)^a$	-0.722^c
			$(0.618, 2.583)^b$	$(-1.664, -0.566)^d$
	6		$(0.512, 2.714)^a$	-0.559^c
			$(0.548, 2.289)^b$	$(-0.798, 0.157)^d$

^{a} The 95% Bayesian confidence interval for θ.

^{b} The 95% classical confidence interval for θ.

^{c} The generated upper kth record $x_{m+1}^{(k)}$.

^{d} The 95% Bayesian prediction interval for $x_{m+1}^{(k)}$.

Comments. As can be seen from Table 1 and Table 2

1. Simulations show the performance of the obtained estimators under many classical and modified loss functions.

2. The Bayes estimations $\widehat{\theta}_{2,BKL}^{(k,m)}$ under Kullback - Leibler divergence type loss function with $r = 2$ are almost equal to the Bayes estimators $\widehat{\theta}_{B}^{(k,m)}$.

3. The MSE of the estimators $\widehat{\theta}_{ML}^{(k,m)}$, $\widehat{\theta}_{MU}^{(k,m)}$ and the Bayes estimators under squared error, squared log error, Kullback - Leibler divergence type and LINEX, General Entropy, Modified General Entropy loss functions decrease when m increases.

4. The MSE of the estimators Bayes estimators under General Entropy , Modified General Entropy loss functions decrease when k increases.

5. The generated kth lower record $x_{m+1}^{(k)}$ belongs to the 95% Bayesian prediction interval for $x_{m+1}^{(k)}$.

References

[1] Ahsanullah, M. (2004). *Record Values - Theory and Applications*. University Press of America, Lanham, Maryland, USA.

[2] Ali Mousa, M. A. M. (2001). Inference and prediction for the Burr type X model based on records. *Statistics*, **35(4)**, 415-425.

[3] Ali Mousa, M. A. M., Jaheen, Z. F. and Ahmed, A. A. (2002). Bayesian estimation, prediction and characterization for the Gumbel model based on records. *Statistics*, **36(1)**, 65-74.

[4] Asgharzadeh, A., (2006). Point and interval estimation for a generalized logistic distribution under progressive type II censoring *Commun. Statist. Theor. Math.*, **32 (4)**, 1685-1702.

[5] Atchison, J., and Dunsmore, I. R. (1975). Statistical Prediction Analysis. *Cambridge University Press*

[6] Calambria, G., and Plucini, G. (1996). Point estimation under asymmetric loss functions for left truncated exponential sample *Commun. Statist. Theor. Math.*, **25 (3)**, 285-600.

[7] Dey, D. K., Ghosh, M. and Srinivasan, C. (1987). Simultaneous estimation of parameters under entropy loss. *J. Statist. Plann. Infer.* **25**, 347-363.

[8] Dziubdziela, W. and Kopociski, B. (1976). Limiting properties of the k-th record values. *Appl. Math.,* **15**, 187-190.

[9] Lin, C. T., Duran, B. S. and Lewis, T. O. (1989). Inverted gamma as a life distribution. *Microelectron. Reliab.,* **29 (4)**, 619-626.

[10] Ferguson, T.S (1967). *Mathematical Statistics*. Academic Press , New York and London, England.

[11] Malinowska, I. and Szynal, D. (2004). On a family of Bayesian estimators and predictors for Gumbel model based on the kth lower records. *Appl. Math.,* **31(1)**, 107-115.

[12] Malinowska, I. and Szynal, D. (2005). Inference and prediction for the Burr type X model based on the kth lower record. *J. Statit. Theor. Appl.* **4(3)**, 107-115.

[13] Pawlas, P. and Szynal, D. (1998). Relations for single and product moments of kth record values from exponential and Gumbel distributions. *J. Appl. Statist. Sci.,* **7(1)**, 53-62.

[14] Soliman, A. A. (2006). Estimators for finite mixture of Rayleight model based on progressively censored date. *Commun. Statist. Theor. Meth.* **32**, 803-820.

[15] Varian, H.R. (1975). A Bayesian approch to real estate assessment. *In Studies in Bayesian Econometrics and statistics in Honer of Leonard J. Savage, North Holland, Amsterdam*, 195-208.

In: New Developments in Applied Statistics
Editor: Mohammad Ahsanullah

ISBN 978-1-61324-648-1

Chapter 11

ON TESTING SEASONALITY OF PERIODIC AUTOREGRESSIVE TIME SERIES

Q. Shao
Department of Mathematics
The University of Toledo

Summary

The number of parameters in a periodic autoregressive model is a multiple of the period. Large sample tests on the parameters of interest may be misleading when the sample size is small and the period is large. The modified profile likelihood proposed in the paper aims at reducing the adverse effect of the nuisance parameters on the inference of seasonality. The likelihood ratio test, the Wald test and the score test are compared by means of a simulation study. The technique is illustrated by the analysis of two real data sets.

Key Words: periodically stationary time series, periodic autoregressive time series model, seasonality, profile likelihood function, parameter orthogonality, likelihood ratio test, Wald test, score test.

1. Introduction

Periodically stationary time series $\{Y_t; t = 0, \pm 1, \pm 2, \cdots\}$ can be found in many fields, such as hydrology (Hipel and McLeod (1994)) and economics (Franses and Paap (2004)). Unlike its stationary counterpart, a periodically stationary time series exhibits a periodic pattern in the first two moments. Specifically, for all integers m and n

$$\mathrm{E}(Y_{n+T}) = \mathrm{E}(Y_n) \text{ and } \mathrm{Cov}(Y_{n+T}, Y_{m+T}) = \mathrm{Cov}(Y_n, Y_m), \tag{1.1}$$

where the period T is the smallest integer that satisfies (1.1). To emphasize periodicity, we often use the subscript $iT + \nu$ instead of t to denote the time series in the νth season ($1 \leq \nu \leq T$) of the ith cycle ($i = 0, \pm 1, \pm 2, \cdots$). From (1.1), it follows that

$\text{Cov}(Y_{iT+\nu}, Y_{iT+\nu-k}) = \text{Cov}(Y_\nu, Y_{\nu-k})$ for any integer k. Therefore, the covariance of such a time series depends on both ν and k.

Periodic autoregressive models are usually applied in modeling such time series; that is,

$$Y_{iT+\nu} - \mu_\nu = \sum_{k=1}^{p} \phi_k(\nu)(Y_{iT+\nu-k} - \mu_{\nu-k}) + \epsilon_{iT+\nu}, \tag{1.2}$$

where $\text{E}(Y_{iT+\nu}) = \mu_\nu$ and $\{\epsilon_{iT+\nu}\}$ is white noise with $\text{E}(\epsilon_{iT+\nu}) = 0$ and $\text{Var}(\epsilon_{iT+\nu}) = \sigma_\nu^2$. A time series generated by a dynamic system is periodically stationary (see Vecchia (1985)) subject to some restriction on the parameters $\{\phi_1(1), \cdots, \phi_p(1), \cdots, \phi_1(T), \cdots, \phi_p(T)\}$. A periodically stationary time series that satisfies the difference equation (1.2) is called a periodic autoregressive time series with order p (PAR(p)).

Although the periodic autoregressive model class works well for many periodic time series, there are two commonly encountered difficulties. One difficulty is that the number of observations for each season is usually very small. For instance, for a monthly data set of ten years, there are only ten observations for each month, whereas the whole data set contains one hundred and twenty observations. However, as is well known, the accuracy of the analysis relies on the number of periods when using the asymptotic distribution of a statistic. The other difficulty is that there are many parameters for a periodic time series with a large period, as the number of parameters in a periodic autoregressive model is a multiple of the period.

The primary interest of this paper is testing seasonality in the first moment for periodic autoregressive time series; that is,

$$H_0 : \mu_1 = \cdots = \mu_T \text{ vs. } H_a : \text{ at least one } \mu \text{ is different.} \tag{1.3}$$

When the number of observations is relatively small and the period is large, the conclusion for the hypotheses by the likelihood ratio test, the Wald test, or the score test is adversely affected by the estimation of the nuisance parameters. The correction of this adverse effect in the presence of a large number of nuisance parameters for a small sample size of a periodic autoregressive time series has not received enough attention and deserves special consideration.

The discussion of bias reduction due to the presence of nuisance parameters has a long history. The parametric approach focuses on the modification of the profile likelihood function in which the nuisance parameters are replaced by the parameters of interest. Barndorff and Cox (1984) applied the so-called Bartlett correction and obtained a modified likelihood function. Based on this function, the likelihood ratio test is approximately a χ^2 distribution with an error of order $O(n^{-2})$. Cox and Reid (1987) attacked this problem from the conditional likelihood approach under the assumption that the parameters of interest and the nuisance parameters are orthogonal in terms of the information. They proposed a pseudo-likelihood function by adding a penalty term to the profile likelihood function. They argued that-due to the orthogonality of the parameters, the likelihood ratio test statistic constructed by this penalized likelihood function achieves the same order of error as $O(n^{-2})$. Severini (1998) provided general conditions to construct pseudo-likelihood functions such that the influence of the estimation of nuisance parameters on the inference of parameters of interest is reduced.

However, in periodic time series analysis, little attention has been paid to bias reduction of nuisance parameters, whereas the need occurs quite often. This paper attempts to remedy this deficiency by extending the idea of Cox and Reid (1987) to the periodic autoregressive model context. The paper is organized as follows: in Section 2, under the assumption that the white noise sequence is independent and identically distributed as $N(0, \sigma_\nu^2)$, we will propose a modified likelihood ratio test statistic for seasonality; in Section 3, we will compare the performance of the proposed likelihood ratio test with the unmodified test by a simulation study; in Section 4, we will illustrate the application of the test statistic by means of the data about quarterly chemical products and quarterly streamflows.

2. Construction of Test Statistic

Suppose that $\boldsymbol{y_n} = \{y_1, y_2, \cdots, y_n\}'$ is a realization of a PAR(p) time series satisfying (1.2). The parameters of primary interest are $\boldsymbol{\mu} = \{\mu_1, \mu_2, \cdots, \mu_T\}'$. In addition to $\boldsymbol{\mu}$, there are nuisance parameters, denoted by $\boldsymbol{\alpha} = \{\phi_1(1), \cdots, \phi_p(1), \sigma_1^2, \cdots, \phi_1(T), \cdots, \phi_p(T), \sigma_T^2\}'$.

Consider the log-likelihood function for this PAR(p) time series as follows:

$$l(\boldsymbol{\mu}, \boldsymbol{\alpha}) = -\frac{nT}{2}\log(2\pi) - \frac{n}{2}\sum_{\nu=1}^{T}\log\sigma_\nu^2 - Q^2, \tag{2.1}$$

where

$$Q^2 = \sum_{i=0}^{n-1}\sum_{\nu=1}^{T}\frac{1}{2\sigma_\nu^2}\left\{(y_{iT+\nu} - \mu_\nu) - \sum_{k=1}^{p}\phi_k(\nu)(y_{iT+\nu-k} - \mu_{\nu-k})\right\}^2, \tag{2.2}$$

and $y_{iT+\nu-k} - \mu_{\nu-k} = 0$ if $iT + \nu - k \leq 0$. The score functions for $1 \leq \nu \leq T$ and $1 \leq j \leq p$ are as follows:

$$\boldsymbol{g_n}(\boldsymbol{\mu}, \boldsymbol{\alpha}) = \begin{cases} \frac{\partial l(\boldsymbol{\mu},\boldsymbol{\alpha})}{\partial \mu_\nu} = \sum_{i=0}^{n-1}\left\{\frac{\epsilon_{iT+\nu}}{\sigma_\nu^2} - \sum_{k=1}^{p}\phi_k(\nu+k)\frac{\epsilon_{iT+\nu+k}}{\sigma_{\nu+k}^2}\right\}, \\ \frac{\partial l(\boldsymbol{\mu},\boldsymbol{\alpha})}{\partial \phi_j(\nu)} = \sum_{i=0}^{n-1}\frac{\epsilon_{iT+\nu}}{\sigma_\nu^2}(y_{iT+\nu-j} - \mu_{\nu-j}), \\ \frac{\partial l(\boldsymbol{\mu},\boldsymbol{\alpha})}{\partial \sigma_\nu^2} = -\frac{n}{2\sigma_\nu^2} + \frac{1}{2\sigma_\nu^4}\sum_{i=0}^{n-1}\epsilon_{iT+\nu}^2. \end{cases} \tag{2.3}$$

Under the assumptions of regularity conditions and the finiteness of the fourth moment of the time series, the maximum likelihood estimators of the parameters are consistent (cf. Basawa and Lund, 2001).

The estimate of $\boldsymbol{\alpha}$ which is the solution to the second and third estimating equations in (2.3) depends on $\boldsymbol{\mu}$ and is denoted by $\hat{\boldsymbol{\alpha}}(\boldsymbol{\mu}) = \{\hat{\phi}_1(1, \boldsymbol{\mu}), \cdots, \hat{\phi}_p(1, \boldsymbol{\mu}), \hat{\sigma}_1^2(\boldsymbol{\mu}), \cdots, \hat{\phi}_1(T, \boldsymbol{\mu}), \cdots, \hat{\phi}_p(T, \boldsymbol{\mu}), \hat{\sigma}_T^2(\boldsymbol{\mu})\}'$. Replace $\boldsymbol{\alpha}$ by $\hat{\boldsymbol{\alpha}}(\boldsymbol{\mu})$, and hence obtain the profile likelihood function as follows:

$$l_p(\boldsymbol{\mu}) = -\frac{nT}{2}\log(2\pi) - \frac{n}{2}\sum_{\nu=1}^{T}\log\hat{\sigma}_\nu^2(\boldsymbol{\mu}) - \frac{nT}{2}. \tag{2.4}$$

The maximum likelihood estimators obtained by the profile likelihood function (2.4) and likelihood function (2.1) are identical. Furthermore, it can be shown that

$$\mathrm{E}\left\{\frac{\partial^2 l(\boldsymbol{\mu},\boldsymbol{\alpha})}{\partial\boldsymbol{\mu}\partial\boldsymbol{\alpha}}\right\} = 0,$$

which means that $\boldsymbol{\mu}$ and $\boldsymbol{\alpha}$ are orthogonal in terms of information. Therefore, the estimators of $\boldsymbol{\mu}$ and $\boldsymbol{\alpha}$ are asymptotically independent. However, the estimating equations on $\boldsymbol{\mu}$ obtained by (2.4) are no longer unbiased.

Following Cox and Reid (1987), we formulate the modified profile likelihood for a PAR(p) time series as follows:

$$l_m(\boldsymbol{\mu}) \;=\; l_p(\boldsymbol{\mu}) - \frac{1}{2}\sum_{\nu=1}^{T}\log|\boldsymbol{B}_\nu(\boldsymbol{\mu},\hat{\boldsymbol{\alpha}}(\boldsymbol{\mu}))|, \tag{2.5}$$

where the $(p+1)\times(p+1)$ matrix $\boldsymbol{B}_\nu(\boldsymbol{\mu},\hat{\boldsymbol{\alpha}}(\boldsymbol{\mu})) = \begin{pmatrix} b_{j,k}\end{pmatrix}$ is the observed information matrix for $\boldsymbol{\alpha}$ in season ν evaluated at $(\boldsymbol{\mu},\hat{\boldsymbol{\alpha}}(\boldsymbol{\mu}))$ where

$$\begin{aligned}
b_{j,k} &= -\frac{\partial^2 l(\boldsymbol{\mu},\boldsymbol{\alpha})}{\partial\phi_j(\nu)\partial\phi_k(\nu)}\bigg|_{(\boldsymbol{\mu},\hat{\boldsymbol{\alpha}}(\boldsymbol{\mu}))}\\
&= \frac{1}{\hat{\sigma}_\nu^2(\boldsymbol{\mu})}\sum_{i=0}^{n-1}(y_{iT+\nu-j}-\mu_{\nu-j})(y_{iT+\nu-k}-\mu_{\nu-k}),\quad 1\le j,k\le p,\\
b_{j,p+1} &= b_{p+1,j} = -\frac{\partial^2 l(\boldsymbol{\mu},\boldsymbol{\alpha})}{\partial\phi_j(\nu)\partial\sigma_\nu^2}\bigg|_{(\boldsymbol{\mu},\hat{\boldsymbol{\alpha}}(\boldsymbol{\mu}))}\\
&= \frac{1}{\hat{\sigma}_\nu^4(\boldsymbol{\mu})}\sum_{i=1}^{n-1}\hat{\epsilon}_{iT+\nu}(\boldsymbol{\mu})(y_{iT+\nu-j}-\mu_{\nu-j}),\quad 1\le j\le p,\\
b_{p+1,p+1} &= -\frac{\partial^2 l(\boldsymbol{\mu},\boldsymbol{\alpha})}{(\partial\sigma_\nu^2)^2}\bigg|_{(\boldsymbol{\mu},\hat{\boldsymbol{\alpha}}(\boldsymbol{\mu}))} = \frac{n}{2\hat{\sigma}_\nu^4(\boldsymbol{\mu})},
\end{aligned}$$

and $\hat{\epsilon}_{iT+\nu}(\boldsymbol{\mu}) = \left\{(y_{iT+\nu}-\mu_\nu) - \sum_{k=1}^{p}\hat{\phi}_k(\nu,\boldsymbol{\mu})(y_{iT+\nu-k}-\mu_{\nu-k})\right\}$.

The hypothesis testing on the equality of means (1.3) can be equivalently written as

$$H_0 : \boldsymbol{C\mu} = 0 \text{ vs. } H_a : \boldsymbol{C\mu} \neq 0, \tag{2.6}$$

where $\boldsymbol{C} = \begin{pmatrix} c_{i,j}\end{pmatrix}$ is a $(T-1)\times T$ matrix with $c_{i,i}=1$, $c_{i,i+1}=-1$, and zero otherwise. The likelihood ratio test Λ, the Wald test W, and the score test S for such composite hypotheses are defined by

$$\begin{aligned}
\Lambda &= -2\{l(\hat{\boldsymbol{\mu}}_0) - l(\hat{\boldsymbol{\mu}})\},\\
W &= n\hat{\boldsymbol{\mu}}'\boldsymbol{C}'\{\boldsymbol{C}\boldsymbol{\Gamma}^{-1}(\hat{\boldsymbol{\mu}})\boldsymbol{C}'\}^{-1}\boldsymbol{C}\hat{\boldsymbol{\mu}},\\
S &= \frac{1}{n}\boldsymbol{g}_n'(\hat{\boldsymbol{\mu}}_0)\boldsymbol{\Gamma}^{-1}(\hat{\boldsymbol{\mu}}_0)\boldsymbol{g}_n(\hat{\boldsymbol{\mu}}_0),
\end{aligned}$$

where $l(\cdot)$ is the log-likelihood; $\hat{\boldsymbol{\mu}}$ is the unrestricted estimate; $\hat{\boldsymbol{\mu}}_0$ is the restricted estimate under H_0; the T-dimensional vector $\{\boldsymbol{g_n}(\hat{\boldsymbol{\mu}}_0)\}$ are the estimating equations in (2.3) for $\boldsymbol{\mu}$ at $\hat{\boldsymbol{\mu}}_0$; the $T \times T$ matrix $\boldsymbol{\Gamma}(\boldsymbol{\mu}) = \left(\gamma_{i,j}\right)$ is the expected information matrix at $\boldsymbol{\mu}$ where

$$
\begin{aligned}
\gamma_{\nu,\nu} &= -\mathrm{E}\left\{\frac{\partial^2 l(\boldsymbol{\mu})}{\partial \mu_\nu^2}\right\} = \sum_{i=0}^{n-1}\left\{\frac{1}{\sigma_\nu^2} + \sum_{k=1}^{p}\frac{\phi_k^2(\nu+k)}{\sigma_{\nu+k}^2}\right\}, \quad 1 \le \nu \le T, \\
\gamma_{\nu,\nu^*} &= \gamma_{\nu^*,\nu} = -\mathrm{E}\left\{\frac{\partial^2 l(\boldsymbol{\mu})}{\partial \mu_\nu \partial \mu_{\nu^*}}\right\} \\
&= \sum_{i=0}^{n-1}\left\{-\frac{\phi_{\nu-\nu^*}(\nu)}{\sigma_\nu^2} + \sum_{k=1}^{p}\frac{\phi_k(\nu+k)\phi_{\nu+k-\nu^*}(\nu+k)}{\sigma_{\nu+k}^2}\right\}, \quad 1 \le \nu^* < \nu \le T.
\end{aligned}
$$

In the above $\phi_{\nu-\nu^*}(\nu) = 0$ if $\nu - \nu^* > p$; $\phi_{\nu+k-\nu^*}(\nu+k) = 0$ if $\nu + k - \nu^* > p$; $\phi_{\nu+k-\nu^*}(\nu+k) = -1$ if $\nu + k - \nu^*$ is a multiple of T.

These three tests are asymptotically equivalent in the sense that they approach the same χ^2 distribution with $T - 1$ degrees of freedom. In the definition of these three tests, we can either use the profile likelihood in (2.4) and its estimates of parameters or the modified likelihood in (2.5) and its estimates. The price paid for the better approximation to the χ^2 distribution is the necessity of calculating the determinants of the information matrices. When deciding between the three tests, the score test only involves calculating the restricted estimates, whereas the likelihood ratio test and the Wald test require both unrestricted and restricted estimates. In the next section, we will compare the performance of these three tests defined by the profile likelihood and modified profile likelihood by a simulation study.

3. Simulation Study

We simulate the PAR(1) time series with $T = 12$ and the true values for $\boldsymbol{\mu}$ equal to 10 in all seasons. The number of periods ranges from 10 to 50. We evaluate $\boldsymbol{\alpha}$ using the functions in Figure 1 to facilitate the discussion and simplify the calculation. Moreover, Fourier series have been successfully applied to modeling parameters when there are slow seasonal changes in parameters, for example, Bloomfield *et al* (1994) and Lund *et al* (2005).

For each PAR(1) time series, we generate five thousand samples. The nominal level of significance used is 0.05. The empirical levels of significance for the three tests are summarized in Table 1. For each test, the first row and the second row are respectively calculated according to (2.4) and (2.5). In terms of the agreement between the empirical levels of significance and nominal level of significance, the modified likelihood ratio tests perform very well even with a number of periods as small as ten, whereas the Wald tests and score tests have an inflated level as large as 0.41 and a deflated level as small as 0.01. Overall the modified likelihood ratio tests are superior to the others. Therefore, only the likelihood ratio tests are reported when analyzing the real data in the next section.

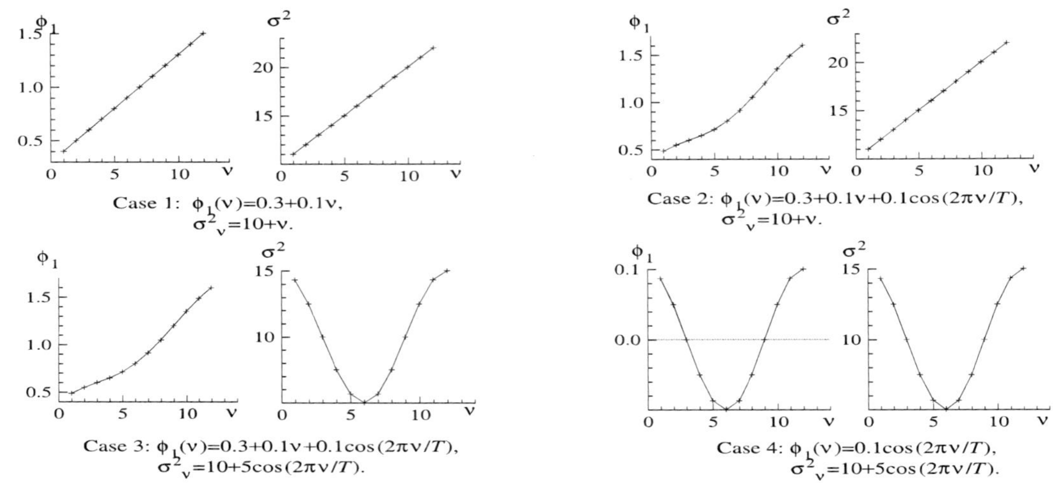

Figure 1. Parameters for Simulations.

4. Applications

4.1. Chemical Products

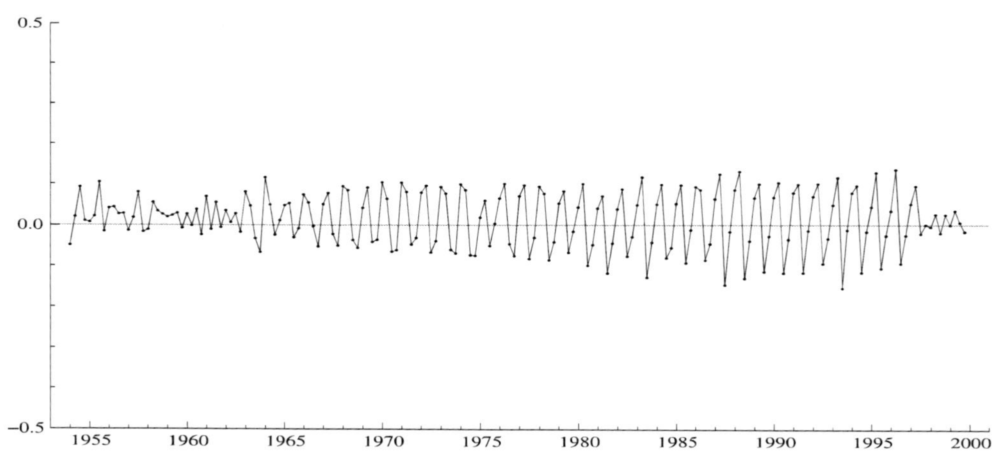

Figure 2. Quarterly Chemical Products.

Franses and Paap (2004) analyzed fourteen quarterly economic time series. They explored many important modeling aspects using the natural logarithm of the original observations. Instead, our interest is whether each quarter has the same mean growth rates. The quarterly growth rates can be approximated by the difference of the logarithmically transformed time series. We will demonstrate the above method with one of the fourteen quarterly time series that concerns chemical products observed from 1954 through 2000.

Table 1. Comparison of Tests for PAR(1) Model

	Number of Periods					
Case 1	10	15	20	30	40	50
Likelihood Ratio Test	0.18	0.12	0.10	0.08	0.07	0.08
	0.08	0.07	0.06	0.06	0.06	0.06
Wald Test	0.15	0.14	0.12	0.11	0.08	0.08
	0.13	0.14	0.12	0.10	0.08	0.08
Score Test	0.37	0.29	0.24	0.18	0.13	0.12
	0.37	0.29	0.24	0.18	0.13	0.12
	Number of Periods					
Case 2	10	15	20	30	40	50
Likelihood Ratio Test	0.18	0.12	0.09	0.08	0.08	0.06
	0.07	0.07	0.06	0.06	0.06	0.05
Wald Test	0.16	0.15	0.12	0.10	0.09	0.07
	0.14	0.15	0.12	0.10	0.09	0.07
Score Test	0.41	0.31	0.25	0.18	0.16	0.14
	0.39	0.31	0.25	0.18	0.16	0.14
	Number of Periods					
Case 3	10	15	20	30	40	50
Likelihood Ratio Test	0.18	0.12	0.10	0.08	0.07	0.07
	0.07	0.07	0.06	0.06	0.06	0.06
Wald Test	0.12	0.14	0.12	0.10	0.09	0.08
	0.10	0.13	0.11	0.10	0.09	0.08
Score Test	0.31	0.24	0.19	0.14	0.11	0.09
	0.30	0.24	0.19	0.14	0.11	0.09
	Number of Periods					
Case 4	10	15	20	30	40	50
Likelihood Ratio Test	0.19	0.13	0.10	0.09	0.07	0.07
	0.06	0.06	0.06	0.06	0.06	0.05
Wald Test	0.02	0.03	0.03	0.04	0.04	0.04
	0.02	0.03	0.03	0.04	0.04	0.04
Score Test	0.01	0.03	0.03	0.04	0.04	0.04
	0.01	0.03	0.03	0.04	0.04	0.04

We transform the original data by taking the natural logarithm and then differencing. The transformed data in the last year are ignored so that we have the 46 full year observations plotted in Figure 2.

The observations $\{y_{iT+\nu}; 0 \le i \le 45, 1 \le \nu \le 4\}$ stand for the approximate quarterly growth rates of each quarter. We implement the maximum likelihood estimation for model fitting, and carry out diagnostic checking using AIC and residuals. The most appropriate model appears to be PAR(3).

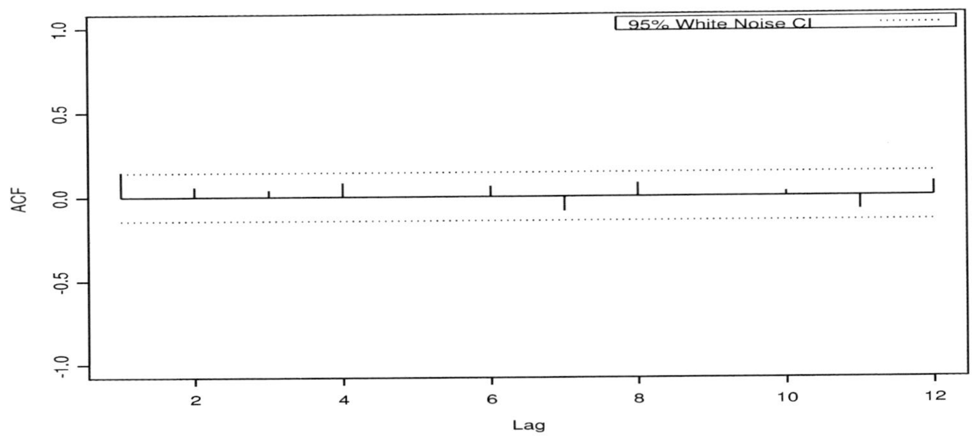

Figure 3. Residual Autocorrelations of Chemical Products.

Figure 3 is the sample autocorrelations of the PAR(3) residuals $\{r_{iT+\nu}; 0 \leq i \leq 45, 1 \leq \nu \leq 4\}$ based on the maximum likelihood estimates given by

$$r_{iT+\nu} = \frac{1}{\hat{\sigma}_\nu^2}\left\{y_{iT+\nu} - \hat{\mu}_\nu - \sum_{k=1}^{p} \hat{\phi}_k(\nu)(y_{iT+\nu-k} - \hat{\mu}_{\nu-k})\right\}.$$

Notice there are only 46 observations for each quarter, giving a total number of observations as 184. In addition, there are 20 parameters in total including 16 nuisance parameters.

The likelihood ratio test is 26.964 with p-value 5.998×10^{-6} and the modified likelihood ratio test is 29.813 with p-value 1.512×10^{-6}. Despite the fairly large number of nuisance parameters, both test statistics lead to the same conclusion providing very strong evidence that the quarterly mean growth rates of the chemical products are different from quarter to quarter.

4.2. Streamflows

Shao and Ni (2004) applied the least squares approach when analyzing streamflow data from Asotin Creek at Asotin in Washington State. There are a total of 11 water years in Figure 4, from the fourth quarter in 1991 to the third quarter in 2002. After determining that PAR(1) is the most appropriate model, they conclude that there is a significant difference among the average streamflows from the approximate F test calculated with the historical information. The modified F test provides a correction for the correlated data to some extent.

We apply the likelihood ratio test to this time series and obtain a value of the likelihood ratio test statistic of 6.739 with p-value 0.008 and a value of the modified likelihood ratio test statistic of 9.444 with p-values 0.002. There is a difference between these test statistics. As a consequence, p-values of the likelihood ratio tests are different as well. However both

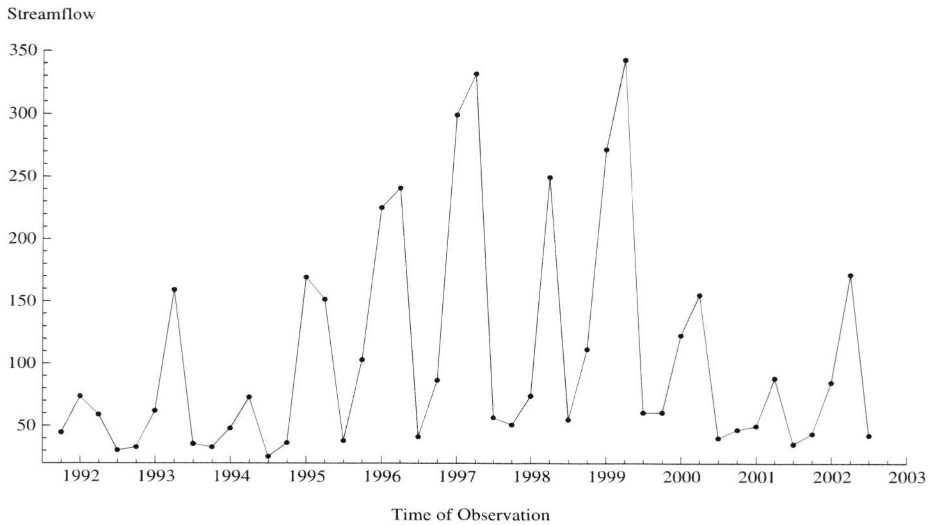

Figure 4. Quarterly Streamflows, Asotin Creek at Asotin, WA.

of these two tests have p-values less than the nominal level 0.05. Hence we reach the same conclusion as in Shao and Ni (2004).

References

Barndorff-Nielsen, O. E. & Cox, D. R. (1984). Bartlett adjustments to the likelihood ratio statistic and the distribution of the maximum likelihood estimator. *Journal of the Royal Statistical Society. Series B*, **46**(3), pp. 483-495.

Basawa, I. V. & Lund, R. B. (2001). Large sample property estimates for periodic ARMA models. *Jouneral of Time Series Analysis*, **22**, pp. 651-663.

Bloomfield, P., Hurd, H. L. & Lund, R. B. (1994). Periodic correlation in stratospheric ozone data, *Journal of Time Series Analysis*, **15**, pp. 127-150.

Cox, D. R. & Reid, N. (1987). Parameter orthogonality and approximate conditional inference. *Journal of the Royal Statistical Society. Series B*, **49**(1), pp. 1-39.

Franses, P. H. & Paap, R (2004). *Periodic Time Series Models*, Oxford University Press, New York.

Hipel, W. K. & McLeod, I. A. (1994). *Time Series Modelling of Water Resources and Environmental Systems*, Elsevier.

Lund, R., Shao, Q., and Basawa, V. I. (2006). Parsimonious periodic time series modeling, *Australian & New Zealand Journal of Statistics*, **48**(1), 33-47.

Severini, A. T. (1998). Likelihood functions for inference in the presence of a nuisance parameter. *Biometrika*, **85**(3), pp. 507-522.

Shao, Q. & Ni, P. P. (2004). Least squares estimation and ANOVA for periodic autoregressive time series, *Statistics & Probability Letters*, **69**, pp. 287-297.

Vecchia, A. V. (1985). Periodic autoregressive-moving average (PARMA) modeling with applications to water resources, *Water Resources Bulletin*, **21**, pp. 721-730.

In: New Developments in Applied Statistics
Editor: Mohammad Ahsanullah
ISBN 978-1-61324-648-1

Chapter 12

BAYESIAN PREDICTION BOUNDS FOR RAYLEIGH AND INVERSE RAYLEIGH LIFETIME MODELS

***Hatem A. Howlader*[1], *Anwar M. Hossain*[2] *and Oleg Makhnin*[2]**
[1]Department of Mathematics and Statistics, University of Winnipeg, Manitoba, Canada R3B 2E9
[1]Department of Mathematics, New Mexico Tech, Socorro, NM 87801

Abstract

Using the first k components of a random sample of n components from Rayleigh and Inverse Rayleigh lifetime models, this paper presents Bayesian prediction bounds for the remaining $(n - k)$ components for each of the above models. The prior belief of the models is represented by its respective natural conjugate prior. It is indicated in this paper that the inverse Rayleigh distribution could be used as an alternative to the lognormal distribution in modelling the survival time of certain diseases. Numerical examples are given to illustrate the results derived.

Keywords and Phrases: Bayes estimator; order statistic; prior/posterior distribution; survival function; prediction interval; highest posterior density; closure property of the natural conjugate prior

1. Introduction

The Rayleigh distribution has been found to provide a useful survival model for experiments that age with time. Dyer and Whisenand (1973) demonstrated the importance of this distribution in communication engineering and Polovko (1968) noted its importance in electrovacuum devices. See Hirano (1986) for a brief history and properties and Siddiqui (1962) for other aspects of this distribution. The prediction problems of the Rayleigh life time models are very important and have been studied, among others Raqab (1992) based on maximum likelihood method. Bhattacharya & Tyagi (1990), Howlader (1985), Sinha (1990) and Howlader and Hossain (1995), Fernandez (2000), Raqab & Madi (2002), Soliman (2000), and Soliman & Al-Aboud (2006) have considered the Bayesian estimation and prediction problems for the Rayleigh distribution. See Aitchison & Dunsmore (1975) for a discussion on predictive distributions. The Rayleigh distribution has probability density function (pdf)

$$f(x|\theta) = \frac{x}{\theta}\, e^{-\frac{x^2}{2\theta}}, \qquad x > 0,\ \theta > 0 \tag{1}$$

survival function

$$S(t) = e^{-\frac{t^2}{2\theta}},\quad t > 0 \tag{2}$$

and linearly increasing failure rate function

$$h(t) = \frac{t}{\theta}.$$

The inverse Rayleigh distribution was introduced in the literature by Treyer (1964) as a model for life testing and reliability. The probability density function (pdf), the reliability function and the failure rate (hazard rate) function of the distribution are, respectively,

$$f(x|\sigma) = \frac{2\sigma}{x^3}\, e^{-\frac{\sigma}{x^2}} \quad x > 0, \quad \sigma > 0 \tag{3}$$

$$S(t) = 1 - e^{-\frac{\sigma}{t^2}}, \quad t > 0 \tag{4}$$

$$h(t) = \frac{2\sigma}{t^3}\left(e^{\sigma/t^2} - 1\right)^{-1}. \tag{5}$$

The failure rate function (5) increases initially to a maximum, and then decreases almost as soon as it passes the median ($\sqrt{\sigma/\ln 2}$) of the distribution, indicating that the inverse Rayleigh distribution is suitable for survival patterns with an initial increasing and then decreasing failure rate, see Fig. 1.

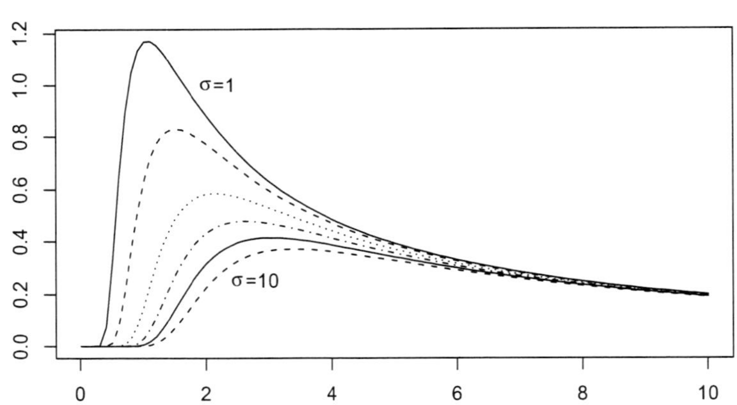

Figure 1. Inverse Rayleigh failure rate function for $\sigma = 1, 2, 4, 6, 8, 10$.

This behaviour of the failure rate function is similar to that of the lognormal distribution. The distribution of survival time of diseases, such as Hodgkin's and chronic leukemia, as well as the distribution of age at onset of Alzheimer's disease is usually modelled by a lognormal distribution (see, for example, Lee (1992)). The above diseases can also be modelled by the inverse Rayleigh distribution. Even though some early work in terms of maximum likelihood, confidence interval, and testing of the parameter of the inverse

Rayleigh was done by Voda (1972) and Illiescu & Voda (1973), very few authors since then have studied this distribution either from Bayesian or classical point of view.

In this study, we assume an inverted gamma, $InGa(\alpha, \beta)$ prior for the Rayleigh parameter θ, with pdf

$$g_1(\theta) = \frac{\beta^\alpha}{\Gamma(\alpha)}\, \theta^{-(\alpha+1)}\, e^{-\beta/\theta} \quad \alpha,\, \beta > 0, \quad \theta > 0 \tag{6}$$

It is the natural conjugate prior (NCP) for θ. A family of priors is conjugate if the choice of the prior in that family generates a posterior that belongs to the same family. In addition to the ease of mathematical manipulation of the posterior generated from an NCP, the prior parameters can be chosen to suit the prior belief of the experimenter in terms of location and variability of the prior distribution.

For the inverse Rayleigh parameter σ, we choose gamma $G(m, p)$ prior having the pdf

$$g_2(\sigma) = \frac{1}{p^m\, \Gamma(m)}\, \sigma^{m-1}\, e^{-\frac{\sigma}{p}} \quad m,\, p > 0, \;\; \sigma > 0 \tag{7}$$

where m and p are the prior parameters. It is also the NCP for σ.

The purpose of this paper is to obtain the highest posterior density (HPD) prediction bounds for the survival times of the remaining $(n-k)$ components of a sample of n components where only the first k components have been observed for both the models. We also derive the predictive survival functions for the above models.

2. Posterior and Predictive Densities

2.1. Rayleigh Distribution

As is quite common in life testing situations that only the first k lifetimes in a sample of n units can be obtained. Let $\mathbf{x} = (x_1, x_2, \ldots x_k)$, be the lifetimes from (1) where x_i is the time of the ith component to fail. Since the remaining $(n-k)$ components have not yet failed and thus have lifetimes greater than x_k, the likelihood function, assuming that the survival times are independent, is

$$L(\mathbf{x}|\theta) = \frac{n!}{(n-k)!}\left[\prod_{i=1}^{k} f(x_i)\right][S(x_k)]^{n-k}$$

$$= \frac{n!}{(n-k)!}\, \theta^{-k}\, exp\left[-\frac{1}{2\theta}\left(\sum_{i=1}^{k} x_i^2 + (n-k)x_k^2\right)\right]. \tag{8}$$

Combining the likelihood function (8) and the prior (6) and restoring the normalizing constant, the posterior density of θ can be shown to take the form:

$$\Pi(\theta|\mathbf{x}) = \frac{(\beta')^{\alpha'}}{\Gamma(\alpha')}\, \theta^{-(\alpha'+1)}\, e^{-\frac{\beta'}{\theta}}, \quad \theta > 0 \tag{9}$$

where $\alpha' = \alpha + k$ and $\beta' = \frac{1}{2}\left(\sum_{i=1}^{k} x_i{}^2 + (n-k)x_k{}^2 + 2\beta\right)$. We note that the posterior pdf (9) is also inverted gamma, $InGa(\alpha', \beta')$, showing that the posterior has the same functional form as the prior, and hence, the inverted gamma priors are closed under sampling.

Let $Y = X_{k+r}$ denote the lifetime of the $(k+r)$-th component, where $r = 1, 2, \ldots, n-k$. Then, the conditional predictive density of Y given **x** is found as

$$f(y|\mathbf{x}, \theta) = \frac{r\binom{n-k}{r}[S(x_k) - S(y)]^{r-1}[S(y)]^{n-k-r}\, f(y)}{[S(x_k)]^{n-k}}, \qquad y > x_k \tag{10}$$

Substituting the Rayleigh pdf $f(y)$ from (1) and $S(t)$ from (2), the equation (10) takes the form

$$f(y|\mathbf{x}, \theta) = r\binom{n-k}{r}\frac{y}{\theta}\sum_{j=0}^{r-1}\binom{r-1}{j}(-1)^j\left[\frac{exp(-y^2/2\theta)}{exp(-x_k^2/2\theta)}\right]^{h_j} \tag{11}$$

where $h_j = n - k - r + j + 1$. Combining (9) and (11), we obtain the joint pdf of y and θ given **x** as

$$f(y, \theta|\mathbf{x}) \propto \theta^{(\alpha'+2)}\, y\sum_{j=0}^{r-1}\binom{r-1}{j}(-1)^j\, exp(-H_j^2/2\theta) \tag{12}$$

where $H_j^2 = h_j y^2 - h_j x_k^2 + 2\beta'$. Integrating out θ from (12), we obtain the predictive density of Y given the observed data **x** as

$$f(y|\mathbf{x}) = c_1\sum_{j=0}^{r-1}\binom{r-1}{j}(-1)^j y(h_j y^2 - h_j x_k^2 + 2\beta')^{-(\alpha'+1)}, y > x_k \tag{13}$$

where $c_1^{-1} = \frac{1}{2\alpha'}(2\beta')^{-\alpha'}\sum_{j=0}^{r-1}\binom{r-1}{j}(-1)^j\, h_j^{-1}$.

2.2. Inverse Rayleigh Distribution

The likelihood function for the inverse Rayleigh distribution, corresponding to (8) is given by

$$L(\mathbf{x}|\sigma) = \frac{n!}{(n-k)!}\left(1 - e^{-\sigma/x_k^2}\right)^{n-k}\prod_{i=1}^{k}\frac{2\sigma}{x_i^3}e^{-\sigma/x_i^2}. \tag{14}$$

Under the gamma prior (7), the posterior pdf of σ is

$$\Pi(\sigma|\mathbf{x}) = c_2\,\sigma^{k+m-1}e^{-\sigma/p'}\left(1 - e^{-\sigma/x_k^2}\right)^{n-k} \tag{15}$$

where $p' = \left[\sum_{i=1}^{k} x_i^{-2} + 1/p\right]^{-1}$, and c_2 is the normalizing constant.

Using the binomial expansion, namely

$$\left(1 - e^{-\sigma/x_k^2}\right)^{n-k} = \sum_{i=0}^{n-k}(-1)^i\binom{n-k}{i}e^{-i\sigma/x_k^2}$$

we have

$$\Pi(\sigma|\mathbf{x}) = c_2\, \sigma^{k+m-1} \sum_{i=0}^{n-k} (-1)^i \binom{n-k}{i} e^{-\sigma/\tau_i}, \quad \sigma > 0 \tag{16}$$

where $\tau_i = (1/p' + i/x_k^2)^{-1}$ and

$$c_2^{-1} = \sum_{i=0}^{n-k} (-1)^i \binom{n-k}{i} \tau_i^{k+m}\, \Gamma(k+m).$$

Thus, the posterior pdf, $\Pi(\sigma|\mathbf{x})$ is a linear combination of gamma densities, but not itself a gamma density, so gamma priors are no longer closed under sampling for inverse Rayleigh.

The equation (10), the conditional predictive density of y given **x** for the inverse Rayleigh pdf (3) takes the form:

$$f(y|\mathbf{x},\sigma) = r\binom{n-k}{r} \left[e^{-\sigma/y^2} - e^{-\sigma/x_k^2}\right]^{r-1} \left[1 - e^{-\sigma/y^2}\right]^{n-k-r}$$

$$\times e^{-\sigma/y^2} \left[1 - e^{-\sigma/x_k^2}\right]^{k-n} \frac{2\sigma}{y^3}. \tag{17}$$

Multiplying (17) and (16) with we obtain the joint posterior of y and σ given **x** as

$$f(y,\sigma|\mathbf{x}) = 2r\, c_2 \binom{n-k}{r} \frac{\sigma^{k+m}}{y^3} e^{-\sigma/p'} \left[e^{-\sigma/y^2} - e^{-\sigma/x_k^2}\right]^{r-1}$$

$$\times \left[1 - e^{-\sigma/y^2}\right]^{n-k-r} e^{-\sigma/y^2}. \tag{18}$$

Using the binomial expansions for $\left[e^{-\sigma/y^2} - e^{-\sigma/x_k^2}\right]^{r-1}$ and $\left[1 - e^{-\sigma/y^2}\right]^{n-k-r}$ we have

$$f(y,\sigma|\mathbf{x}) = 2r\, c_2 \binom{n-k}{r} \frac{\sigma^{k+m}}{y^3} \sum_{i=0}^{r-1} \sum_{j=0}^{n-k-r} (-1)^{i+j} \binom{r-1}{i} \binom{n-k-r}{j} e^{-\sigma\tau_{ij}} \tag{19}$$

where $\tau_{ij} = \dfrac{1}{p'} + \dfrac{i}{x_k^2} + \dfrac{r-i+j}{y^2}$.

Integrating out σ, the predictive density of y is

$$f(y|\mathbf{x}) = 2r\, c_2 \binom{n-k}{r} \frac{\Gamma(k+m+1)}{y^3}$$

$$\times \sum_{i=0}^{r-1} \sum_{j=0}^{n-k-r} (-1)^{i+j} \binom{r-1}{i} \binom{n-k-r}{j} \tau_{ij}^{-(k+m+1)}, \quad y > x_k \tag{20}$$

3. Predictive Survival Function

3.1. Rayleigh Distribution

The predictive survival function, $S_R^*(t)$ from (13) is given by

$$S_R^*(t) = P(Y \geq t) = \int_t^\infty c_1 \sum_{j=0}^{r-1} \binom{r-1}{j} (-1)^j \, y(h_j y^2 - h_j x_k^2 + 2\beta')^{-\alpha'-1} dy$$

$$= \frac{\sum_{j=0}^{r-1} \binom{r-1}{j} (-1)^j \, h_j^{-1} \, (2\beta' + h_j t^2 - h_j {x_k}^2)^{-\alpha'}}{(2\beta')^{-\alpha'} \sum_{j=0}^{r-1} \binom{r-1}{j} (-1)^j \, h_j^{-1}} \tag{21}$$

Special cases of (21) are

(i) For r=1, i.e., predicting the failure time $Y = X_{k+1}$ of the first future component, the equation (21) reduces to

$$S_R^*(t) = P(Y > t) = \left(1 + \frac{(n-k)(t^2 - {x_k}^2)}{2\beta'}\right)^{-\alpha'} \tag{22}$$

(ii) For r=2, the equation (21) reduces to

$$S_R^*(t) = P(Y \geq t) = \frac{{h_0}^{-1}(2\beta' + h_0 t^2 - h_0 {x_k}^2)^{-\alpha'} - {h_1}^{-1}(2\beta' + h_1 t^2 - h_1 {x_k}^2)^{-\alpha'}}{{h_0}^{-1}(2\beta')^{-\alpha'} - {h_1}^{-1}(2\beta')^{-\alpha'}} \tag{23}$$

where $h_j = n - k - r + j + 1, h_0 = n - k - 1$ and $h_1 = n - k$

(iii) For $r = n - k$, i.e., predicting the failure time x_n, of the last future component, the equation (21) reduces to

$$S_R^*(t) = P(Y \geq t) = \frac{\sum_{i=1}^{n-k} \binom{n-k-1}{i-1} (-1)^{i-1} (i)^{-1} \, (2\beta' + it^2 - i{x_k}^2)^{-\alpha'}}{\sum_{i=1}^{n-k} \binom{n-k-1}{i-1} (-1)^{i-1} (i)^{-1} \, (2\beta')^{-\alpha'}} \tag{24}$$

3.2. Inverse Rayleigh Distribution

The predictive survival function $S_I^*(t)$ from (19) for inverse Rayleigh distribution is

$$S_I^*(t) = \int_t^\infty f(y|\mathbf{x}) \, dy = rc_2 \Gamma(k+m) \binom{n-k}{r}$$

$$\times \sum_{i=0}^{r-1} \sum_{j=0}^{n-k-r} (-1)^{i+j} \binom{r-1}{i} \binom{n-k-r}{j} (r-i+j)^{-1} \left[\tau_i^{-(k+m)} - \left(\tau_i + \frac{r-i+j}{t^2} \right)^{-(k+m)} \right] \tag{25}$$

where $\tau_i = \dfrac{1}{p'} + \dfrac{i}{x_k^2}$ and c_2 as given in (16) . We note that for r=1, 2, etc, the equation (25) does not reduce significantly.

4. HPD Prediction Bounds

A highest posterior density (HPD) interval is one in which the posterior density for every point inside the interval is greater than that for every point outside of it so that the interval includes the more probable values of the parameter and excludes the less probable ones. For a uni-modal pdf, a γ- HPD interval (h_1, h_2) must satisfy the following two equations simultaneously:

$$\int_{h_1}^{h_2} \Pi(\theta|\mathbf{x})\, d\theta = \gamma \tag{26}$$

$$\Pi(h_1|\mathbf{x}) = \Pi(h_2|\mathbf{x}) \tag{27}$$

4.1. Rayleigh Distribution

For r=1, the predictive density of x_{k+1} from (13) takes the form

$$f(x_{k+1}|\mathbf{x}) = \frac{x_{k+1}\left[(n-k){x^2}_{k+1} - (n-k)x_k^2 + 2\beta'\right]^{-(\alpha'+1)}}{\frac{1}{2\alpha'}[2\beta']^{-\alpha'}\frac{1}{n-k}}$$

$$= 2\alpha'(2\beta')^{\alpha'}(n-k)x_{k+1}\left[(n-k){x^2}_{k+1} - (n-k)x_k^2 + 2\beta'\right]^{-(\alpha'+1)}, \quad x_{k+1} > x_k$$

which is uni-modal. Hence the γ-HPD prediction interval (I_1, I_2) for x_{k+1} is the simultaneous solution of the following equations (28) and (29):

$$S_R^*(I_1) - S_R^*(I_2) = \gamma \tag{28}$$

and

$$\left[\frac{(n-k)I_2^2 - (n-k)x_k^2 + 2\beta'}{(n-k)I_1^2 - (n-k)x_k^2 + 2\beta'}\right]^{\alpha'+1} = \frac{I_2}{I_1} \tag{29}$$

where $S_R^*(.)$ is defined in (21).

Similarly, for $r = 2, 3, \ldots$

4.2. Inverse Rayleigh Distribution

For r=1, the predictive density of x_{k+1} from (20) is

$$f(x_{k+1}|\mathbf{x}) = 2c_2(n-k)\frac{\Gamma(k+m-1)}{x_{k+1}^3}\sum_{i=0}^{n-k-1}(-1)^i\binom{n-k-1}{i}\left(\frac{1}{p'} + \frac{i+1}{x_{k+1}^2}\right)^{-(k+m+1)}$$

$$x_{k+1} > x_k$$

which is uni-modal. The γ-HPD prediction interval (d_1, d_2) for a x_{k+1} is the simultaneous solution of the following equations:

$$S_I^*(d_1) - S_I^*(d_2) = \gamma \tag{30}$$

and

$$\frac{\sum_{i=0}^{n-k-1}(-1)^i\binom{n-k-1}{i}\left(\frac{1}{p'}+\frac{i+1}{d_1^2}\right)^{-(k+m+1)}}{\sum_{i=0}^{n-k-1}(-1)^i\binom{n-k-1}{i}\left(\frac{1}{p'}+\frac{i+1}{d_2^2}\right)^{-(k+m+1)}}=\left(\frac{d_1}{d_2}\right)^3 \quad (31)$$

where $S_I^*(.)$ is defined in (25) with r=1.

Similarly, for $r = 2, 3, \ldots$

5. Numerical Examples

5.1. Rayleigh Distribution

For illustrative purposes, we consider the numerical example given in Howlader & Hossain (1995). A sample of $n = 20$ obervations was taken, with 20% censoring, so the remaining 16 observations are

```
0.506, 0.549, 0.679, 1.127, 1.209, 1.212, 1.418, 1.469,
1.701, 1.805, 1.859, 2.231, 2.471, 2.758, 2.813, 2.917
```

Using the above data and the choice of prior parameters $\alpha = 2, \beta = 2$, we plot (Fig. 2) the predictive density $f(y|\mathbf{x})$ for $r = 1, 2, 3, 4$. Fig. 3 shows the corresponding 95% HPD posterior interval (2.928,4.057) for $r = 2$, i.e. the 18th order statistic.

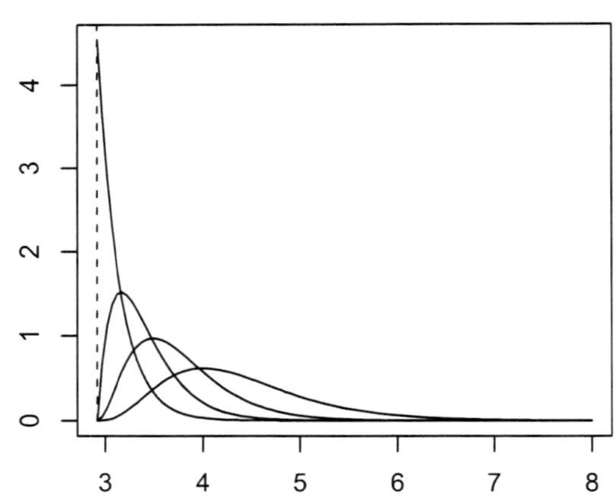

Figure 2. Predictive densities $f(y|\mathbf{x})$.

For different values of t we compute and present the predictive survival function $S_R^*(t)$ for $r = 2$ in Table 1. Assuming different values of the prior parameters α and β, we also compute and present 95% HPD posterior intervals for future observations for $r = 1, 2, 3, 4$ in Table 2. To compute the intervals, Nelder-Mead (1965) method was used, and there were no computational difficulties.

Table 2 shows that the length of the HPD prediction interval increases as r increases, implying that the density function of a smaller order statistic has a smaller variance.

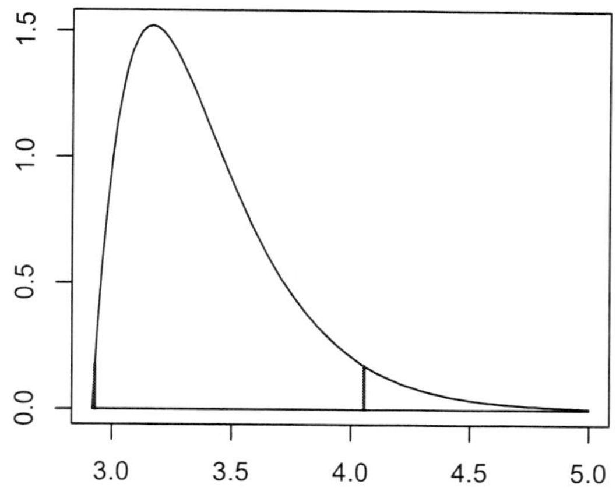

Figure 3. Predictive density for $r = 2$ and 95% HPD interval.

Table 1. Predictive survival function $S_R^*(t)$ for Rayleigh distribution, $r = 2$

t	3.0	3.1	3.2	3.3	3.4	3.5
$S_R^*(t)$	0.958	0.841	0.698	0.559	0.436	0.334
t	3.6	3.7	3.8	4.0	4.2	4.4
$S_R^*(t)$	0.251	0.187	0.137	0.073	0.038	0.019

Table 2. HPD prediction intervals for future observations from Rayleigh

(α, β) \ r	1	2	3	4
(0.5, 0.5)	2.917 3.598	2.928 4.121	3.010 4.803	3.183 6.013
(1,1)	2.917 3.585	2.928 4.098	3.008 4.770	3.179 5.961
(1,5)	2.917 3.638	2.929 4.187	3.017 4.902	3.204 6.162
(5,1)	2.917 3.459	2.926 3.883	2.991 4.446	3.132 5.463
(2,2)	2.917 3.561	2.928 4.057	3.006 4.708	3.171 5.867
(5,5)	2.917 3.503	2.927 3.957	2.999 4.557	3.153 5.636

5.2. Inverse Rayleigh Distribution

The following data represent noise levels (in degrees Kelvin) in cryogenic microwave receivers, shifted by 16 to achieve a reasonable fit with inverse Rayleigh distribution. The data are supplied by Darrell Hicks (National Radio Astronomy Observatory, Socorro, NM, personal communication).

1.3	1.6	1.7	1.8	1.8	2.0	2.0	2.0	2.2	2.2	2.3	2.3	2.3
2.4	2.4	2.5	2.5	2.6	2.6	2.6	2.7	2.7	2.8	2.8	2.9	2.9
2.9	2.9	2.9	2.9	3.0	3.0	3.2	3.3	3.3	3.4	3.4	3.4	3.4
3.4	3.5	3.5	3.6	3.6	3.7	3.8	3.8	3.8	3.8	3.9	4.0	4.1
4.2	4.2	4.2	4.2	4.2	4.3	4.5	4.6	5.0	5.8	6.0	6.0	6.1

```
6.2  6.5  6.5  6.6  6.9  7.6  7.9  8.3  8.5  9.0  9.3  9.7 10.
11.4 14.9 16.1 16.8 19.0 19.8 21.5 24.4
```

Using the Kolmogorov-Smirnov statistic, the inverse Rayleigh distribution was found to fit the data very well ($p > 0.20$). Here, $n = 86$ and we omit the largest 10 observations. Thus, the largest uncensored x_k is 9.3.

The plot (Fig. 4) of the empirical CDF and the true inverse Rayleigh CDF also shows a very good match between the data and inverse Rayleigh distribution. Note that the MLE for σ (= 9.9) was used in the plot.

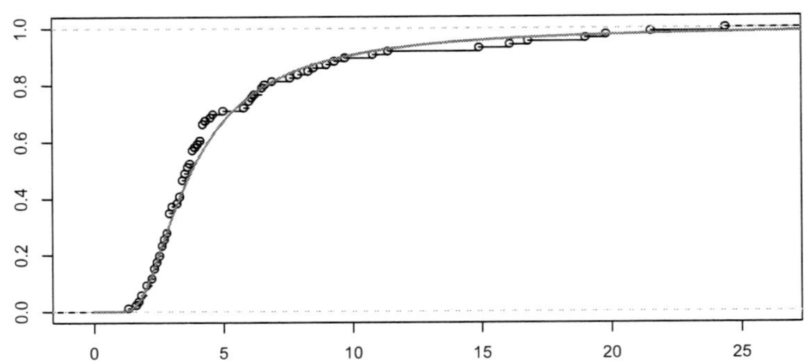

Figure 4. Inverse Rayleigh distribution. The dotted line indicates the empirical CDF and the solid line indicates the true CDF.

Using the above noise level data, we plot (Fig. 5) the predictive density $f(y|\mathbf{x})$ for $r = 1, 2, ..., 10$ and for the choice of prior parameters $m = 2, p = 2$. Fig. 6 shows the corresponding 95% HPD posterior interval (9.658,15.182) for $r = 4$, i.e. the future 80th order statistic.

The predictive survival function for $r = 4$ and for different values of t is presented in Table 3. We also present the HPD prediction intervals for the future observations for $r = 1, 2, ..., 5$, and some selected values of the prior parameters m and p in Table 4.

Table 3. Predictive survival function $S_I^*(t)$ for inverse Rayleigh, $r = 4$

t	9.5	10	10.5	11	11.5	12
$S_I^*(t)$	1	0.969	0.866	0.715	0.556	0.415
t	12.5	13	13.5	14.5	15.5	16.5
$S_I^*(t)$	0.301	0.214	0.151	0.073	0.034	0.016

Table 4 also shows that the length of the HPD prediction interval increases as r increases implying that the pdf of a smaller order statistic is concentrated on a shorter domain, i.e, has a smaller variance; and that a larger order statistic is concentrated on a larger domain.

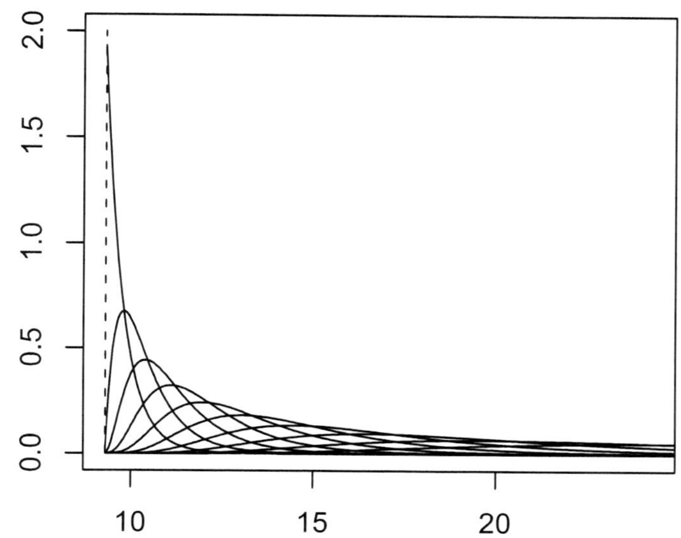

Figure 5. Predictive densities $f(y|\mathbf{x})$ for inverse Rayleigh.

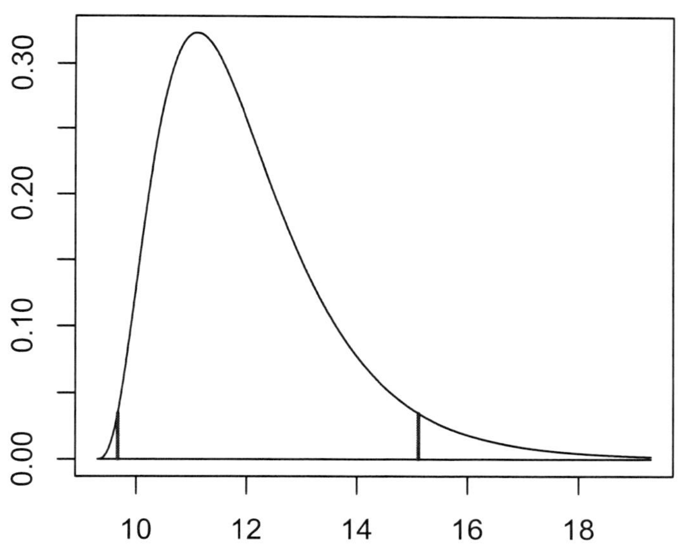

Figure 6. Predictive density for $r = 4$ and 95% HPD interval.

6. Summary

In this paper, the Bayesian prediction of the survival data arising from Rayleigh and inverse Rayleigh distributions is carried out based on the first k lifetimes in a sample of n lifetimes. As known in the literature, the posterior pdf of the Rayleigh parameter under its inverted gamma natural conjugate prior possesses the closure property under sampling. We observe in this paper that the posterior pdf of the inverse Rayleigh parameter under its gamma natural conjugate prior does not possess the above property; the pdf rather is a linear

Table 4. HPD prediction intervals for future observations from inverse Rayleigh

(m, p) \ r	1	2	3	4	5
(1, 1)	9.300 10.867	9.317 12.083	9.439 13.471	9.657 15.165	9.956 17.346
(2, 2)	9.300 10.872	9.317 12.091	9.440 13.483	9.658 15.182	9.959 17.368
(2, 5)	9.300 10.875	9.317 12.096	9.440 13.490	9.659 15.191	9.960 17.379
(10, 2)	9.300 10.879	9.317 12.104	9.441 13.501	9.660 15.206	9.962 17.399

combination of gamma densities. There are similarities between lognormal and inverse Rayleigh distributions in terms of pdf and failure rate functions. The inverse Rayleigh distribution thus has a potential to be a competitor to the lognormal distribution in the field of survival analysis, especially in the Bayesian arena.

References

[1] Aitchison, J. and Dunsmore, I. R. (1975). *Statistical Prediction Analysis,* Cambridge University Press.

[2] Bhattacharya, S.K. & Tyagi, R.K. (1990). *Bayesian survival analysis based on the Rayleigh model. Trobajos de Estadistica,* **5**(1), 81-92.

[3] Dyer, D.D. and Whisenand (1973). Best linear unbiased estimator of the parameter of the Rayleigh distribution. *IEEE Transactions on Reliability,* **R-22**, 27-34; and 455-466.

[4] Fernandez, A. J.(2000). Bayesian Inference from type II doubly censored Rayleigh data. *Statistics and Probability Letters,* **48**, 393-399.

[5] Hirano, K.(1986). "Rayleigh distribution". In Kotz, S. Johnson, N. L., and Reed, C. B. (Eds.), *Encyclopedia of Statistical sciences*, pp 647-649, John Wiley and Sons Inc., New York.

[6] Howlader, H. A.(1985). HPD Prediction Intervals for Rayleigh Distribution. *IEEE Transactions on Reliability*, **R-34**, 1985 Jan, 121-123.

[7] Howlader, H. A. and Hossain, A. M. (1995). On Bayesian estimation and Prediction from Rayleigh Based on Type II censored data. *Commun. Statist-Theory Meth.*, **24**(9), 2249-2259.

[8] Iliescu, D.V., and Voda, V. Gh (1973). *Studiul Inversei Unei Variable Aleatoare De Tip Rayleigh. Studii si Cercetari Matematica,* Academia Republica Sociate Romania, Bucharest, 25, 1507-1521.

[9] Lee, E.T. (1992). *Statistical methods for survival data analysis,* pp. 141-42, John Wiley and Sons Inc., New York.

[10] Nelder, J.A. and Mead, R. (1965). A simplex method for function optimization. *Computer Journal,* **7**, 308-313.

[11] Polovko, A. M. (1968). *Fundamentals of reliability Theory,* Academic Press, New York.

[12] Raqab, M.A. (1992). *Predictors of Future Order Statistics from Type II Censored Samples*. Ph.D Thesis, Ohio State University, Columbus.

[13] Raqab, M.A.and Madi, M.T. (2002). Bayesian Prediction of the total time on test using doubly censored Rayleigh data. *Journal of Statistical Computation and Simulation* **72**, 781-789.

[14] Siddiqui, M. M. (1962). Some problems connected with Rayleigh distribution. *J. of Res Nat. Bur. Stand*-**66D**, 167-174.

[15] Sinha, S. K. (1990). On the prediction limits for Rayleigh distribution. *Calcutta Statistical Association Bulletin,* **39**, 104-109.

[16] Soliman, A.A. (2000). Comparison of linex and quadratic Bayes estimators for the Rayleigh distribution. *Commun.Statist.-Theory Meth.*, **2**, 95-107.

[17] Soliman, A.A. and Al-Aboud, F.M. (2008). Bayesian inference using record values from Rayleigh model with applications. *European Journal of Operational Research*, **185**, 659-672.

[18] Treyer, V.N. (1964). Закон распределения случайных величин для расчёта надёжности изнашивающихся элементов машин и приборов (A distribution law of random variables for the reliability computations for wearable parts of machines and devices.) In: *Doklady Acad. Nauk Belorus. SSR (In Russian)*, **8**, 47-50.

[19] Voda, V. Gh. (1972). On the inverse Rayleigh distributed random variable. *Reports in Statistical Applied Research JUSE*, **19**, 13-21.

In: New Developments in Applied Statistics
Editor: Mohammad Ahsanullah
ISBN 978-1-61324-648-1

Chapter 13

Ranked Set Sampling Technique to Estimate a Time-to-Failure Distribution of a Linear Degradation Model

***M.T. Alodat*[*] *and Mohammed Al-Haj Ebrahem*[†]**
Department of Statistics, Yarmouk University
Irbid, Jordan

Abstract

In degradation analysis, the degradation measurements are often taken over time, and these measurements can be used to get information about the product's reliability. The time-to-failure is defined as the time when the degradation of a unit reaches a critical level. In this paper, we used a Ranked Set Sampling (RSS) technique to estimate the parameters of a time-to-failure distribution of the linear degradation model. We obtained the maximum likelihood estimates (MLE's) of the parameters based on a RSS and based on a simple random sample (SRS). The performance of these estimators are compared in terms of their efficiencies. Simulation results showed that the MLE's based on a RSS are more efficient than the MLE's based on a SRS. Application to a real data set is given.

Keywords: Reliability; Degradation; Mixed-effect; Ranked set sampling; Simple random sample; Maximum likelihood.

MSC 2000: 62N02

1. Introduction

With today's high technology many products are designed to work without failures for years. This results in very few or no failures, so it is very difficult to analyze life time data with traditional reliability analysis. A recent approach to assess products reliability using degradation measurements of product performance has to pre-specify a level of degradation

[*]E-mail address: malodat@yu.edu.jo
[†]E-mail address: hassanb@hotmail.com

and obtain measurements at different times. Thus the time to failure is defined as the time when the degradation of a unit reaches a critical level. Degradation test has many applications in many different areas, examples are loss of tread on rubber tires and degradation of the active ingredient of a drug as a result of chemical reactions with oxygen and water, and by microbial growth. In the literature, Nelson (1990) provided a general background on degradation models. Lu and Meeker (1993) considered a nonlinear mixed-effect model and used a two-stage method to obtain estimates of percentiles of the failure time distribution. Meeker and Hamda (1995) gave a statistical tools and concepts that are used in designing a high reliability product using degradation data. Meeker, Escolar and Lu (1998) described accelerated degradation models that are related to physical failure mechanisms. Tang and chang (1995) modeled nondestructive accelerated degradation data from power supply anis. Lu, Park and Yang (1997) proposed a model with regression coefficients and standard deviation function for analyzing linear degradation data from semiconductors. For more references see Meeker and Escobar (1998).

In this paper, in order to save the time and the cost of a degradation test, a *Ranked Set Sampling* is introduced to estimate the parameters of the time-to-failure. The MLE's of the parameters based on a RSS and a SRS will be compared in terms of their efficiencies. In both cases there will be no closed form for these estimators, so the efficiencies will be computed numerically. The ranked set sampling and the description of the of the problem will be introduced in Sec. 2.. The linear degradation model and the failure time distribution are given in Sec. 3. . MLE's based on a SRS and MLE's based on a RSS are discussed in Sec. 4. and Sec. 5., respectively. Simulation study is presented in Sec. 6.. Application to a real data set is given in Sec.7.. Our conclusions are given in Sec. 8..

2. Ranked Set Sampling and Description of the Problem

The ranked set sampling is a sampling technique was first introduced by McIntyre (1952) for estimating mean pasture yield. Takahasi and Wakimoto (1968) showed that the sample mean of a RSS is more efficient than the sample mean of a SRS of the same size. The RSS technique can be performed as follows:

- Select a random sample of size m^2 from the population of interest.
- Divide the m^2 units into m subsets of m units each.
- Visually, rank the m units in each subset based on the characteristic of interest.
- From the first set, quantify the minimum, from the second set quantify the second smallest, and continue this until the largest observation in the last set is chosen.
- Repeat the last 4 steps r cycles in order to get a random sample of size $n = mr$.

This has to be done under ranking the units without errors. Let $\{Y_1, Y_2, \ldots, Y_n\}$ be a ranked set sample with mean $\bar{Y}$ and $\{X_1, X_2, \ldots, X_n\}$ be a simple random sample with mean $\bar{X}$. Takahasi and Wakimoto (1968) established the following inequality,

$$1 \leq \mathit{eff}(\bar{X}, \bar{Y}) = \frac{Var(\bar{X})}{Var(\bar{Y})} \leq \frac{m+1}{2}.$$

It was noted that the efficiency of a RSS is increasing in m and the difficulty in ranking is increasing in m. So we use, in practice, $m = 2$ or 3. By using a RSS of size n we can achieve the same accuracy which needs more units to be achieved if we use a SRS, i.e., we save more units. Many authors studied the MLE from a RSS. A good reference about the MLE from a RSS is Zehua, Zhidong and Bimal (2004).

Assume that the sample path of the units is monotone and convex. In order to obtain a RSS of size $3n$, a simple random sample of size 3 is to be chosen and the units are to be tested. From the earlier degradation measurements, we can know which one of these three units will produce the minimum failure time (i.e., which unit touches a critical level of degradation D first), so we stop testing the other two units and we continue to measure the minimum failure time. If another simple random sample of size 3 is tested, we can know which unit will produce the median failure time, so we stop testing the other two units and we continue to measure the median failure time. Again, if another simple random sample of size 3 is tested we can know which unit will produce the maximum failure time, so we stop testing the other two units and we continue to measure maximum failure time. By this plan we can get a ranked set sample of size 3. This plan can be repeated n times to obtain a RSS of size $3n$. The resulting sample will be used to obtain the MLE's of the parameters of the time-to-failure distribution. These estimators will be compared to their MLE's obtained using a SRS.

3. Model and and Failure Time Distribution

The general form of the degradation model commonly used in the reliability analysis is of the form

$$y_{ij} = \eta(t_{ij}, \Theta_j) + \epsilon_{ij}, \quad j = 1, 2, \ldots, n \;\; i = 1, 2, \ldots, m_j,$$

where y_{ij} is the observed degradation amount of the jth unit at the time t_{ij}, $\eta(.)$ is the actual degradation amount of the unit under study and is a function of Θ_j and t_{ij}. Θ_j is a set of parameters and t_{ij} is the time of the ith measurement for the jth unit. $\epsilon'_{ij}s$ are iid random errors with mean 0 and constant variance σ^2

In this paper, we will consider the following linear degradation model, which is a special case of the above model

$$y_{ij} = \phi + \theta_j t_i + \epsilon_{ij}, \quad j = 1, 2, \ldots, n, \;\; i = 1, 2, \ldots, m_j$$

where ϕ is a fixed effect parameter and $\theta'_j s$ are random effects. In this case the degradation function is $\eta(t) = \phi + \theta t$, where ϕ represents the common initial amount of degradation of all test units at the beginning of the test, i.e., $\eta(0) = \phi$. θ represents degradation rate and varies from unit to unit according to a Weibull(α, β) distribution, i.e.,

$$G(\theta) = P(\theta \leq x) = 1 - \exp\{-(\frac{x}{\alpha})^{\beta}\}.$$

Note that θ may follow any other distribution like lognormal, normal, etc. Also we will assume that the units degrade monotonically in time and η is increasing function so $P(\theta > 0) = 1$. θ_j and $\epsilon'_{ij}s$ are independent distributed.

For critical level D, we can write $D = \phi + \theta T$. Then $T = (D - \phi)/\theta$. The distribution function of T, the failure time, is

$$
\begin{aligned}
F_T(t) = P(T \leq t) &= P(\frac{D-\phi}{\theta} \leq t), \\
&= P(\theta \geq \frac{D-\phi}{t}), \\
&= \begin{cases} 0, & t < 0, \\ \exp\{-(\frac{D-\phi}{t\alpha})^{\beta}\}, & t \geq 0. \end{cases}
\end{aligned}
\tag{1}
$$

The corresponding probability density function is

$$
f_T(t) = \begin{cases} \frac{\beta}{t}(\frac{D-\phi}{t\alpha})^{\beta} \exp\{-(\frac{D-\phi}{t\alpha})^{\beta}\}, & t > 0, \\ 0, & t \leq 0. \end{cases}
$$

Thus, in order to estimate the parameters of the failure time distribution, we need to estimate the parameters ϕ, α and β.

4. MLE Based on a SRS

Let $\mathbf{t} = (t_1, t_2, \ldots, t_{3n})$ be the observed values of a random sample of size $3n$ from $F_T(t)$. The likelihood function of $\mathbf{t}$ is given by

$$
\begin{aligned}
L(\mathbf{t}, \alpha, \beta, \phi) &= \prod_{j=1}^{3n} f_T(t_j) \\
&= \prod_{j=1}^{3n} \frac{\beta}{t_j}(\frac{D-\phi}{t_j\alpha})^{\beta} \exp\{-(\frac{D-\phi}{t_j\alpha})^{\beta}\}.
\end{aligned}
$$

The MLE's of α, β, and ϕ are the values of these parameters that maximizes the $L(\mathbf{t}, \alpha, \beta, \phi)$ or equivalently $\ln L(\mathbf{t}, \alpha, \beta, \phi)$. So we need to find the solution of the following system of equations

$$
\begin{aligned}
\frac{\partial}{\partial_{\alpha}} \ln L(\mathbf{t}, \alpha, \beta, \phi) = 0, \\
\frac{\partial}{\partial_{\beta}} \ln L(\mathbf{t}, \alpha, \beta, \phi) = 0, \\
\frac{\partial}{\partial_{\phi}} \ln L(\mathbf{t}, \alpha, \beta, \phi) = 0,
\end{aligned}
$$

where

$$
\ln L(\mathbf{t}, \alpha, \beta, \phi) = 3n \ln \beta + 3n\beta \ln(D-\phi) - \beta\sum_{j=1}^{3n} \ln \alpha - \beta\sum_{j=1}^{3n} \ln t_j - \sum_{j=1}^{3n} \ln t_j - \sum_{j=1}^{3n}(\frac{D-\phi}{t_j\alpha})^{\beta}
$$

Clearly, there is no closed form of the estimators, hence finding the solution of the above system requires numerical calculations.

5. MLE Based on a RSS

Let $\{T_{11}, T_{12}, T_{13}\}, \{T_{21}, T_{22}, T_{23}\}$ and $\{T_{31}, T_{32}, T_{33}\}$ be the failure times of three SRS's each of size 3. The ranked set sample of size 3 is $\{S_1, S_2, S_3\}$, where

$$S_1 = \min\{T_{11}, T_{12}, T_{13}\}, \; S_2 = \text{med}\{T_{21}, T_{22}, T_{23}\} \text{ and } S_3 = \max\{T_{31}, T_{32}, T_{33}\}.$$

It is easy to verify that the distribution functions of S_1, S_2, and S_3, respectively, are

$$\begin{aligned}
F_{S_1}(s_1) &= 1 - \{1 - \exp(-(\frac{D-\phi}{s_1\alpha})^{\beta})\}^3, \\
F_{S_2}(s_2) &= 3\exp(-2(\frac{D-\phi}{s_2\alpha})^{\beta}) - 2\exp(-3(\frac{D-\phi}{s_2\alpha})^{\beta}), \\
F_{S_3}(s_3) &= \exp(-3(\frac{D-\phi}{s_3\alpha})^{\beta}).
\end{aligned}$$

The corresponding densities functions are

$$\begin{aligned}
f_{S_1}(s_1) &= \frac{3\beta}{s_1}(\frac{D-\phi}{s_1\alpha})^{\beta}\exp(-3(\frac{D-\phi}{s_1\alpha})^{\beta})\big(\exp((\frac{D-\phi}{s_1\alpha})^{\beta}) - 1\big)^2, \\
f_{S_2}(s_2) &= \frac{6\beta}{s_2}(\frac{D-\phi}{s_2\alpha})^{\beta}\exp(-3(\frac{D-\phi}{s_2\alpha})^{\beta})\big(\exp((\frac{D-\phi}{s_2\alpha})^{\beta}) - 1\big), \\
f_{S_3}(s_3) &= \frac{3\beta}{s_3}(\frac{D-\phi}{s_3\alpha})^{\beta}\exp(-3(\frac{D-\phi}{s_3\alpha})^{\beta}).
\end{aligned}$$

Let $s_{11}, s_{12}, \ldots, s_{1n}$ be the observed values of an i.i.d sample from $f_{S_1}(s_1)$, $s_{21}, s_{22}, \ldots, s_{2n}$ be the observed values of an i.i.d sample from $f_{S_2}(s_2)$ and $s_{31}, s_{32}, \ldots, s_{3n}$ be the observed values of an i.i.d sample from $f_{S_3}(s_3)$. Let $\mathbf{s} = (s_{11}, \ldots, s_{1n}, s_{21}, \ldots, s_{2n}, s_{31}, \ldots, s_{3n})$. The likelihood function of $\mathbf{s}$ is

$$L(\mathbf{s}, \alpha, \beta, \phi) = \prod_{j=1}^{n} f_{S_1}(s_{1j}) \prod_{j=1}^{n} f_{S_2}(s_{2j}) \prod_{j=1}^{n} f_{S_3}(s_{3j}).$$

To find the MLE's of α, β, and ϕ we need to find the maximizers of $L(\mathbf{s}, \alpha, \beta, \phi)$ which will be obtained numerically.

6. Simulation

In order to compare the MLE's of ϕ, α and β obtained using a RSS with the MLE's obtained using a SRS, we simulated 5000 RSS and SRS from (1). At each realization we computed the MLE's numerically. Let $\hat{\phi}^{SRS}$ and $\hat{\phi}^{RSS}$ denote the MLE of ϕ obtained using a SRS and a RSS, respectively. The efficiency of $\hat{\phi}^{RSS}$ with respect to $\hat{\phi}^{SRS}$ is defined as follows:

$$\textit{eff}\,\hat{\phi} = \textit{eff}(\hat{\phi}^{SRS}, \hat{\phi}^{RSS}) = \frac{MSE(\hat{\phi}^{SRS})}{MSE(\hat{\phi}^{RSS})},$$

where $MSE(\hat{\phi}^{SRS})$ and $MSE(\hat{\phi}^{RSS})$ are the mean squared errors of $\hat{\phi}^{SRS}$ and $\hat{\phi}^{RSS}$, respectively. Similarly we can define the efficiencies of $\hat{\alpha}$ and $\hat{\beta}$. In this simulation study,

Table 1. Efficiencies of RSS MLE's of ϕ, α and β w.r.t their SRS MLE's.

n	ϕ	α	β	*eff* $\hat{\phi}$	*eff* $\hat{\alpha}$	*eff* $\hat{\beta}$
5	1	2	3	1.9345	1.9309	1.7123
10	1	2	3	1.8337	1.8410	1.4897
15	1	2	3	2.0571	2.0600	1.3692
5	1	2	1.5	2.0827	2.0903	1.5751
10	1	2	1.5	2.0137	2.0247	1.4404
15	1	2	1.5	1.8793	1.9007	1.4471
5	1	1	1.5	1.9749	1.98427	1.6426
10	1	1	1.5	1.9518	1.9559	1.4729
15	1	1	1.5	1.8727	1.8793	1.3649
5	.5	1	1.5	1.9567	1.9755	1.6561
10	.5	1	1.5	1.8540	1.8528	1.4313
15	.5	1	1.5	1.8835	1.8867	1.2636

the value of the critical level of degradation is equal to 2. Table 1 contains the values of *eff* $\hat{\phi}$, *eff* $\hat{\alpha}$ and *eff* $\hat{\beta}$ for different values of the parameters. The entries of this table show that efficiencies are greater than 1, which means that the MLE obtained using a RSS is more efficient than the MLE obtained using a SRS.

7. Application

We will consider a real data set from Meeker and Escobar (1998) Table C.17, page 642. The data consists of 15 units. For each unit the percent increase (PI) in operating current for GaAs laser tested at 80°**C** is given. The measurements are taken at times range from 250 to 4000 with step equal to 250. In our analysis, we will scale these times by dividing them by 100, also we used $D = 5$. Figure 1 shows that these data follow a linear degradation path. The estimates of ϕ, α and β and their MSE's are presented in Table 2. Clearly, form the results the MLE obtained from a RSS is more efficient comparing with the MLE obtained from a SRS. The MSE's of the MLE's are obtained using parametric Bootstrap.A goodness-of-fit test analysis of this real data set shows that the slopes $\theta_j^{\prime}s$ follow the Weibull distribution.

8. Conclusion

In this paper, we used a ranked set sampling technique to estimate the parameters of the time-to-failure distribution when the degradation function is linear. Simulation results show that the MLE's of the parameters obtained using a RSS are more efficient than the corresponding MLE's obtained using a SRS. Thus using a RSS in degradation analysis not only gives more efficient estimators, but also saves more units if it is used in stead of a simple random sample of the same size.

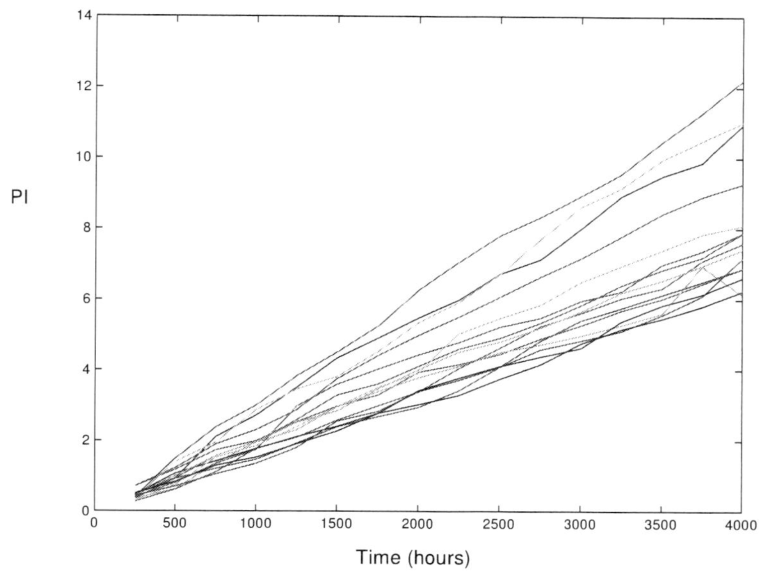

Figure 1. The percent increase in operating current for GaAs Laser tested at 80°**C**.

Table 2. The values of $\hat{\phi}$, $\hat{\alpha}$ and $\hat{\beta}$ and their MSE's.

Estimate	RSS	SRS
$\hat{\phi}$	.28996	.28316
$MSE(\hat{\phi})$	.0255	.12714
$\hat{\alpha}$	2.1267	2.111
$MSE(\hat{\alpha})$	.03152	.071325
$\hat{\beta}$	4.7813	5.2622
$MSE(\hat{\beta})$	119.48	205.322

References

[1] Lu, J. and Meeker, W. (1993). Using Degradation meausres to estimate a Time-to-Failure Distriburion. *Technometrics,* **35**, 161-174.

[2] Lu, J., Park, J. and Yang, Q. (1997). Statistical inference of a Time-to-Failure Distriburion Derived from Linear Degradation Data. *Technometrics,* **39**, 391-400.

[3] McIntyre, G. A. (1952). A method for unbiased selective sampling, using ranked sets. *Austral. J. Agric. rees.,* **3**,385-390.

[4] Meeker, W. and Hamada, M. (1995). Statistical Tools for the Rapid Development & Evaluation of High-Reliability Products. *IEEE Transations on Reliability,* **44**, 187-198.

[5] Meeker, W., Escobar, L. and Lu, J. (1998). Accelerated Degradations Tests: Modeling and Analysis. *Technometrics,* **40**, 89-99.

[6] Meeker, W. and Escobar, L. (1998). *Statistical Methods for Reliability Data.* John Wiley and Sons, Inc.

[7] Nelson, W. (1990). *Accelerated Testing: Statistical Models, Test plans, and Data Analysis.* John Wiley and Sons, Inc.

[8] Takahasi, K., Wakimoto, K. (1968). On unbiased estimates of the population mean based on the sample stratified by means of ordering. *Ann. Ins. Statist. Math.* **20**, 1-30

[9] Tang, L. and Chang, D. (1995). Reliability Prediction Using Nondestructive Accelerated-Degradation Data: Case Study on Power Supplies. *IEEE Transactions on Reliability,* **46**, 562-566.

[10] Zehua C., Zhidong, B. and Bimal, K. S. (2004). *Ranked Set Sampling: Theory and Applications.* Springer-Verlag New York, Inc.

In: New Developments in Applied Statistics
Editor: Mohammad Ahsanullah

ISBN: 978-1-61324-648-1

Chapter 14

REVIEW OF "RELIABILITY AND RISK: A BAYESIAN PERSPECTIVE", NOZER D. SINGPURWALLA, WILEY, NEW YORK (2006)

Nader Ebrahimi
Northern Illinois University
Ehsan S. Soofi
University of Wisconsin-Milwaukee

Over the past few decades, a substantial number of books have been published on the topics of reliability and risk analysis. This book serves as an introduction to the mathematical and statistical aspects of this field from the Bayesian viewpoint. It is aimed at advanced graduate students and statistical researchers in the above topics.

The book's main distinguishing feature is the author's deliberate attempt to go beyond the popular notions such as hazard function, survival function and so on as the benchmark for reliability and risk analysis. Instead, the second and third chapters are devoted entirely to justifying why probability, utility and exchangeability and indifference are the fundamental ideas behind reliability and risk.

The crux of the book comes in Chapters 4 to 10 that offers a comprehensive review of the material in the field. Chapter 4 is devoted largely to the univariate and multivariate failure rate functions. This chapter also includes nice discussions of hazard potential, copula, several bivariate models, and probabilistic causality.

Parametric failure data analysis using Bayesian methods is addressed in Chapter 5. The focal topic of this chapter is predictive models for reliability and risk analysis in the absence and presence of data. Predictive distributions of various reliability models under various priors for the models' parameters are presented. Methods for prior elicitation and construction of prior distributions without elicitation (non-informative, default priors) are presented. This chapter also presents a rather extensive discussion of topics related to design and life-testing under various models. Application of information theory and utility theory are presented in the context of learning from the data and optimal design of life-testing experiments.

Chapter 6 deals with component reliability and signatures. Chapter 7 focuses on survival in a dynamic environment. Chapter 8 deals with recurrent events. Chapter 9 focuses on non-parametric methods in reliability and Chapter 10 discusses the survivability of Co-operative, Competing and Vague system. Finally, the last chapter introduces some of the risk notions to the finance world.

In summary, this book is one of a kind in providing "a clear account of the subject that will be of value to both theoreticians and practitioners" (Dennis Lindley), "in its coverage of reliability and risk concepts and applications" (A.F.M. Smith), and in being "invaluable to those who wish to understand reliability and survival analysis from a modern Bayesian perspective" (Joseph Kadane); quotations are from the back cover of the book. This book is an excellent source of reference for researchers in reliability and risk analysis, particularly useful for students writing theses that involve reliability and risk topics. It also will be an excellent book for reliability courses intended for advanced students in statistics, biostatistics, operations research, decision analysis, and engineering, if supplemented by computational algorithms and exercise problems.

In: New Developments in Applied Statistics
Editor: Mohammad Ahsanullah
ISBN: 978-1-61324-648-1

Chapter 15

MINIMUM PERPENDICULAR DISTANCE SQUARED METHOD ESTIMATION

Forhad Hossain [1] ***and Ghadban Khalaf*** [2]
[1]Department of Statistics, Jahangirnagar University, Savar, Dhaka, Bangladesh
[2]Department of Mathematics, King Khalid University, Abha, Saudi Arabia

Abstract

A new approach 'minimum perpendicular distance squared method' has been suggested to estimate the regression coefficient and constant, instead of traditionally least squared method to fit a regression line, with an aim to get a better fitted regression line, so that the estimated line will be most closest one to the observed points. Mathematical form of the estimator for the parameters have been derived and a logical stand has been placed behind the relationship between the slope parameters of the lines $y = \alpha x + \beta$ and $x = \alpha' y + \beta'$. Finally, a simulation study has been conducted to justify the validity of the suggested new method against the traditionally least squared method.

Keywords: Estimation, Least squares, Simulation.

1. Introduction

Study on inter relationship between two variables or the dependency of one variable on other is a vital area of statistics. In real life problem many situation arises where we need to find the relationship between two variables, when they are interdependent or one is dependent on other. If a mathematical functional relationship can be established then this functional form can be used for future prediction purpose. The functional relationship between two variables may be linear or even sometimes non-linear. The linear relationship between two variables occupied an important area in the field of regression analysis, which is an integrated part of statistics. If the relationship between two variables can be expressed by a linear equation then this linear equation will be much useful for future prediction purpose. Linear regression has a long history in its way of development from the very beginning of eighteen century to up to date. A lot of literatures are available in this area, these literatures involve the estimation of regression coefficient and constant by least squared method i.e. by minimizing the squared of the vertical distance between the observed points and assumed regression line, and estimate the regression coefficient traditionally known as least squared estimation procedure.

In this paper we have noticed (/mention) the basic deflection of the assumption of difference from the observed points to the assumed regression line that are minimized during the use of traditionally least squared method to estimate the regression coefficients (parameters) and have suggested a new idea to estimate the regression coefficient by minimizing the real difference from the observed point to the assumed regression line, we called it actual deviation from the points to the line or perpendicular deviation from the points to the line. Hence, here our main effort is to estimate the regression coefficient by minimizing the squared of the perpendicular distances of the line from the observed points, and comparison with this fitted regression line to the traditionally least squared estimated fitted regression line.

2. Simplle Linear Model

The relation between two variables x and y may be represented by a simple linear equation $y = ax + b$, where a be the slope of the line and b be the intercept for set of pairs (x, y); x be the independent variable and y be the dependent variable. A probabilistic model that may be obtained as an approximate generalization of this deterministic model, where we assume that the expected value of Y is a linear function of x , but for fixed value of x , the value of Y differs from its expected value by a random amount.

2.1. The Simple Regression Model

The model $y = ax + b + e$ where the response variable y is related the explanatory variable x through the regression coefficient a constant intercept b and the random error e . Assuming that the random error $e \sim N(0, \sigma^2)$. If we have a set of n pairs $(x_1, y_1), (x_2, y_2), \ldots, (x_n, y_n)$ of observations and if we construct a scatter plot then for each value of x_i we will find a value of y_i falling exactly on the line $y = ax + b$, call the true regression line. But due to the presence of error term e_i the pair (x, y) fall either above the regression line (if e_i>0) or below the line (if e_i<0). The points (x_1, y_1), $(x_2, y_2), \ldots \ldots, (x_n, y_n)$ resulting from n independent observations will then be scattered about the true regression line.

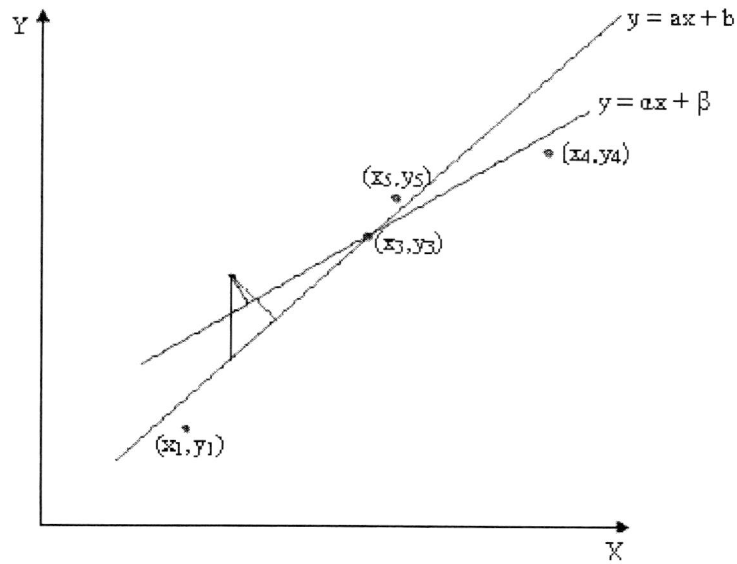

Figure 1.

3. Estimation of Regression Coefficient by Least Squared Method

Having idea about the relationship between the predictor variable x and dependent variable y from the scattered diagram, we are interested to have a line $y = ax + b$ that provides in some sense a best fit to the observed data points, so that the random error (e_i) for the data point from the line will be minimized. This is what motivates the principle of least squares, which can be tracked back to the German Mathematician Gauss (1777-1855). According to this principle, a line is said to be a best fitted line for the data set if the vertical distances (deviations) from the observed points to the line are minimum (*i.e.* e_i 's are minimum which is shown in Figure - 1).

3.1. Principle of Least Square

There is no corroborative exact history who first use the least square method; Carl Friedrich Gauss or Adrien Marie Legendre. It may be noted that Carl Friedrich Gauss started using it before 1803 and by Legendre in 1805. Plackett (1972), Eisenhart (1964), Stigler (1974) and Stigler (1986) etc. discuss in detail about the history of least square.

According to the principle of ordinary least square or traditional least square estimation procedure, we consider the vertical deviation of the points (x_i, y_i) from the line $y = ax + b$ is the vertical distance of the point – vertical distance of the line = $y_i - (ax_i + b)$ and the sum of squared of the vertical deviation from the points $(x_1, y_1), (x_2, y_2), \dots, (x_n, y_n)$ to the line is $\sum_{i=1}^{n} e_i^2 = \sum_{i=1}^{n} [y_i - (ax + b)]^2$, differentiating both sides with respect to a and b, setting equal to zero we get the following two normal equations:

$$a\sum_{i=1}^{n} x_i + nb = \sum_{i=1}^{n} y_i$$

$$a\sum_{i=1}^{n} x_i^2 + b\sum_{i=1}^{n} x_i = \sum_{i=1}^{n} x_i y_i$$

solving these two normal equations we get the traditional (vertical) least square estimator of a and b as follows:

$$\hat{a} = \frac{\sum_{i=1}^{n}(x_i - \bar{x})(y_i - \bar{y})}{\sum_{i=1}^{n}(x_i - \bar{x})^2} = \frac{SPxy}{SSx} \text{ and } \hat{b} = \bar{y} - \hat{a}\bar{x} = \bar{y} - \bar{x}\frac{SP_{xy}}{SS_x}, \tag{3.1}$$

But if we consider the line $x = a'y + b'$, and we are interested to have the least square estimate of a' and b', we could proceed in the similar way as above and we will get the following results:

$$\hat{\alpha}' = \frac{\sum_{i=1}^{n}(x_i - \bar{x})(y_i - \bar{y})}{\sum_{i=}^{n}(y_i - \bar{y})^2} = \frac{SP_{xy}}{SS_y}, \quad \hat{b} = \bar{x} - \hat{\alpha}\bar{y} = \bar{x} - \bar{y}\frac{SP_{xy}}{SS_y}, \tag{3.2}$$

4. Parameter Estimation Based on Minimum Perpendiculat Distance Square Method

The real distance of the assumed regression model $y = \alpha x + \beta$ from the points $(x_1, y_1), (x_2, y_2), \ldots, (x_n, y_n)$ are not the vertical distances or height of the point – height of regression i.e. $y_i - (\alpha x_i + \beta)$. Therefore, estimating regression parameters (regression coefficient and intercept constant) α and β by minimizing the sum of squared of these vertical deviations will not produce a model that will be most closest fitted regression line from the observed points.

In fact the actual distances from the line $y = \alpha x + \beta$ to the points $(x_1, y_2), (x_2, y_2), \ldots, (x_n, y_n)$ are the perpendicular distances e_i' 's (as indicated in the above Figure-1) from the points to the line. These perpendicular distances would also be positive and negative according to (x_i, y_i) is above the line $(e_i' > 0)$ or below the line $(e_i' < 0)$. Also assuming that $e_i's \sim N(0, \sigma^2)$ (it could also be shown from the assumption of section 2.1 that similar assumption condition is true for perpendicular distances). Hence estimating α and β by minimizing the squared of these perpendicular distances will produce the most closest fitted regression line from the points $(x_i, y_i), i = 1, 2, \ldots, n$; and any estimation within the range of the independent variable based on the regression line using these latest estimated parameter will gives us the minimum error due to regression.

4.1. Perpendicular Distance from the Points to the Line $Y = \alpha X + \beta$

Let the fitted regression line be $y = \alpha x + \beta$, then the perpendicular line from the point

(x_i, y_i) to the line $y = \alpha x + \beta$ at the point (x, y) be

$$\frac{y - y_i}{x - x_i} = -\frac{1}{\alpha}$$

$$\Rightarrow \alpha y = x_i - x + \alpha y_i$$

$$\Rightarrow \alpha(\alpha x + \beta) = x_i - x + \alpha y_i$$

$$\Rightarrow x = \frac{\alpha y_i + x_i - \alpha\beta}{\alpha^2 + 1}$$

Hence the perpendicular distance e_i' from the points (x_i, y_i) to the line $y = \alpha x + \beta$ is

$$\begin{aligned}
e_i' &= \sqrt{(x_i - x)^2 + (y_i - y)^2}\ , \\
&= \sqrt{(x_i - \frac{\alpha y_i + x_i - \alpha\beta}{\alpha^2 + 1})^2 + (y_i - \left[\frac{\alpha(\alpha x + x_i - \alpha\beta)}{\alpha^2 + 1} + \beta\right])^2}, \\
&= \sqrt{(\frac{\alpha^2 x_i + x_i - \alpha y_i - x_i + \alpha\beta}{\alpha^2 + 1})^2 + (\frac{\alpha^2 y_i + y_i - \alpha^2 y_i - \alpha x_i + \alpha^2\beta - \alpha^2\beta - \beta}{\alpha^2 + 1})^2}, \\
&= \sqrt{\alpha^2 (\frac{y_i - \alpha x_i - \beta}{\alpha\ \ + 1})^2 + (\frac{y_i - \alpha x_i - \beta}{\alpha^2 + 1})^2}\ , \\
&= \sqrt{(\alpha^2 + 1)(\frac{y_i - \alpha x_i - \beta}{\alpha^2 + 1})^2}\ , \\
&= \frac{(y_i - \alpha x_i - \beta)}{\pm\sqrt{\alpha^2 + 1}} \qquad (4.1)
\end{aligned}$$

Similar type of result was indicated by Spiegel and John (1999).

4.2. Minimum Perendicular Distance Squared Method Estimation

To estimate regression coefficient α and regression constant β by minimizing e_i' 's square, the perpendicular distance from the line $y = \alpha x + \beta$ to the points $(x_i, y_i), i = 1, 2, \ldots, n$; we do the following steps:

$$\sum_{i=1}^{n}(e_i')^2 = \sum_{i=1}^{n}\frac{(y_i - \alpha x_i - \beta)^2}{\alpha^2 + 1} \tag{4.2}$$

differentiating both sides with respect to α and putting equal to zero we get

$$\frac{d\sum_{i=1}^{n}(e_i')^2}{d\alpha} = \frac{(\alpha^2+1)\{\sum_{i=1}^{n}(y_i - \alpha x_i - \beta)(-2x_i)\} - \{\sum_{i=1}^{n}(y_i - \alpha x_i - \beta)^2(2\alpha)\}}{(\alpha^2+1)^2} = 0$$

$$\Rightarrow \quad (\alpha^2+1)\{\sum_{i=1}^{n}(-x_i y_i + \alpha x_i^2 + \beta x_i)\} - \{\sum_{i=1}^{n}(y_i - \alpha x_i - \beta)^2 \alpha\} = 0$$

$$\Rightarrow -\sum_{i=1}^{n}x_i y_i + \alpha\sum_{i=1}^{n}x_i^2 + \beta\sum_{i=1}^{n}x_i - \alpha^2\sum_{i=1}^{n}x_i y_i + \alpha^3\sum_{i=1}^{n}x_i^2 + \alpha^2\beta\sum_{i=1}^{n}x_i$$

$$-a\sum_{i=1}^{n}\{y_i^2 + \alpha^2 x_i^2 + \beta^2 - 2\alpha x_i y_i - 2\beta y_i + 2\alpha\beta x_i\} = 0$$

$$\Rightarrow -\sum_{i=1}^{n}x_i y_i + \alpha^2\sum_{i=1}^{n}x_i y_i + \alpha\sum_{i=1}^{n}x_i^2 + \beta\sum_{i=1}^{n}x_i - \alpha\sum_{i=1}^{n}y_i^2 - n\alpha\beta^2 + 2\alpha\beta\sum_{i=1}^{n}y_i - \alpha^2\beta\sum_{i=1}^{n}x_i = 0 \tag{4.3}$$

Again differentiating equation (4.2) with respect to β and equating zero we get

$$\frac{d\sum_{i=1}^{n}(e_i')^2}{d\beta} = \frac{\sum_{i=1}^{n}(y_i - \alpha x_i - \beta)(-2)}{\alpha^2+1} = 0$$

$$\Rightarrow \quad \sum_{i=1}^{n}y_i - \alpha\sum_{i=1}^{n}x_i - n\beta = 0$$

$$\Rightarrow \beta = \frac{\sum_{i=1}^{n}y_i}{n} - \alpha\frac{\sum_{i=1}^{n}x_i}{n}, \tag{4.4}$$

using equation (4.4) in equation (4.3) we get

$$-\sum_{i=1}^{n} x_i y_i + \alpha^2 \sum_{i=1}^{n} x_i y_i + \alpha \sum_{i=1}^{n} x_i^2 + \left(\frac{\sum_{i=1}^{n} y_i}{n} - \alpha \frac{\sum_{i=1}^{n} x_i}{n}\right) \sum_{i=1}^{n} x_i - \alpha \sum_{i=1}^{n} y_i$$

$$-\alpha\left(\frac{\sum_{i=1}^{n} y_i}{n} - \alpha \frac{\sum_{i=1}^{n} x_i}{n}\right)^2 + 2\alpha\left(\frac{\sum_{i=1}^{n} y_i}{n} - \alpha \frac{\sum_{i=1}^{n} x_i}{n}\right) \sum_{i=1}^{n} y_i - \alpha^2\left(\frac{\sum_{i=1}^{n} y_i}{n} - \alpha \frac{\sum_{i=1}^{n} x_i}{n}\right) \sum_{i=1}^{n} x_i = 0 \text{ or}$$

$$\alpha^2 \left\{\sum_{i=1}^{n} x_i y_i - \frac{(\sum_{i=1}^{n} x_i)(\sum_{i=1}^{n} y_i)}{n}\right\} + \alpha\left[\left\{\sum_{i=1}^{n} x_i^2 - \frac{(\sum_{i=1}^{n} x_i)^2}{n}\right\} - \left\{\sum_{i=1}^{n} y_i^2 - \frac{(\sum_{i=1}^{n} y_i)^2}{n}\right\}\right]$$

$$-\left\{\sum_{i=1}^{n} x_i y_i - \frac{(\sum_{i=1}^{n} x_i)(\sum_{i=1}^{n} y_i)}{n}\right\} = 0$$

or

$$\alpha^2 SPxy + \alpha(SSx - SSy) - SPxy = 0$$

where, $$SPxy = \sum_{i=1}^{n} x_i y_i - \frac{(\sum_{i=1}^{n} x_i)(\sum_{i=1}^{n} y_i)}{n}; \quad SSx = \sum_{i=1}^{n} x_i^2 - \frac{(\sum_{i=1}^{n} x_i)^2}{n}$$

$$SSy = \sum_{i=1}^{n} y_i^2 - \frac{(\sum_{i=1}^{n} y_i)^2}{n}.$$

or $$\alpha = \frac{-(SSx - SSy) \pm \sqrt{(SSx - SSy)^2 + 4(SPxy)^2}}{2SPxy}, \text{ hence}$$

$$\hat{\alpha}_1 = \frac{-(SSx - SSy) + \sqrt{(SSx - SSy)^2 + 4(SPxy)^2}}{2SPxy} \text{ or}$$

$$\hat{\alpha}_2 = \frac{-(SS_x - SS_y) - \sqrt{(SS_x - SS_y)^2 + 4(SP_{xy})^2}}{2SP_{xy}} \qquad (4.5)$$

Using this result in equation (4.4) we can estimate β. And hence

$$\hat{\beta} = \bar{y} - \hat{\alpha}\bar{x}. \qquad (4.6)$$

In this method we get two regression coefficients, we suggest the reader to use $\hat{\alpha}_1$ as the regression coefficient and accordingly the regression constant β could be estimated by using $\hat{\alpha}_1$ in equation (4.6) to fit the regression line y on x i.e. $y = \hat{\alpha}x + \hat{\beta}$. To check the relative efficiency of the present model, i.e. regression line $y = \hat{\alpha}_1 x + \hat{\beta}$ based on estimated parameter using minimum perpendicular distance squared method with the traditional least squared method, a simulation study was conducted and the results are shown in the table's appeared in the Appendix.

4.3. Estimation of Regression Coefficient by Using Minimum Perpendicular Distance Squared Method for the Model $X = \alpha' Y + \beta'$

To estimate regression coefficient α' and regression constant β' by minimizing sum of squared of the error term c_i's (assumed) the perpendicular distances from the line $x = \alpha' y + \beta'$ to the points $(x_i, y_i), i = 1, 2, \ldots, n$; we do the similar steps as in section 4.2:

$$\sum_{i=1}^{n}(c_i)^2 = \sum_{i=1}^{n}\frac{(x_i - \alpha' y_i - \beta')^2}{(\alpha')^2 + 1}, \tag{4.7}$$

differentiating both sides with respect to α' and putting equal to zero we get

$$\frac{d\sum_{i=1}^{n}(c_i)^2}{d\alpha'} = \frac{((\alpha')^2+1)\{\sum_{i=1}^{n}(x_i - \alpha' y_i - \beta')(-2y_i)\} - \{\sum_{i=1}^{n}(x_i - \alpha' y_i - \beta')^2(2\alpha')\}}{((\alpha')^2+1)^2} = 0$$

$$\Rightarrow \ ((\alpha')^2+1)\{\sum_{i=1}^{n}(-x_i y_i + \alpha' y_i^2 + \beta' y_i)\} - \{\sum_{i=1}^{n}(x_i - \alpha' y_i - \beta')^2 \alpha'\} = 0$$

$$\Rightarrow -\sum_{i=1}^{n}x_i y_i + \alpha'\sum_{i=1}^{n}y_i^2 + \beta'\sum_{i=1}^{n}y_i - (\alpha')^2\sum_{i=1}^{n}x_i y_i + (\alpha')^3\sum_{i=1}^{n}y_i^2 + (\alpha')^2\beta'\sum_{i=1}^{n}y_i$$

$$-a'\sum_{i=1}^{n}\{x_i^2 + (\alpha')^2 y_i^2 + (\beta')^2 - 2\alpha' x_i y_i - 2\beta' x_i + 2\alpha'\beta' y_i\} = 0$$

$$\Rightarrow -\sum_{i=1}^{n}x_i y_i + (\alpha')^2\sum_{i=1}^{n}x_i y_i + \alpha'\sum_{i=1}^{n}y_i^2 + \beta'\sum_{i=1}^{n}y_i - \alpha'\sum_{i=1}^{n}x_i^2$$

$$-n\alpha'(\beta')^2 + 2\alpha'\beta'\sum_{i=1}^{n}x_i - (\alpha')^2\beta'\sum_{i=1}^{n}y_i = 0 \quad . \tag{4.8}$$

Again differentiating equation (4.7) with respect to β' and equating to zero we get

$$\frac{d\sum_{i=1}^{n}(c_i)^2}{d\beta'} = \frac{\sum_{i=1}^{n}(x_i - \alpha' y_i - \beta')(-2)}{(\alpha')^2 + 1} = 0$$

$$\Rightarrow \quad \sum_{i=1}^{n} x_i - \alpha' \sum_{i=1}^{n} y_i - n\beta' = 0$$

$$\Rightarrow \quad \beta' = \frac{\sum_{i=1}^{n} x_i}{n} - \alpha' \frac{\sum_{i=1}^{n} y_i}{n}, \tag{4.9}$$

using equation (4.9) in equation (4.8) we have

$$-\sum_{i=1}^{n} x_i y_i + (\alpha')^2 \sum_{i=1}^{n} x_i y_i + \alpha' \sum_{i=1}^{n} y_i^{\,2} + (\frac{\sum_{i=1}^{n} x_i}{n} - \alpha' \frac{\sum_{i=1}^{n} y_i}{n}) \sum_{i=1}^{n} y_i - \alpha' \sum_{i=1}^{n} x_i$$

$$-\alpha'(\frac{\sum_{i=1}^{n} x_i}{n} - \alpha' \frac{\sum_{i=1}^{n} y_i}{n})^2 + 2\alpha'(\frac{\sum_{i=1}^{n} x_i}{n} - \alpha' \frac{\sum_{i=1}^{n} y_i}{n}) \sum_{i=1}^{n} x_i - (\alpha')^2 (\frac{\sum_{i=1}^{n} x_i}{n} - \alpha' \frac{\sum_{i=1}^{n} y_i}{n}) \sum_{i=1}^{n} y_i = 0$$

or $$(\alpha')^2 \{\sum_{i=1}^{n} x_i y_i - \frac{(\sum_{i=1}^{n} x_i)(\sum_{i=1}^{n} y_i)}{n}\} + \alpha'[\{\sum_{i=1}^{n} y_i^{\,2} - \frac{(\sum_{i=1}^{n} y_i)^2}{n}\} - \{\sum_{i=1}^{n} x_i^{\,2} - \frac{(\sum_{i=1}^{n} x_i)^2}{n}\}]$$

$$-\{\sum_{i=1}^{n} x_i y_i - \frac{(\sum_{i=1}^{n} x_i)(\sum_{i=1}^{n} y_i)}{n}\} = 0$$

or $$(\alpha')^2 SPxy + \alpha'(SSy - SSx) - SPxy = 0$$

where, $$SPxy = \sum_{i=1}^{n} x_i y_i - \frac{(\sum_{i=1}^{n} x_i)(\sum_{i=1}^{n} y_i)}{n}; \quad SSx = \sum_{i=1}^{n} x_i^{\,2} - \frac{(\sum_{i=1}^{n} x_i)^2}{n}$$

$$SSy = \sum_{i=1}^{n} y_i^{\,2} - \frac{(\sum_{i=1}^{n} y_i)^2}{n}.$$

Or $$\alpha' = \frac{-(SSy - SSx) \pm \sqrt{(SSy - SSx)^2 + 4(SPxy)^2}}{2SPxy},$$ hence

$$\hat{\alpha}_1' = \frac{-(SSy - SSx) + \sqrt{(SSy - SSx)^2 + 4(SPxy)^2}}{2SPxy}$$ or

$$\hat{\alpha}_2' = \frac{-(SS_y - SS_x) - \sqrt{(SS_y - SS_x)^2 + 4(SP_{xy})^2}}{2SP_{xy}}. \tag{4.10}$$

Using this result in equation (4.9) we can estimate β'. And hence

$$\hat{\beta}' = \overline{x} - \hat{\alpha}'\overline{y} \tag{4.11}$$

In this method we get two regression coefficients and we will suggest the reader to use $\hat{\alpha}_1'$ as regression coefficient and accordingly the estimation of $\hat{\beta}'$ may be obtained to fit the regression line x on y i.e.

$$x = \hat{\alpha}'y + \hat{\beta}'.$$

4.3.1. Relationship Between Regression Coefficients

In Section 3.1 we use the principle of traditional least square method to estimate regression coefficient y on x $a = \dfrac{SPxy}{SSx}$ by minimizing the error term e_i, hence the fitted regression line will be $y = ax + b$. And the regression coefficient x on y is $a' = \dfrac{SPxy}{SSx}$, and hence the fitted regression line will be $x = a'y + b'$. Now if really the error terms are minimized during the estimation of the regression coefficients a or a', we will get two straight line $y = ax + b$ and $x = a'y + b'$, and hence a should be proportional to a'. In strict sense a should be equal to $\dfrac{1}{a'}$, or $aa' = 1$. But in reality we get the relationship as follows $aa' = \dfrac{SPxy}{SSx}\dfrac{SPxy}{SSy}$ which is nothing but the squared of the Pearson's Correlation coefficient (r^2), and this will be unity only when x and y have one to one correlation.

But on the other hand, if we consider the minimum perpendicular distance squared method to estimate the regression coefficients α and α' as we have indicate in section 4 by minimizing the error term e_i' (the perpendicular distances from the line to the observed points), we get $\hat{\alpha}_1 = \dfrac{-(SSx - SSy) + \sqrt{(SSx - SSy)^2 + 4(SPxy)^2}}{2SPxy}$ for the line $y = \hat{\alpha}x + \hat{\beta}$ and $\hat{\alpha}_1' = \dfrac{-(SSy - SSx) + \sqrt{(SSy - SSx)^2 + 4(SPxy)^2}}{2SPxy}$ for the line $x = \hat{\alpha}'y + \hat{\beta}'$, we see that $\hat{\alpha}_1$ is proportional to $\hat{\alpha}_1'$ i.e. $\hat{\alpha}_1 = \dfrac{1}{\hat{\alpha}_1'}$ or $\hat{\alpha}_1\hat{\alpha}_1' = 1$, which indicate that during estimating regression coefficient by using minimum perpendicular distance squared method, error term is minimized. This is a new angle to advocate the advantage of our newly suggested minimum perpendicular distance squared method to estimate the regression coefficients.

5. Simylated Results and Conclusions

Simulation study was conducted based on 10,000 times repetition and the results are shown in the table in Appendix, where σ_x and σ_y represents the standard deviations of the variables X and Y respectively, r represents the sample correlation coefficient and m indicates the sample sizes.

In this simulation study we have compared our newly suggested model i.e. fitted regression line based on the estimated regression coefficient and constant using minimum perpendicular distance squared method with the fitted regression line using traditionally used least squared method estimator.

In the result Tables: first column indicate the different values of σ_x and σ_y. In the second column of the table, it is indicating that out of 10000 times repetitions how many times the sum of vertically absolute deviations from the observed points to our newly suggested line is smaller (better) than the sum of vertically absolute deviations from the same points to the traditionally least squared fitted regression line.

In the third column it is indicating that out of total 10000 times repetitions how many times the sum of absolute perpendicular distances from the observed points to our newly suggested fitted regression line is smaller (better) than the sum of absolute perpendicular distances from the same observed points to the traditionally least squared estimated fitted regression line. And in the fourth column it is indicating that out of 10000 times repetitions, how many times both third column and fourth column simultaneously have smaller (better) values incase of sum of vertically absolute deviations from the observed points and sum of absolute perpendicular distances from the same points for our newly suggested model than the traditionally least squared estimated fitted regression line.

It has been observed that for highly correlated data ($r = 0.99$) if $\sigma_x = 5$ and $\sigma_y = 5$, about 48 percent cases our newly suggested method produces smaller absolute deviations (better result) from observed data to the fitted regression line, in case of perpendicular absolute deviation as well as vertically absolute deviations and also their common cases; and as the variability of the dependent variable y, i.e. , σ_x increases these percent increases (better cases) even up to about 90% cases, it is evident that these percent is not significantly changing with increasing sample sizes. But as the relationship between x (the independent variable) and y (the dependent variable) become weaker and weaker such as $r = 0.82, r = 0.53$ or even $r = 0.22$ the percentage of vertically absolute deviation decreases significantly. On the other hand percentage of perpendicular absolute deviations increases significantly even up to 100% (better result), sometimes as sample size m increases as well as variability of the variable x keeping fixed ($i.e., \sigma_x = 5$) but variability of the dependent variable y increases from $\sigma_y = 5$ to $\sigma_y = 10, \sigma_y = 20$ or even $\sigma_y = 30$.

From the table (page 21 and on word up to page 24) it is evident that for different values of r $(i.e., r = 0.99, r = 0.82, r = 0.53$ and or $r = 0.22)$ if we fix up the value of $\sigma_y = 5$ and change the value of σ_x $(i.e., \sigma_x = 10, \sigma_x = 20$ and/or $\sigma_x = 30)$, 30% to 50% cases our newly suggested model gives smaller values (better result) for the absolute vertical deviation in comparison with the traditionally least squared estimated fitted regression line for different sample sizes. On the other hand 30% to up to 71% cases the absolute perpendicular deviation from the observed points to our newly suggested regression line gives smaller value (better results) than that of traditionally least squared estimated fitted regression model, these variation happen with different values of σ_x's and sample sizes m's. It may be noted here that in this situation for both cases (absolute vertical deviation and absolute perpendicular deviations) as the strength of relationship between x and y become weaker and weaker, our newly suggested model fit better and better than the traditionally least squared estimated fitted regression line.

From the table (in page 26 and on ward) it is evident that for $r = 0.99$, if the variability of x is maximum in this case say $\sigma_x = 30$ and the variability of y varies from $\sigma_y = 5$, to up to $\sigma_y = 30$ and, for different sample sizes , 29% to 52% cases the vertically absolute deviation as well as absolute perpendicular distances of our newly suggested model gives smaller values (better results) than that of traditionally least squared estimated fitted regression line.

But for $r = 0.82$ or smaller and, for similar variability in the independent variable ($\sigma_x = 30$) and dependent variable ($\sigma_y = 5$ to up to $\sigma_y = 30$), percentage of vertically absolute deviations of our newly suggested model varies up to 50% cases. These percentage decreases as variability of the dependent variable increases as well as sample sizes increases. This indicate that if both independent and dependent variable are highly dispersed then use of vertically deviation procedure to detect the suitability of a model, our newly suggested model is not showing a significant cases better result.

On the other hand if we consider the perpendicular deviation procedure (in which we are mainly focusing this paper) we see that 42% cases to up to 98% cases our newly suggested model gives better result depending on the strength of relationship between x and y, sample sizes, variability within variable x and y, and the variability between these variables x and y. It could be mentioned here that irrespective of the value of r (varies from $r = 0.82$, or $r = 0.53$, or even $r = 0.22$) if the variability of x and y is very high e.g. ($\sigma_x = 30$ and $\sigma_y = 30$), and the sample sizes are increasing, our newly suggested method produces smaller perpendicular deviations (better results) between the observed points and the fitted regression line(it varies from 69% to up to 97% cases) than the traditionally least squared fitted regression line.

In this paper our main focus is on the minimum perpendicular deviations, so we want to draw attention to the reader towards perpendicular deviation results. Even though we have seen there are some significant cases where our newly suggested method produce better result if we consider vertically deviation to measure the deflection. It is notable that every cases irrespective of high or poor relationship between dependent and independent variables x and y, and higher or smaller variability of both the variables x and/or y; perpendicular deviation of our newly suggested model gives considerable cases smaller deviations(better result, varying from 48% to 100% depending on different values of $r's$ and, $\sigma_x's$ and $\sigma_y's$) in compared with the traditionally least squared estimated fitted regression model. But at the same time we have tried to show that even there are some situation when the vertical deviation of our newly suggested model gives smaller values (better result) than the traditionally least squared estimated fitted regression line (varies from 30% to up to 50% cases in general, depending on the strength of relationship between x and y, as well as the different variability of x and y).

Acknowledgement

The authors would like to thank Prof. Yousif Atique and Dr. Munsur Abdulkhalek of the department of Mathematics, King Khalid University, Abha, Saudi Arabia, for their valuable suggestions and help during preparing this paper.

Appendix

Table

The model for $y = \alpha x + \beta$

If $r = 0.99$

$m = 10$

The standard deviations	Absolute deviation	Perpendicular deviation	Common
$\sigma_x = 5, \quad \sigma_y = 5$	4767	4781	4766
$\sigma_x = 5, \quad \sigma_y = 10$	6859	6861	6858
$\sigma_x = 5, \quad \sigma_y = 20$	8481	8484	8487
$\sigma_x = 5, \quad \sigma_y = 30$	8941	8943	8941

$m = 20$

The standard deviations	Absolute deviation	Perpendicular deviation	Common
$\sigma_x = 5, \quad \sigma_y = 5$	4735	4741	4732
$\sigma_x = 5, \quad \sigma_y = 10$	6924	6923	6923
$\sigma_x = 5, \quad \sigma_y = 20$	8523	8523	8523
$\sigma_x = 5, \quad \sigma_y = 30$	8939	8939	8939

If $r = 0.99$

$m = 30$

The standard deviations	Absolute deviation	Perpendicular deviation	Common
$\sigma_x = 5, \quad \sigma_y = 5$	5056	5064	5050
$\sigma_x = 5, \quad \sigma_y = 10$	6900	6901	6900
$\sigma_x = 5, \quad \sigma_y = 20$	8643	8645	8643
$\sigma_x = 5, \quad \sigma_y = 30$	9094	9094	9094

$m = 40$

The standard deviations	Absolute deviation	Perpendicular deviation	Common
$\sigma_x = 5, \quad \sigma_y = 5$	4669	4672	4661
$\sigma_x = 5, \quad \sigma_y = 10$	6750	6755	6755
$\sigma_x = 5, \quad \sigma_y = 20$	8602	8601	8601
$\sigma_x = 5, \quad \sigma_y = 30$	8958	8957	8957

$m = 80$

The standard deviations	Absolute deviation	Perpendicular deviation	Common
$\sigma_x = 5, \quad \sigma_y = 5$	4657	4662	4650
$\sigma_x = 5, \quad \sigma_y = 10$	6831	6831	6830
$\sigma_x = 5, \quad \sigma_y = 20$	8627	8627	8627
$\sigma_x = 5, \quad \sigma_y = 30$	8937	8937	8937

$m = 100$

The standard deviations	Absolute deviation	Perpendicular deviation	Common
$\sigma_x = 5, \quad \sigma_y = 5$	5246	5244	5238
$\sigma_x = 5, \quad \sigma_y = 10$	6848	6853	6847
$\sigma_x = 5, \quad \sigma_y = 20$	8569	8569	8569
$\sigma_x = 5, \quad \sigma_y = 30$	9034	9034	9034

If $r = 0.82$

$m = 10$

The standard deviations	Absolute deviation	Perpendicular deviation	Common
$\sigma_x = 5, \quad \sigma_y = 5$	3586	6919	3580
$\sigma_x = 5, \quad \sigma_y = 10$	1945	8722	1945
$\sigma_x = 5, \quad \sigma_y = 20$	1086	9516	1086
$\sigma_x = 5, \quad \sigma_y = 30$	922	9628	922

$m = 20$

The standard deviations	Absolute deviation	Perpendicular deviation	Common
$\sigma_x = 5, \quad \sigma_y = 5$	3234	7791	3234
$\sigma_x = 5, \quad \sigma_y = 10$	1178	9663	1178
$\sigma_x = 5, \quad \sigma_y = 20$	470	9942	470
$\sigma_x = 5, \quad \sigma_y = 30$	373	9970	373

$m = 30$

The standard deviations	Absolute deviation	Perpendicular deviation	Common
$\sigma_x = 5, \quad \sigma_y = 5$	2776	8364	2776
$\sigma_x = 5, \quad \sigma_y = 10$	749	9878	749
$\sigma_x = 5, \quad \sigma_y = 20$	203	9992	203
$\sigma_x = 5, \quad \sigma_y = 30$	165	9998	165

$m = 40$

The standard deviations	Absolute deviation	Perpendicular deviation	Common
$\sigma_x = 5, \quad \sigma_y = 5$	2544	8714	2544
$\sigma_x = 5, \quad \sigma_y = 10$	485	9957	485
$\sigma_x = 5, \quad \sigma_y = 20$	108	9997	108
$\sigma_x = 5, \quad \sigma_y = 30$	72	9999	72

$$If \quad r = 0.82$$

$$m = 80$$

The standard deviations	Absolute deviation	Perpendicular deviation	Common
$\sigma_x = 5, \quad \sigma_y = 5$	1714	9435	1714
$\sigma_x = 5, \quad \sigma_y = 10$	106	10000	106
$\sigma_x = 5, \quad \sigma_y = 20$	5	10000	5
$\sigma_x = 5, \quad \sigma_y = 30$	8	10000	8

$$m = 100$$

The standard deviations	Absolute deviation	Perpendicular deviation	Common
$\sigma_x = 5, \quad \sigma_y = 5$	1406	9639	1406
$\sigma_x = 5, \quad \sigma_y = 10$	33	10000	33
$\sigma_x = 5, \quad \sigma_y = 20$	2	10000	2
$\sigma_x = 5, \quad \sigma_y = 30$	1	10000	1

$$If \quad r = 0.53$$

$$m = 10$$

The standard deviations	Absolute deviation	Perpendicular deviation	Common
$\sigma_x = 5, \quad \sigma_y = 5$	2811	6985	2811
$\sigma_x = 5, \quad \sigma_y = 10$	518	9239	516
$\sigma_x = 5, \quad \sigma_y = 20$	175	9648	143
$\sigma_x = 5, \quad \sigma_y = 30$	144	9689	144

$$m = 20$$

The standard deviations	Absolute deviation	Perpendicular deviation	Common
$\sigma_x = 5, \quad \sigma_y = 5$	2218	7996	2218
$\sigma_x = 5, \quad \sigma_y = 5$	90	9872	90
$\sigma_x = 5, \quad \sigma_y = 5$	6	9979	5
$\sigma_x = 5, \quad \sigma_y = 30$	8	9975	6

$$If \quad r = 0.53$$

$$m = 30$$

The standard deviations	Absolute deviation	Perpendicular deviation	Common
$\sigma_x = 5, \quad \sigma_y = 5$	1856	8565	1856
$\sigma_x = 5, \quad \sigma_y = 10$	19	9979	19
$\sigma_x = 5, \quad \sigma_y = 20$	1	9997	1
$\sigma_x = 5, \quad \sigma_y = 30$	1	9997	1

$$m = 40$$

The standard deviations	Absolute deviation	Perpendicular deviation	Common
$\sigma_x = 5, \quad \sigma_y = 5$	1489	8947	1489
$\sigma_x = 5, \quad \sigma_y = 10$	2	9995	2
$\sigma_x = 5, \quad \sigma_y = 20$	0	10000	0
$\sigma_x = 5, \quad \sigma_y = 30$	0	10000	0

$$m = 80$$

The standard deviations	Absolute deviation	Perpendicular deviation	Common
$\sigma_x = 5, \quad \sigma_y = 5$	222	9640	722
$\sigma_x = 5, \quad \sigma_y = 10$	0	10000	0
$\sigma_x = 5, \quad \sigma_y = 20$	0	10000	0
$\sigma_x = 5, \quad \sigma_y = 30$	0	10000	0

$$m = 100$$

The standard deviations	Absolute deviation	Perpendicular deviation	Common
$\sigma_x = 5, \quad \sigma_y = 5$	505	9778	505
$\sigma_x = 5, \quad \sigma_y = 10$	0	10000	0
$\sigma_x = 5, \quad \sigma_y = 20$	0	10000	0
$\sigma_x = 5, \quad \sigma_y = 30$	0	10000	0

$If \quad r = 0.22$

$m = 10$

The standard deviations	Absolute deviation	Perpendicular deviation	Common
$\sigma_x = 5, \quad \sigma_y = 5$	1810	6021	1810
$\sigma_x = 5, \quad \sigma_y = 5$	365	7667	293
$\sigma_x = 5, \quad \sigma_y = 20$	435	7828	112
$\sigma_x = 5, \quad \sigma_y = 30$	455	7691	52

$m = 20$

The standard deviations	Absolute deviation	Perpendicular deviation	Common
$\sigma_x = 5, \quad \sigma_y = 5$	1449	7091	1449
$\sigma_x = 5, \quad \sigma_y = 10$	166	8667	126
$\sigma_x = 5, \quad \sigma_y = 20$	196	8740	49
$\sigma_x = 5, \quad \sigma_y = 30$	232	8506	16

$m = 30$

The standard deviations	Absolute deviation	Perpendicular deviation	Common
$\sigma_x = 5, \quad \sigma_y = 5$	1134	7621	1134
$\sigma_x = 5, \quad \sigma_y = 10$	89	9136	72
$\sigma_x = 5, \quad \sigma_y = 20$	115	9144	23
$\sigma_x = 5, \quad \sigma_y = 30$	120	9080	14

$m = 40$

The standard deviations	Absolute deviation	Perpendicular deviation	Common
$\sigma_x = 5, \quad \sigma_y = 5$	864	7961	864
$\sigma_x = 5, \quad \sigma_y = 10$	62	9425	47
$\sigma_x = 5, \quad \sigma_y = 20$	58	9405	18
$\sigma_x = 5, \quad \sigma_y = 30$	60	9405	5

If $r = 0.22$

$m = 80$

The standard deviations	Absolute deviation	Perpendicular deviation	Common
$\sigma_x = 5, \quad \sigma_y = 5$	316	8910	316
$\sigma_x = 5, \quad \sigma_y = 10$	5	9868	4
$\sigma_x = 5, \quad \sigma_y = 20$	12	9873	4
$\sigma_x = 5, \quad \sigma_y = 30$	7	9862	1

$m = 100$

The standard deviations	Absolute deviation	Perpendicular deviation	Common
$\sigma_x = 5, \quad \sigma_y = 5$	191	9167	191
$\sigma_x = 5, \quad \sigma_y = 10$	4	9933	3
$\sigma_x = 5, \quad \sigma_y = 20$	5	9923	1
$\sigma_x = 5, \quad \sigma_y = 30$	1	9927	0

If $r = 0.99$

$m = 10$

Common	Perpendicular deviation	Absolute deviation	The standard deviations
3417	3417	3416	$\sigma_y = 5, \quad \sigma_x = 10$
3113	3128	3113	$\sigma_y = 5, \quad \sigma_x = 20$
3001	3040	3001	$\sigma_y = 5, \quad \sigma_x = 30$

$m = 20$

Common	Perpendicular deviation	Absolute deviation	The standard deviations
3370	3370	3370	$\sigma_y = 5, \quad \sigma_x = 10$
3055	3068	3055	$\sigma_y = 5, \quad \sigma_x = 20$
3020	3025	3020	$\sigma_y = 5, \quad \sigma_x = 30$

$If\ r = 0.99$

$m = 30$

Common	Perpendicular deviation	Absolute deviation	The standard deviations
3571	3571	3571	$\sigma_y = 5, \quad \sigma_x = 10$
3444	3450	3444	$\sigma_y = 5, \quad \sigma_x = 20$
3348	3352	3348	$\sigma_y = 5, \quad \sigma_x = 30$

$m = 40$

Common	Perpendicular deviation	Absolute deviation	The standard deviations
3300	3357	3355	$\sigma_y = 5, \quad \sigma_x = 10$
3114	3116	3114	$\sigma_y = 5, \quad \sigma_x = 20$
3079	3082	3079	$\sigma_y = 5, \quad \sigma_x = 30$

$m = 80$

Common	Perpendicular deviation	Absolute deviation	The standard deviations
3330	3330	3330	$\sigma_y = 5, \quad \sigma_x = 10$
3071	3071	3071	$\sigma_y = 5, \quad \sigma_x = 20$
3089	3090	3089	$\sigma_y = 5, \quad \sigma_x = 30$

$m = 100$

Common	Perpendicular deviation	Absolute deviation	The standard deviations
3702	3702	3702	$\sigma_y = 5, \quad \sigma_x = 10$
3522	3522	3522	$\sigma_y = 5, \quad \sigma_x = 20$
3419	3420	3419	$\sigma_y = 5, \quad \sigma_x = 30$

$If \quad r = 0.82$

$m = 10$

Common	Perpendicular deviation	Absolute deviation	The standard deviations
4561	5525	4561	$\sigma_y = 5, \quad \sigma_x = 10$
4912	5159	4912	$\sigma_y = 5, \quad \sigma_x = 20$
4904	5029	4904	$\sigma_y = 5, \quad \sigma_x = 30$

If $r = 0.82$

$m = 20$

Common	Perpendicular deviation	Absolute deviation	The standard deviations
4469	5916	4469	$\sigma_y = 5, \quad \sigma_x = 10$
4851	5218	4851	$\sigma_y = 5, \quad \sigma_x = 20$
5039	5227	5039	$\sigma_y = 5, \quad \sigma_x = 30$

$m = 30$

Common	Perpendicular deviation	Absolute deviation	The standard deviations
4454	6216	4454	$\sigma_y = 5, \quad \sigma_x = 10$
4814	5279	4814	$\sigma_y = 5, \quad \sigma_x = 20$
4888	5104	4888	$\sigma_y = 5, \quad \sigma_x = 30$

$m = 40$

Common	Perpendicular deviation	Absolute deviation	The standard deviations
4269	6330	4269	$\sigma_y = 5, \quad \sigma_x = 10$
4739	5317	4739	$\sigma_y = 5, \quad \sigma_x = 20$
4972	5220	4972	$\sigma_y = 5, \quad \sigma_x = 30$

$m = 80$

Common	Perpendicular deviation	Absolute deviation	The standard deviations
4000	6937	4000	$\sigma_y = 5, \quad \sigma_x = 10$
4699	5469	4699	$\sigma_y = 5, \quad \sigma_x = 20$
4912	5267	4912	$\sigma_y = 5, \quad \sigma_x = 30$

$m = 100$

Common	Perpendicular deviation	Absolute deviation	The standard deviations
3775	7086	3775	$\sigma_y = 5, \quad \sigma_x = 10$
4689	5598	4689	$\sigma_y = 5, \quad \sigma_x = 20$
4817	5219	4817	$\sigma_y = 5, \quad \sigma_x = 30$

If $r = 0.53$

$m = 10$

Common	Perpendicular deviation	Absolute deviation	The standard deviations
4427	5501	4427	$\sigma_y = 5, \quad \sigma_x = 10$
4809	5073	4809	$\sigma_y = 5, \quad \sigma_x = 20$
4818	4952	4818	$\sigma_y = 5, \quad \sigma_x = 30$

$m = 20$

Common	Perpendicular deviation	Absolute deviation	The standard deviations
4397	5931	4397	$\sigma_y = 5, \quad \sigma_x = 10$
4780	5165	4780	$\sigma_y = 5, \quad \sigma_x = 20$
4893	5088	4893	$\sigma_y = 5, \quad \sigma_x = 30$

$m = 30$

Common	Perpendicular deviation	Absolute deviation	The standard deviations
4347	6292	4347	$\sigma_y = 5, \quad \sigma_x = 10$
4839	5302	4839	$\sigma_y = 5, \quad \sigma_x = 20$
4883	5098	4883	$\sigma_y = 5, \quad \sigma_x = 30$

$m = 40$

Common	Perpendicular deviation	Absolute deviation	The standard deviations
4141	6352	4141	$\sigma_y = 5, \quad \sigma_x = 10$
4862	5458	4862	$\sigma_y = 5, \quad \sigma_x = 20$
4918	5145	4918	$\sigma_y = 5, \quad \sigma_x = 30$

$m = 80$

Common	Perpendicular deviation	Absolute deviation	The standard deviations
3814	6912	3814	$\sigma_y = 5, \quad \sigma_x = 10$
4663	5438	4663	$\sigma_y = 5, \quad \sigma_x = 20$
4875	5244	4875	$\sigma_y = 5, \quad \sigma_x = 30$

If $r = 0.53$

$m = 100$

Common	Perpendicular deviation	Absolute deviation	The standard deviations
3669	7135	3669	$\sigma_y = 5, \quad \sigma_x = 10$
4697	5543	4697	$\sigma_y = 5, \quad \sigma_x = 20$
4913	5308	4913	$\sigma_y = 5, \quad \sigma_x = 30$

If $r = 0.22$

$m = 10$

Common	Perpendicular deviation	Absolute deviation	The standard deviations
3829	4723	3829	$\sigma_y = 5, \quad \sigma_x = 10$
4128	4333	4128	$\sigma_y = 5, \quad \sigma_x = 20$
4166	4278	4166	$\sigma_y = 5, \quad \sigma_x = 30$

$m = 20$

Common	Perpendicular deviation	Absolute deviation	The standard deviations
4153	5283	4153	$\sigma_y = 5, \quad \sigma_x = 10$
4530	4802	4530	$\sigma_y = 5, \quad \sigma_x = 20$
4562	4682	4562	$\sigma_y = 5, \quad \sigma_x = 30$

$m = 30$

Common	Perpendicular deviation	Absolute deviation	The standard deviations
4270	5629	4270	$\sigma_y = 5, \quad \sigma_x = 10$
4739	5042	4739	$\sigma_y = 5, \quad \sigma_x = 20$
4755	4892	4755	$\sigma_y = 5, \quad \sigma_x = 30$

$m = 40$

Common	Perpendicular deviation	Absolute deviation	The standard deviations
4211	5752	4221	$\sigma_y = 5, \quad \sigma_x = 10$
4790	5114	4790	$\sigma_y = 5, \quad \sigma_x = 20$
4931	5105	4931	$\sigma_y = 5, \quad \sigma_x = 30$

$If \quad r = 0.22$

$m = 80$

Common	Perpendicular deviation	Absolute deviation	The standard deviations
4143	6357	4149	$\sigma_y = 5, \quad \sigma_x = 10$
4849	5406	4849	$\sigma_y = 5, \quad \sigma_x = 20$
4996	5212	4996	$\sigma_y = 5, \quad \sigma_x = 30$

$m = 100$

Common	Perpendicular deviation	Absolute deviation	The standard deviations
3976	6447	3976	$\sigma_y = 5, \quad \sigma_x = 10$
4794	5380	4794	$\sigma_y = 5, \quad \sigma_x = 20$
4951	5169	4951	$\sigma_y = 5, \quad \sigma_x = 30$

$If \; r = 0.99$

$m = 10$

The standard deviations	Absolute deviation	Perpendicular deviation	Common
$\sigma_x = 30, \quad \sigma_y = 5$	2919	2925	2919
$\sigma_x = 30, \quad \sigma_y = 10$	3079	3092	3079
$\sigma_x = 30, \quad \sigma_y = 20$	3704	3729	3704
$\sigma_x = 30, \quad \sigma_y = 30$	4755	4772	4755

$m = 20$

The standard deviations	Absolute deviation	Perpendicular deviation	Common
$\sigma_x = 30, \quad \sigma_y = 5$	3021	3032	3021
$\sigma_x = 30, \quad \sigma_y = 10$	3119	3125	3119
$\sigma_x = 30, \quad \sigma_y = 20$	3726	3731	3726
$\sigma_x = 30, \quad \sigma_y = 30$	4711	4715	4711

$m = 30$

The standard deviations	Absolute deviation	Perpendicular deviation	Common
$\sigma_x = 30, \quad \sigma_y = 5$	3369	3376	3369
$\sigma_x = 30, \quad \sigma_y = 10$	3434	3436	3434
$\sigma_x = 30, \quad \sigma_y = 20$	3963	3967	3963
$\sigma_x = 30, \quad \sigma_y = 30$	4835	4839	4835

If $r = 0.99$

$m = 40$

The standard deviations	Absolute deviation	Perpendicular deviation	Common
$\sigma_x = 30, \quad \sigma_y = 5$	3074	3075	3074
$\sigma_x = 30, \quad \sigma_y = 10$	3165	3169	3165
$\sigma_x = 30, \quad \sigma_y = 20$	3827	3827	3827
$\sigma_x = 30, \quad \sigma_y = 30$	4680	4681	4680

$m = 80$

The standard deviations	Absolute deviation	Perpendicular deviation	Common
$\sigma_x = 30, \quad \sigma_y = 5$	3104	3105	3104
$\sigma_x = 30, \quad \sigma_y = 10$	3118	3118	3118
$\sigma_x = 30, \quad \sigma_y = 20$	3847	3848	3847
$\sigma_x = 30, \quad \sigma_y = 30$	4683	4683	4683

$m = 100$

The standard deviations	Absolute deviation	Perpendicular deviation	Common
$\sigma_x = 30, \quad \sigma_y = 5$	3461	3462	3461
$\sigma_x = 30, \quad \sigma_y = 10$	3459	3460	3459
$\sigma_x = 30, \quad \sigma_y = 20$	4022	4023	4022
$\sigma_x = 30, \quad \sigma_y = 30$	5189	5189	5189

If $r = 0.82$

$m = 10$

The standard deviations	Absolute deviation	Perpendicular deviation	Common
$\sigma_x = 30, \quad \sigma_y = 5$	4937	5079	4937
$\sigma_x = 30, \quad \sigma_y = 10$	4820	5311	4820
$\sigma_x = 30, \quad \sigma_y = 20$	4242	5974	4242
$\sigma_x = 30, \quad \sigma_y = 30$	3576	6870	3576

If $r = 0.82$

$m = 20$

The standard deviations	Absolute deviation	Perpendicular deviation	Common
$\sigma_x = 30, \quad \sigma_y = 5$	5013	5792	5013
$\sigma_x = 30, \quad \sigma_y = 10$	4767	5471	4717
$\sigma_x = 30, \quad \sigma_y = 20$	3959	6477	3959
$\sigma_x = 30, \quad \sigma_y = 30$	3068	7769	3068

$m = 30$

The standard deviations	Absolute deviation	Perpendicular deviation	Common
$\sigma_x = 30, \quad \sigma_y = 5$	4984	5196	4984
$\sigma_x = 30, \quad \sigma_y = 10$	4717	5513	4717
$\sigma_x = 30, \quad \sigma_y = 20$	3903	6890	3903
$\sigma_x = 30, \quad \sigma_y = 30$	2840	8328	2840

$m = 40$

The standard deviations	Absolute deviation	Perpendicular deviation	Common
$\sigma_x = 30, \quad \sigma_y = 5$	4878	5099	4878
$\sigma_x = 30, \quad \sigma_y = 10$	4638	5614	4638
$\sigma_x = 30, \quad \sigma_y = 20$	3845	7245	3845
$\sigma_x = 30, \quad \sigma_y = 30$	2523	8734	2523

$m = 80$

The standard deviations	Absolute deviation	Perpendicular deviation	Common
$\sigma_x = 30, \quad \sigma_y = 5$	4950	5303	4950
$\sigma_x = 30, \quad \sigma_y = 10$	4486	5821	4486
$\sigma_x = 30, \quad \sigma_y = 20$	3286	7943	3286
$\sigma_x = 30, \quad \sigma_y = 30$	1695	9471	1965

If $r = 0.82$

$m = 100$

The standard deviations	Absolute deviation	Perpendicular deviation	Common
$\sigma_x = 30, \quad \sigma_y = 5$	4845	5236	4845
$\sigma_x = 30, \quad \sigma_y = 10$	4533	6086	4533
$\sigma_x = 30, \quad \sigma_y = 20$	3096	8198	3096
$\sigma_x = 30, \quad \sigma_y = 30$	1418	9637	1418

If $r = 0.53$

$m = 10$

The standard deviations	Absolute deviation	Perpendicular deviation	Common
$\sigma_x = 30, \quad \sigma_y = 5$	4801	4924	4801
$\sigma_x = 30, \quad \sigma_y = 10$	4657	5166	4657
$\sigma_x = 30, \quad \sigma_y = 20$	3978	5822	3978
$\sigma_x = 30, \quad \sigma_y = 30$	2808	7071	2808

$m = 20$

The standard deviations	Absolute deviation	Perpendicular deviation	Common
$\sigma_x = 30, \quad \sigma_y = 5$	4844	5012	4844
$\sigma_x = 30, \quad \sigma_y = 10$	4777	5466	4777
$\sigma_x = 30, \quad \sigma_y = 20$	3852	6555	3852
$\sigma_x = 30, \quad \sigma_y = 30$	2191	7959	2191

$m = 30$

The standard deviations	Absolute deviation	Perpendicular deviation	Common
$\sigma_x = 30, \quad \sigma_y = 5$	4828	5057	4828
$\sigma_x = 30, \quad \sigma_y = 10$	4626	5466	4626
$\sigma_x = 30, \quad \sigma_y = 20$	3705	6999	3705
$\sigma_x = 30, \quad \sigma_y = 30$	1893	8587	1893

If $r = 0.53$

$m = 40$

The standard deviations	Absolute deviation	Perpendicular deviation	Common
$\sigma_x = 30, \quad \sigma_y = 5$	4831	5072	4831
$\sigma_x = 30, \quad \sigma_y = 10$	4625	5584	4625
$\sigma_x = 30, \quad \sigma_y = 20$	3439	7322	3439
$\sigma_x = 30, \quad \sigma_y = 30$	1455	8948	1455

$m = 80$

The standard deviations	Absolute deviation	Perpendicular deviation	Common
$\sigma_x = 30, \quad \sigma_y = 5$	4898	5247	4898
$\sigma_x = 30, \quad \sigma_y = 10$	4448	5782	4448
$\sigma_x = 30, \quad \sigma_y = 20$	2840	8093	2840
$\sigma_x = 30, \quad \sigma_y = 30$	675	9638	675

$m = 100$

The standard deviations	Absolute deviation	Perpendicular deviation	Common
$\sigma_x = 30, \quad \sigma_y = 5$	4900	5296	4900
$\sigma_x = 30, \quad \sigma_y = 10$	4449	5972	4449
$\sigma_x = 30, \quad \sigma_y = 20$	2681	8360	2681
$\sigma_x = 30, \quad \sigma_y = 30$	503	9804	503

If $r = 0.22$

$m = 10$

The standard deviations	Absolute deviation	Perpendicular deviation	Common
$\sigma_x = 30, \quad \sigma_y = 5$	4171	4271	4171
$\sigma_x = 30, \quad \sigma_y = 10$	4022	4396	4022
$\sigma_x = 30, \quad \sigma_y = 20$	3267	5042	3267
$\sigma_x = 30, \quad \sigma_y = 30$	1812	6126	1812

If $r = 0.22$

$m = 20$

The standard deviations	Absolute deviation	Perpendicular deviation	Common
$\sigma_x = 30, \quad \sigma_y = 5$	4562	4672	4562
$\sigma_x = 30, \quad \sigma_y = 10$	4445	4917	4445
$\sigma_x = 30, \quad \sigma_y = 20$	3605	5787	3605
$\sigma_x = 30, \quad \sigma_y = 30$	1458	7050	1458

$m = 30$

The standard deviations	Absolute deviation	Perpendicular deviation	Common
$\sigma_x = 30, \quad \sigma_y = 5$	4772	4923	4772
$\sigma_x = 30, \quad \sigma_y = 10$	4569	5174	4569
$\sigma_x = 30, \quad \sigma_y = 20$	3675	6270	3675
$\sigma_x = 30, \quad \sigma_y = 30$	1201	7695	1201

$m = 40$

The standard deviations	Absolute deviation	Perpendicular deviation	Common
$\sigma_x = 30, \quad \sigma_y = 5$	4861	5040	4861
$\sigma_x = 30, \quad \sigma_y = 10$	4650	5361	4650
$\sigma_x = 30, \quad \sigma_y = 20$	3526	6489	3526
$\sigma_x = 30, \quad \sigma_y = 30$	847	8004	847

$m = 80$

The standard deviations	Absolute deviation	Perpendicular deviation	Common
$\sigma_x = 30, \quad \sigma_y = 5$	4874	5130	4874
$\sigma_x = 30, \quad \sigma_y = 10$	4671	5619	4671
$\sigma_x = 30, \quad \sigma_y = 20$	3094	7184	3094
$\sigma_x = 30, \quad \sigma_y = 30$	320	8870	320

$$If \quad r = 0.22$$
$$m = 100$$

The standard deviations	Absolute deviation	Perpendicular deviation	Common
$\sigma_x = 30, \quad \sigma_y = 5$	4873	5137	4873
$\sigma_x = 30, \quad \sigma_y = 10$	4576	5698	4576
$\sigma_x = 30, \quad \sigma_y = 20$	2882	7469	2882
$\sigma_x = 30, \quad \sigma_y = 30$	157	9157	157

References

Eisenhart, C. (1964). The Meaning of 'Least' in Least Squares. *Journal of the Washington Academy of Science*, Vol. 54, 24 - 33.

Plackett, R. L. (1972). Studies in the History of Probability and Statistics. XXIX. The Discovery of the Method of Least Squares. *Biometrika*, Vol. 59, 239-251.

Stigler, S. M. (1974): "Gergonne's 1815 Paper on the Design and Analysis of Polynomial Regression Experiments", *Historia Mathematica,* Vol. 1, 431-447.

Stigler, S. M. (1986). T*he History of Statistics*. Belknap Press of Harvard University.

Spiegel, M. R. & John Lin (1999). Mathematical Handbook of Formulas and Tables. Second edition, *Schaum's Outline Series*, Mcgraw - Hill.

In: New Developments in Applied Statistics
Editor: Mohammad Ahsanullah

ISBN 978-1-61324-648-1

Chapter 16

BAYESIAN ESTIMATION USING EM ALGORITHM IN BIVARIATE EXPONENTIAL DISTRIBUTIONS BASED ON CENSORED SAMPLES

David D. Hanagal* and K.A. Ahmadi
Department of Statistics, University of Pune, Pune 411007, India

Abstract

In this paper four bivariate exponential (BVE) distributions with subject to right censoring samples are presented. We have opted gamma distribution as a prior for the parameters of the model on which the hyper parameters have been estimated based on the method of moments. A simulation was used to compute the mean and variance of the moment estimators. Furthermore, it is observed from the simulation study that the posterior mode of parameters obtained by moment method are very close to the true values. We estimate the parameters of posterior pdf using EM algorithm in BVE models. The estimations and standard errors of parameters have also been obtained. We compare the Bayesian estimates with MLEs.

Key words: Bivariate exponential, Bayesian estimation, Censored samples, EM algorithm, Missing data, Simulation study.

1. Introduction

In this paper bivariate exponential (BVE) models have been modeled for the life time of two components in which two components are interdependent in a parallel system. In medical studies, consider, a paired organs like kidneys, eyes, lungs, or any other paired organs in an individual as a two components system which work under interdependency circumstance. We assume that life-time of an individual is independent of the life time of the paired organs. Failure of an individual may censor failure of either one of the paired organ or both. This scheme of censoring is right censoring.

Furthermore in engineering and reliability study, consider, a sub-system having two components and their life times are independent of the life time of the main system. Failure

*E-mail address: ddh@stats.unipune.ernet.in

of main system may censor failure of either one component or both. Hanagal (1992a,1992b) has obtained MLE's of the parameters when univariate right censoring is taken into account.

Censoring may also occur in other ways. Patients may be lost to follow up during the study, the patient may decide to move elsewhere therefore the experimenter may not follow him or her again, or the patients may become non-cooperative which is due to some bad side effects of the therapy. In such a case it is called withdrawal from the study. Patient with censored data contribute valuable information and they should not be omitted from the analysis.

In censoring scheme, data is incomplete therefore the estimations are not precise, to overcome this problem we suggest Bayesian EM algorithm.

Prior density in Bayesian analysis has advantage to effect on estimation of the parameters so that hyper parameters of prior should be controlled by reasonable way. Hence we suggest to estimate the parameters of prior through complete-pass data then reform the posterior distribution along with incomplete data (likelihood part). These amalgam data (complete-pass data) and incomplete data (censored data at hand) are used to improve the censored part. EM algorithm is an intensive computational iterative method to update the censored part. Hanagal and Ahmadi (2008) obtained MLE's using EM algorithm for BVE of Marshall-Olkin (1967). Now we pursue to solve the right censoring problem via Bayesian EM algorithm. Four BVE models are presented in this article have been proposed by Marshall-Olkin(1967), Block-Basu(1974), Freund(1961) and Proschan-Sullo(1974).

Marshall-Olkin(M-O)

Marshall-Olkin(1967) proposed BVE distribution with many interesting properties like marginal exponential and bivariate LMP (loss of memory property). Marshall-Olkin(1967) proposed BVE with pdf given by

$$f(x,y) = \begin{cases} \lambda_1(\lambda_2+\lambda_3)\exp(-\lambda_1 x-(\lambda_2+\lambda_3)y) & 0<x<y \\ \lambda_2(\lambda_1+\lambda_3)exp(-\lambda_2 y-(\lambda_1+\lambda_3)x) & 0<y<x \\ \lambda_3\exp(-\lambda x) & 0<x=y \end{cases} \tag{1}$$

where

$$\lambda = \lambda_1+\lambda_2+\lambda_3$$

and $x, y > 0$, $\lambda_1, \lambda_2 > 0$ and $\lambda_3 \geq 0$.

The above BVE model is not absolutely continuous with respect to Lebesque measure in R^2. It has singularity on the diagonal $X = Y$. Here the parameter λ_3 corresponds to the dependence between the two life-times of components and is also related to the simultaneous failures of the two components. When $\lambda_3 = 0$, the life-times X and Y are independent. The probability of simultaneous failure, that is, $P[X = Y] = \lambda_3/(\lambda_1+\lambda_2+\lambda_3)$, which is also equal to the correlation between X and Y.

The marginal distribution of X and Y are exponential with failure rates $(\lambda_1+\lambda_3)$ and $(\lambda_2+\lambda_3)$ and $\min(X,Y)$ is also exponential with failure rate λ, where $\lambda = \lambda_1+\lambda_2+\lambda_3$.

The distribution of $\min(X, Y)$ and the conditional distributions of X and Y are given by

$$\begin{array}{lll} \min(X,Y) & \sim & \exp(\lambda) \\ X - Y|X > Y & \sim & \exp(\lambda_1 + \lambda_3) \\ Y - X|Y > X & \sim & \exp(\lambda_2 + \lambda_3). \end{array} \tag{2}$$

where $\exp(\lambda)$ denotes exponential distribution with failure rate λ.

Block-Basu(B-B)

Block-Basu(1974) proposed an absolutely continuous BVE as an alternative model to Marshall-Olkin (1967) which has a singular component on the diagonal arising out of possible simultaneous failures. Marshall-Olkin is not absolutely continuous with respect to Lebesque measure in R^2, there are some situations when this model is not appropriate. The pdf of BVE of Block-Basu (1974) is given by

$$f(x,y) = \begin{cases} \frac{\lambda\lambda_1(\lambda_2+\lambda_3)}{(\lambda_1+\lambda_2)} \exp(-\lambda_1 x - (\lambda_2 + \lambda_3)y) & 0 < x < y \\ \frac{\lambda\lambda_2(\lambda_1+\lambda_3)}{(\lambda_1+\lambda_2)} \exp(-\lambda_2 y - (\lambda_1 + \lambda_3)x) & 0 < y < x \end{cases} \tag{3}$$

where $\lambda_1, \lambda_2, \lambda_3 > 0$, $\lambda = \lambda_1 + \lambda_2 + \lambda_3$.

This distribution is the absolutely continuous part of BVE of Marshall-Olkin(1967) and it satisfies bivariate LMP. Here the marginal distribution of BVE of Block-Basu(1974) are now not exponential but a weighted combination of two exponentials with scales $(\lambda_i + \lambda_3)$ and λ, i=1,2 with weights $1 + \lambda_3/(\lambda_1 + \lambda_2)$ and $-\lambda_3/(\lambda_1 + \lambda_2)$. The distribution of $\min(X, Y)$ is also exponential with failure rate λ, where $\lambda = \lambda_1 + \lambda_2 + \lambda_3$. The distribution of $\min(X, Y)$ and the conditional distributions of X and Y are given by

$$\begin{array}{lll} \min(X,Y) & \sim & \exp(\lambda) \\ X - Y|X > Y & \sim & \exp(\lambda_1 + \lambda_3) \\ Y - X|Y > X & \sim & \exp(\lambda_2 + \lambda_3). \end{array} \tag{4}$$

Freund(F)

Freund (1961) proposed BVE as a model for failure time distribution of a system with life-times (X, Y) operating in the following manner. Initially X and Y are independent exponential with failure rates θ_1 and θ_2 respectively, θ_1, θ_2, > 0. The interdependence of the components is such that failure of a component changes the failure rate of other component from θ_1 to θ_1' (θ_2 to θ_2'). The BVE of Freund (1961) with its joint pdf is given by

$$f(x,y) = \begin{cases} \theta_1\theta_2' \exp(-\theta_2' y - (\theta_1 + \theta_2 - \theta_2')x) & 0 < x < y \\ \theta_2\theta_1' \exp(-\theta_1' x - (\theta_1 + \theta_2 - \theta_1')y) & 0 < y < x \end{cases} \tag{5}$$

Provided $\theta_1' \neq \theta_1 + \theta_2 \neq \theta_2'$ The distribution of $\min(X, Y)$ and the conditional distributions of X and Y are given by

$$\begin{array}{lll} \min(X,Y) & \sim & \exp(\theta_1 + \theta_2) \\ X - Y|X > Y & \sim & \exp(\theta_1') \\ Y - X|Y > X & \sim & \exp(\theta_2'). \end{array} \tag{6}$$

Proschan-Sullo(P-S)

Proschan-Sullo (1974) proposed BVE which is the combination of both Marshall-Olkin and Freund models and the two component system operate in the following manner. Initially X and Y follow BVE of Marshall-Olkin (1967). Here the failure of a component changes the failure rate of other component from $\theta_1 + \theta_3$ to $\theta_1' + \theta_3$ ($\theta_2 + \theta_3$ to $\theta_2' + \theta_3$). The BVE of Proschan-Sullo (1974) with its pdf is given by

$$f(x,y) = \begin{cases} \theta_1 \eta_2' \exp(-\eta_2' y - (\theta - \eta_2')x) & 0 < x < y \\ \theta_2 \eta_1' \exp(-\eta_1' x - (\theta - \eta_1')y) & 0 < y < x \\ \theta_3 \exp(-\theta x) & 0 < x = y \end{cases} \tag{7}$$

where

$$\theta = \theta_1 + \theta_2 + \theta_3, \eta_1' = \theta_1' + \theta_3, \eta_2' = \theta_2' + \theta_3$$

Provided $\eta_1' \neq \theta \neq \eta_2'$. We call the above model as BVE of P-S(1). The distribution of $\min(X,Y)$ and the conditional distributions of X and Y are given by

$$\begin{array}{lcl} \min(X,Y) & \sim & \exp(\theta) \\ X - Y|X > Y & \sim & \exp(\eta_1') \\ Y - X|Y > X & \sim & \exp(\eta_2'). \end{array} \tag{8}$$

In this case if $\theta_3 = 0$ the Proschan-Sullo reduces to Freund model. When $\eta_1' = \theta = \eta_2'$ the corresponding pdf is given by

$$f(x,y) = \begin{cases} \theta_1 \theta \exp(-\theta y) & 0 < x < y \\ \theta_2 \theta \exp(-\theta x) & 0 < y < x \\ \theta_3 \exp(-\theta x) & x = y \end{cases} \tag{9}$$

We call the above model as BVE of P-S(2). The distribution of $\min(X,Y)$ and the conditional distributions of X and Y are given by

$$\begin{array}{lcl} \min(X,Y) & \sim & \exp(\theta) \\ X - Y|X > Y & \sim & \exp(\theta) \\ Y - X|Y > X & \sim & \exp(\theta). \end{array} \tag{10}$$

The aim of this paper is to obtain the estimation of parameters by Bayesian EM algorithm in four BVE models discussed above. In Section 2, we deal with the Bayesian EM algorithm for Marshall-Olkin's BVE model. In Sections 3 and 4, we extend the parameter estimation using Bayesian EM algorithm in BVE of Block-Basu(1974) and Proschan-Sullo(1974) models respectively. Finally in Section 5, we present a simulation study.

2. BVE of Marshall-Olkin(M-O)

The marginal distributions are obtained from the pdf (1.1)

$$f(X) = (\lambda_1 + \lambda_3)\exp(-(\lambda_1 + \lambda_3)x) \quad x > 0$$
$$f(Y) = (\lambda_2 + \lambda_3)\exp(-(\lambda_2 + \lambda_3)y) \quad y > 0$$

The moments can easily be obtained by $E(X) = \frac{1}{\lambda_1+\lambda_3}$, $E(Y) = \frac{1}{\lambda_2+\lambda_3}$ and $E(\min(X,Y)) = \frac{1}{\lambda}$ therefore, the consistent estimators are given by

$$\frac{n}{\sum_{i=1}^{n}\min(x_i,y_i)} - \frac{n}{\sum_{i=1}^{n} y_i} \xrightarrow{p} \lambda_1,$$
$$\frac{n}{\sum_{i=1}^{n}\min(x_i,y_i)} - \frac{n}{\sum_{i=1}^{n} x_i} \xrightarrow{p} \lambda_2$$

and

$$\frac{n}{\sum_{i=1}^{n} x_i} + \frac{n}{\sum_{i=1}^{n} y_i} - \frac{n}{\sum_{i=1}^{n}\min(x_i,y_i)} \xrightarrow{p} \lambda_3$$

It is recalled that above consistence estimators are used to estimate hyper parameters of prior. Now we continue the rest of this section to estimate the parameters via computational Bayesian EM algorithm.

For the bivariate life-time distribution, we use univariate censoring scheme given by Hanagal (1992a, 1992b) because the individuals do not enter at the same time and withdrawal or death of an individual or termination of the study will censor both life times of the components. Here the censoring time is independent of the life times of both components. This is the standard univariate right censoring for both failure times X and Y.

Suppose that there are n independent pairs of components, for example, paired kidneys, lungs, eyes, ears in an individual under study and i-th pair of the components have life times (x_i, y_i) and a censoring time (t_i). The life times associated with i-th pair of the components are given by

$$\begin{aligned}(X_i, Y_i) &= (x_i, y_i), && \max(x_i, y_i) < t_i\\ &= (x_i, t_{iy}), && x_i < t_i < t_{iy} = y_i\\ &= (t_{ix}, y_i), && y_i < t_i < t_{ix} = x_i\\ &= (t_{ixy}, t_{ixy}), && t_i < t_{ixy} = \min(x_i, y_i)\end{aligned}$$

where t_{ix}, t_{iy} and t_{ixy} are the missing data or unobserved random observations.

Now the conditional likelihood of the sample of size n is given by

$$L_c(\underline{\lambda}, t_{ix}, t_{iy}, t_{ixy}|x_i, y_i, t_i) = \prod_{j=1}^{6}\prod_{i=1}^{n_j} f_j(x_i, y_i) \tag{11}$$

where

$$\begin{aligned}f_1(x_i,y_i) &= \lambda_1(\lambda_2+\lambda_3)e^{-\lambda_1 x_i-(\lambda_2+\lambda_3)y_i}, \quad 0<x_i<y_i<t_i\\ f_2(x_i,y_i) &= \lambda_2(\lambda_1+\lambda_3)e^{-\lambda_2 y_i-(\lambda_1+\lambda_3)x_i}, \quad 0<y_i<x_i<t_i\\ f_3(x_i,y_i) &= \lambda_3 e^{-\lambda x}, \quad 0<x_i=y_i<t_i\\ f_4(x_i,y_i) &= \lambda_1(\lambda_2+\lambda_3)e^{-\lambda_1 x_i-(\lambda_2+\lambda_3)t_{iy}}, \quad 0<x_i<t_i<t_{iy}=y_i\\ f_5(x_i,y_i) &= \lambda_2(\lambda_1+\lambda_3)e^{-\lambda_2 y_i-(\lambda_1+\lambda_3)t_{ix}}, \quad 0<y_i<t_i<t_{ix}=x_i\\ f_6(x_i,y_i) &= f(t_{ixy}) = \lambda e^{-\lambda t_{ixy}}, \quad t_i<t_{ixy}=\min(x_i,y_i)\end{aligned}$$

$\lambda = (\lambda_1 + \lambda_2 + \lambda_3)$, n_1, n_2, n_3, n_4, n_5 and n_6 are number of observations observed to fall in the range corresponding to $f_1(x_i, y_i)$, $f_2(x_i, y_i)$, $f_3(x_i, y_i)$, $f_4(x_i, y_i)$, $f_5(x_i, y_i)$ and $f_6(x_i, y_i)$ respectively. $f_1(x_i, y_i)$, $f_2(x_i, y_i)$, $f_4(x_i, y_i)$ and f_{5i} are the p.d.f. with respect to Lebesque measure in R^2, $f_3(x_i, y_i)$ and $f_6(x_i, y_i)$ are the p.d.f. with respect to Lebesque measure in R^1.

Let the range space corresponding to $f_j(x_i, y_i)$ be A_j, $j = 1, 2, ..., 6$. $A_1, ..., A_6$ are disjoint sets or regions and letting $B_1 = A_1+A_2+A_3+A_4$, $B_2 = A_1+A_2+A_3+A_5$, and $B_3 = A_1+A_2+A_3$, the expression for log of the posterior density is $\log p(\underline{\lambda}, t_{ix}, t_{iy}, t_{ixy}|x_i, y_i, t_i)$ can be written as follows

$$\begin{aligned}\log p(\underline{\lambda}, t_{ix}, t_{iy}, t_{ixy}|x_i, y_i, t_i) &= (n_1 + n_4)\log[\lambda_1(\lambda_2 + \lambda_3)] + (n_2 + n_5)\log[\lambda_2(\lambda_1 + \lambda_3)] \\ &+ n_3 \log \lambda_3 + n_6 \log \lambda - \lambda_1 \left(\sum_{i \in B_1} x_i + \sum_{i \in A_5} t_{ix} + \sum_{i \in A_6} t_{ixy}\right) - \lambda_2 \left(\sum_{i \in B_2} y_i + \sum_{i \in A_4} t_{iy} + \sum_{i \in A_6} t_{ixy}\right) \\ &- \lambda_3 \left(\sum_{i \in B_3} \max(x_i, y_i) + \sum_{i \in A_5} t_{ix} + \sum_{i \in A_4} t_{iy} + \sum_{i \in A_6} t_{ixy}\right) + \sum_{i=1}^{3}(\alpha_i - 1)\log \lambda_i - \sum_{i=1}^{3} \beta_i \lambda_i \end{aligned} \quad (12)$$

E-step:

The unobserved random variables, $T_x|T_x > Y$, $T_y|T_y > X$, and T_{xy} follow exponential distribution with failure rates, $(\lambda_1 + \lambda_3)$, $(\lambda_2 + \lambda_3)$, and λ respectively as stated in equation (1.2). By loss of memory property(LMP) of exponential distribution, the conditional distributions of $T_x|T_x > t > Y$, $T_y|T_y > t > X$, and $T_{xy}|T_{xy} > t$ still follow exponential distribution with failure rates, $(\lambda_1 + \lambda_3)$, $(\lambda_2 + \lambda_3)$, and λ and the corresponding p.d.f. of these unobserved random variables are

$$\begin{aligned} f(t_x|T_x > t > Y) &= (\lambda_1 + \lambda_3)e^{-(\lambda_1+\lambda_3)(t_x - t)} \\ f(t_y|T_y > t > X) &= (\lambda_2 + \lambda_3)e^{-(\lambda_2+\lambda_3)(t_y - t)} \\ f(t_{xy}|T_{xy} > t) &= \lambda e^{-\lambda(t_{xy} - t)} \end{aligned} \quad (13)$$

The expected values of the conditional unobserved random variables, $T_x|T_x > t > Y$, $T_y|T_y > t > X$, and $T_{xy}|T_{xy} > t$ are as follows.

$$\begin{aligned} E(T_x|T_x > t > Y) &= t + \frac{1}{\lambda_1 + \lambda_3} = t + a \ (say) \\ E(T_y|T_y > t > X) &= t + \frac{1}{\lambda_2 + \lambda_3} = t + b \ (say) \\ E(T_{xy}|T_{xy} > t) &= t + \frac{1}{\lambda_1 + \lambda_2 + \lambda_3} = t + c \ (say) \end{aligned} \quad (14)$$

Now $Q(\underline{\lambda}|\underline{\lambda}^{(k)})$, [the equation (2.2) with expectations given in (2.4)] can be obtained as follows

$$Q(\underline{\lambda}|\underline{\lambda}^{(k)}) = (n_1+n_4+\alpha_1-1)log(\lambda_1) + (n_1+n_4)log(\lambda_2+\lambda_3) + (n_2+n_5+\alpha_2-1)log(\lambda_2) + (n_2+n_5)log(\lambda_1+\lambda_3) + (n_3+\alpha_3-1)log(\lambda_3) + n_6\log\lambda - \lambda_1(\sum_{i\in B_1} x_i + \sum_{i\in A_5}(t_i+a^{(k)}) + \sum_{i\in A_6}(t_i+c^{(k)}) + \beta_1)$$
$$- \lambda_2(\sum_{i\in B_2} y_i + \sum_{i\in A_4}(t_i+b^{(k)}) + \sum_{i\in A_6}(t_i+c^{(k)}) + \beta_2) - \lambda_3(\sum_{i\in B_3}\max(x_i,y_i) + \sum_{i\in A_5}(t_i+a^{(k)}) + \sum_{i\in A_4}(t_i+b^{(k)}) + \sum_{i\in A_6}(t_i+c^{(k)}) + \beta_3) \quad (15)$$

where $a^{(k)}$,$b^{(k)}$,$c^{(k)}$ are the estimated values at k-th iteration and λ,λ_1,λ_2,λ_3 are replaced by their estimated values at(up dated values) k-th iteration, i.e., $\lambda^{(k)}$,$\lambda_1^{(k)}$,$\lambda_2^{(k)}$,$\lambda_3^{(k)}$.

M-step:

To accomplish this part we required to maximize $Q(\underline{\lambda}, \underline{\lambda}^{(k)})$ which leads solving the partial derivatives of $Q(\underline{\lambda}, \underline{\lambda}^{(k)})$ and equating to zero. We use Newton-Raphson procedure for the purpose of maximization of posterior density.

3. BVE of Block-Basu(B-B)

The marginal pdf are given by

$$f(x) = \frac{\lambda(\lambda_1+\lambda_3)}{\lambda_1+\lambda_2}\exp(-(\lambda_1+\lambda_3)x) - \frac{\lambda\lambda_3}{\lambda_1+\lambda_2}\exp(-\lambda x) \quad x > 0$$
$$f(y) = \frac{\lambda(\lambda_2+\lambda_3)}{\lambda_1+\lambda_2}\exp(-(\lambda_2+\lambda_3)y) - \frac{\lambda\lambda_3}{\lambda_1+\lambda_2}\exp(-\lambda y) \quad y > 0$$

The moment values are:

$$E(X) = \frac{1}{\lambda_1+\lambda_3} + \frac{\lambda_2\lambda_3}{\lambda(\lambda_1+\lambda_2)(\lambda_1+\lambda_3)}, \qquad E(Y) = \frac{1}{\lambda_2+\lambda_3} + \frac{\lambda_1\lambda_3}{\lambda(\lambda_1+\lambda_2)(\lambda_2+\lambda_3)}$$

and

$$E(\min(X,Y)) = \frac{1}{\lambda}, \qquad \frac{n_1}{n} \xrightarrow{p} p(X<Y) = \frac{\lambda_1}{\lambda_1+\lambda_2}, \qquad \frac{n_2}{n} \xrightarrow{p} p(Y<X) = \frac{\lambda_2}{\lambda_1+\lambda_2}$$

Hanagal(1992a) has obtained following consistent estimators

$$\frac{n}{\sum_{i=1}^n \min(x_i,y_i)} + \frac{n_1}{\sum_{i=1}^n \min(x_i,y_i) - \sum_{i=1}^n y_i} \xrightarrow{p} \lambda_1,$$

$$\frac{n}{\sum_{i=1}^n \min(x_i,y_i)} + \frac{n_2}{\sum_{i=1}^n \min(x_i,y_i) - \sum_{i=1}^n x_i} \xrightarrow{p} \lambda_2,$$

$$\frac{n_1}{\sum_{i=1}^n y_i - \sum_{i=1}^n \min(x_i,y_i)} + \frac{n_2}{\sum_{i=1}^n x_i - \sum_{i=1}^n \min(x_i,y_i)} - \frac{n}{\sum_{i=1}^n \min(x_i,y_i)} \xrightarrow{p} \lambda_3$$

where $n_1(n_2)$ is the number of observations in the region $X < Y(X > Y)$ in the sample of size n.

The conditional likelihood of the sample size n is given by

$$L_c(\underline{\lambda}, t_{ix}, t_{iy}, t_{ixy}|x_i, y_i, t_i) = \prod_{j=1}^{5}\prod_{i=1}^{n_j} f_j(x_i, y_i) \tag{16}$$

where

$$\begin{aligned}
f_1(x_i, y_i) &= \frac{\lambda\lambda_1(\lambda_2+\lambda_3)}{(\lambda_1+\lambda_2)}\exp(-\lambda_1 x_i - (\lambda_2+\lambda_3)y_i), \quad 0 < x_i < y_i < t_i \\
f_2(x_i, y_i) &= \frac{\lambda\lambda_2(\lambda_1+\lambda_3)}{(\lambda_1+\lambda_2)}\exp(-\lambda_2 y_i - (\lambda_1+\lambda_3)x_i), \quad 0 < y_i < x_i < t_i \\
f_3(x_i, y_i) &= \frac{\lambda\lambda_1(\lambda_2+\lambda_3)}{(\lambda_1+\lambda_2)}\exp(-\lambda_1 x_i - (\lambda_2+\lambda_3)t_{iy}), \quad 0 < x_i < t_i < t_{iy} = y_i \\
f_4(x_i, y_i) &= \frac{\lambda\lambda_2(\lambda_1+\lambda_3)}{(\lambda_1+\lambda_2)}\exp(-\lambda_2 y_i - (\lambda_1+\lambda_3)t_{ix}), \quad 0 < y_i < t_i < t_{ix} = x_i \\
f_5(x_i, y_i) &= f(t_{ixy}) = \lambda\exp(-\lambda t_{ixy}), \quad t_i < t_{ixy} = \min(x_i, y_i)
\end{aligned}$$

n_1, n_2, n_3, n_4, and n_5 are the number of observations that fall in the range corresponding to $f_1(x_i, y_i)$, $f_2(x_i, y_i)$, $f_3(x_i, y_i)$, $f_4(x_i, y_i)$, and $f_5(x_i, y_i)$ respectively. $f_1(x_i, y_i)$, $f_2(x_i, y_i)$, $f_3(x_i, y_i)$, and $f_4(x_i, y_i)$ are the p.d.f. with respect to Lebesque measure in R^2, $f_5(x_i, y_i)$ is the p.d.f. with respect to Lebesque measure in R^1. t_{ix}, t_{iy} and t_{ixy} are the missing data or unobserved random observations as defined earlier.

Let the range space corresponding to $f_j(x_i, y_i)$ be A_j, $j = 1, 2, ..., 5$. $A_1, ..., A_5$ are disjoint sets or regions and letting $B_1 = A_1 + A_2 + A_3$, $B_2 = A_1 + A_2 + A_4$, and $B_3 = A_1 + A_2$. The posterior pdf $p(\underline{\lambda}, t_{ix}, t_{iy}, t_{ixy}|x_i, y_i, t_i)$ is given as follows

$$\begin{aligned}
p(\underline{\lambda}, t_{ix}, t_{iy}, t_{ixy}|x_i, y_i, t_i) &= \lambda^{n-n_5}[\lambda_1(\lambda_2+\lambda_3)]^{n_1+n_3}[\lambda_2(\lambda_1+\lambda_3)]^{n_2+n_4}(\lambda_1+\lambda_2)^{-(n-n_5)} \\
&\exp[-\lambda_1\left(\sum_{i\in B_1} x_i + \sum_{i\in A_4} t_{ix} + \sum_{i\in A_5} t_{ixy}\right) - \lambda_2\left(\sum_{i\in B_2} y_i + \sum_{i\in A_3} t_{iy} + \sum_{i\in A_5} t_{ixy}\right) \\
&- \lambda_3\left(\sum_{i\in B_3}\max(x_i, y_i) + \sum_{i\in A_4} t_{ix} + \sum_{i\in A_3} t_{iy} + \sum_{i\in A_5} t_{ixy}\right) \times \prod_{i=1}^{3}\lambda_i^{\alpha_i-1}\exp(-\beta_i\lambda_i)
\end{aligned} \tag{17}$$

E-step:

The unobserved random variables, $T_x|T_x > Y$, $T_y|T_y > X$, and T_{xy} follow exponential distribution with failure rates, $(\lambda_1 + \lambda_3)$, $(\lambda_2 + \lambda_3)$, and λ respectively as stated in equation (1.4). By loss of memory property(LMP) of exponential distribution, the conditional distributions of $T_x|T_x > t > Y$, $T_y|T_y > t > X$, and $T_{xy}|T_{xy} > t$ still follow exponential distribution with failure rates, $(\lambda_1 + \lambda_3)$, $(\lambda_2 + \lambda_3)$, and λ and the corresponding p.d.f. of these unobserved random variables are

$$\begin{aligned}
f(t_x|T_x > t > Y) &= (\lambda_1+\lambda_3)e^{-(\lambda_1+\lambda_3)(t_x-t)} \\
f(t_y|T_y > t > X) &= (\lambda_2+\lambda_3)e^{-(\lambda_2+\lambda_3)(t_y-t)} \\
f(t_{xy}|T_{xy} > t) &= \lambda e^{-\lambda(t_{xy}-t)}
\end{aligned} \tag{18}$$

The expected values of unobserved random variables can be obtained in a similar way as given in (2.4) of the BVE of Marshall-Olkin model. Now $Q(\underline{\lambda}|\underline{\lambda}^{(k)})$ [the equation (3.2) with expectations given in (2.4)] can be obtained as follows

$$Q(\underline{\lambda}|\underline{\lambda}^{(k)}) = E_{\underline{\lambda}^{(k)}} log(p(\psi|(x,y)) = (n-n_5)log\lambda + (n_1+n_3+\alpha_1-1)log\lambda_1 + (n_1+n_3)log(\lambda_2+\lambda_3)+ (n_2+n_4+\alpha_2-1)log\lambda_2 + (n_2+n_4)log(\lambda_1+\lambda_3) - (n-n_5)log(\lambda_1+\lambda_2) + (\alpha_3-1)log(\lambda_3) + n_5\log\lambda - \lambda_1(\sum_{i\in B_1} x_i + \sum_{i\in A_4}(t_i+a^{(k)}) + \sum_{i\in A_5}(t_i+c^{(k)}) + \beta_1) - \lambda_2(\sum_{i\in B_2} y_i + \sum_{i\in A_3}(t_i+b^{(k)}) + \sum_{i\in A_5}(t_i+c^{(k)}) + \beta_2) - \lambda_3(\sum_{i\in B_3}\max(x_i,y_i) + \sum_{i\in A_4}(t_i+a^{(k)}) + \sum_{i\in A_3}(t_i+b^{(k)}) + \sum_{i\in A_5}(t_i+c^{(k)}) + \beta_3). \quad (19)$$

M-step:
We maximize $Q(\underline{\lambda}|\underline{\lambda}^{(k)})$ using Newton-Raphson procedure until the convergence is reached.

4. BVE of Proschan-Sullo(P-S)

Case (i) When $\eta_1' \neq \theta \neq \eta_2'$: The marginal pdfs are given

$$f(x) = \theta_1\exp(-\theta x) + \frac{\theta_2\eta_1'}{\eta_1'-\theta}(\exp(-\theta x) - exp(-\eta_1' x)) + \theta_3\exp(-\theta x) \qquad x>0$$

$$f(y) = \theta_2\exp(-\theta y) + \frac{\theta_1\eta_2'}{\eta_2'-\theta}(\exp(-\theta y) - \exp(-\eta_2' y)) + \theta_3\exp(-\theta y) \qquad y>0$$

Also the moment values are given $E(X) = \frac{1}{\theta} + \frac{\theta_2}{\eta_1'\theta}$, $E(Y) = \frac{1}{\theta} + \frac{\theta_1}{\eta_2'\theta}$ and $E(min(X,y)) = \frac{1}{\theta}$ also $\frac{n_1}{n} \xrightarrow{p} p(X<Y) = \frac{\theta_1}{\theta}$, $\frac{n_2}{n} \xrightarrow{p} p(Y<X) = \frac{\theta_2}{\theta}$ and $\frac{n_3}{n} \xrightarrow{p} p(X=Y) = \frac{\theta_3}{\theta}$ where $n_3 = n - n_1 - n_2$. Using above expression the consistent estimators are given by

$$\frac{n_1}{\sum_{i=1}^n \min(x_i,y_i)} \xrightarrow{p} \theta_1, \quad \frac{n_2}{\sum_{i=1}^n \min(x_i,y_i)} \xrightarrow{p} \theta_2, \quad \frac{n_3}{\sum_{i=1}^n \min(x_i,y_i)} \xrightarrow{p} \theta_3,$$

$$\frac{n_2}{\sum_{i=1}^n (x_i-y_i)I(x_i>y_i)} \xrightarrow{p} \eta_1', \quad \frac{n_1}{\sum_{i=1}^n (y_i-x_i)I(y_i>x_i)} \xrightarrow{p} \eta_2'$$

where $I(.)$ is indicator function.

The Likelihood function of BVE of Proschan-Sullo(1) model is given by

$$L_c(\underline{\lambda}, t_{ix}, t_{iy}, t_{ixy}|x_i, y_i, t_i) = \prod_{j=1}^{6}\prod_{i=1}^{n_j} f_j(x_i,y_i) \quad (20)$$

where

$$\begin{aligned}
f_1(x_i,y_i) &= \theta_1\eta_2'\exp(-\eta_2' y_i - (\theta-\eta_2')x_i), \quad 0<x_i<y_i<t_i\\
f_2(x_i,y_i) &= \theta_2\eta_1'\exp(-\eta_1' x_i - (\theta-\eta_1')y_i), \quad 0<y_i<x_i<t_i\\
f_3(x_i,y_i) &= \theta_3\exp(-\theta x_i), \quad 0<x_i=y_i<t_i\\
f_4(x_i,y_i) &= \theta_1\eta_2'\exp(-\eta_2' t_{iy} - (\theta-\eta_2')x_i), \quad 0<x_i<t_i<y_{iy}=y_i\\
f_5(x_i,y_i) &= \theta_2\eta_1'\exp(-\eta_1' t_{ix} - (\theta-\eta_1')y_i), \quad 0<y_i<t_i<t_{ix}=x_i\\
f_6(x_i,y_i) &= f(t_{ixy}) = \theta\exp(-\theta t_{ixy}), \quad t_i<t_{ixy}=\min(x_i,y_i)
\end{aligned}$$

n_1, n_2, n_3, n_4, n_5 and n_6 are number of observations observed to fall in the range corresponding to $f_1(x_i,y_i)$, $f_2(x_i,y_i)$, $f_3(x_i,y_i)$, $f_4(x_i,y_i)$, $f_5(x_i,y_i)$ and $f_6(x_i,y_i)$ respectively. $f_1(x_i,y_i)$, $f_2(x_i,y_i)$, $f_4(x_i,y_i)$ and f_{5i} are the pdf with respect to Lebesque measure in R^2, $f_3(x_i,y_i)$ and $f_6(x_i,y_i)$ are the pdf with respect to Lebesque measure in R^1. t_{ix}, t_{iy} and t_{ixy} are the missing data or unobserved data defined earlier.

Let the range space corresponding to $f_j(x_i,y_i)$ be A_j, $j = 1,2,...,5$. $A_1,...,A_5$ are disjoint sets or regions and letting $B = A_1 + A_2 + A_3 + A_4 + A_5$. The posterior pdf $p(\underline{\lambda}, t_{ix}, t_{iy}, t_{ixy}|x_i, y_i, t_i)$ is

$$p(\underline{\lambda}, t_{ix}, t_{iy}, t_{ixy}|x_i, y_i, t_i) = \theta_1^{n_1+n_4} \times (\eta_2')^{n_1+n_4} \times \theta_2^{n_2+n_5} \times (\eta_1')^{n_2+n_5} \times \theta_3^{n_3}\theta^{n_6}$$
$$\exp[-\eta_2'(\sum_{i\in A_1} y_i - \sum_{i\in A_1+A_4} x_i + \sum_{i\in A_4} t_{iy}) - \eta_1'(\sum_{i\in A_2} x_i + \sum_{i\in A_5} t_i - \sum_{i\in A_2+A_5} y_i + \sum_{i\in A_5} t_{ix})$$
$$- \theta(\sum_{i\in B}\min(x_i,y_i) + \sum_{i\in A_6}(t_i + t_{ixy}))] \times \prod_{i=1}^{3}\theta_i^{\alpha_i-1}\exp(-\beta_i\theta_i) \times \prod_{j=1}^{2}\prod_{i=4}^{5}(\eta_j')_i^{\alpha_i-1}\exp(-\beta_i\eta_j') \tag{21}$$

E-step:
The unobserved random variables, $T_x|T_x > Y$, $T_y|T_y > X$, and T_{xy} follow exponential distribution with failure rates, η_1', η_2', and θ respectively as stated in equation (1.8). By loss of memory property(LMP) of exponential distribution, the conditional distributions of $T_x|T_x > t > Y, T_y|T_y > t > X$, and $T_{xy}|T_{xy} > t$ still follow exponential distribution with failure rates, η_1', η_2', and θ respectively and the corresponding p.d.f. of these unobserved random variables are

$$\begin{aligned} f(t_x|T_x > t > Y) &= \eta_1' e^{-\eta_1'(t_x - t)} \\ f(t_y|T_y > t > X) &= \eta_2' e^{-\eta_2'(t_y - t)} \\ f(t_{xy}|T_{xy} > t) &= \theta e^{-\theta(t_{xy} - t)} \end{aligned} \tag{22}$$

The expected values of the conditional unobserved random variables, $T_x|T_x > t > Y$, $T_y|T_y > t > X$, and $T_{xy}|T_{xy} > t$ are as follows.

$$\begin{aligned} E(T_x|T_x > t > Y) = t + \frac{1}{\eta_1'} &= t + a \ (say) \\ E(T_y|T_y > t > X) = t + \frac{1}{\eta_2'} &= t + b \ (say) \\ E(T_{xy}|T_{xy} > t) = t + \frac{1}{\lambda_1 + \lambda_2 + \lambda_3} &= t + c \ (say) \end{aligned} \tag{23}$$

Now $Q(\underline{\lambda}|\underline{\lambda}^{(k)})$ [the equation (4.2) with expectations given in (4.4)] can be computed as follows

$$Q(\underline{\lambda}|\underline{\lambda}^(k)) = (n_1 + n_4 + \alpha_1 - 1)log\theta_1 + (n_1 + n_4 + \alpha_5 - 1)log\eta_2' + (n_2 + n_5 + \alpha_2 - 1)log\theta_2$$
$$+ (n_2 + n_5 + \alpha_4 - 1)log\eta_1' + (n_3 + \alpha_3 - 1)log\theta_3 + n_6\log\theta - \eta_2'(\sum_{i\in A_1} y_i + \sum_{i\in A_4}(t_i + b^{(k)}) - \sum_{i\in A_1+A_4}(x_i + \beta_5)$$
$$- \eta_1'(\sum_{i\in A_2} x_i + \sum_{i\in A_5}(t_i + a^{(k)}) - \sum_{i\in A_2+A_5} y_i + \beta_4) - \sum_{j=1}^{3}\theta_j(\sum_{i\in B}\min(x_i,y_i) + \sum_{i\in A_6}(t_i + c^{(k)}) + \beta_j) \tag{24}$$

M-step:
We maximize $Q(\underline{\lambda}|\underline{\lambda}^{(k)})$ using Newton-Raphson procedure until the convergence is reached.
Case (ii) When $\eta_1' = \theta = \eta_2'$:
The marginal pdfs are given by

$$f(x) = \theta_1 \exp(-\theta x) + \theta_2 \theta x \exp(-\theta x) + \theta_3 \exp(-\theta x) \quad x > 0$$
$$f(y) = \theta_2 \exp(-\theta y) + \theta_1 \theta y \exp(-\theta y) + \theta_3 \exp(-\theta y) \quad y > 0$$

$E(X) = \frac{\theta_2}{\theta^2} + \frac{1}{\theta}$, $E(Y) = \frac{\theta_1}{\theta^2} + \frac{1}{\theta}$ and $E(\min(X,Y)) = \frac{1}{\theta}$
Therefore the consistent estimators are given below

$$\frac{n\sum_{i=1}^n y_i}{(\sum_{i=1}^n \min(x_i,y_i))^2} - \frac{n}{\sum_{i=1}^n \min(x_i,y_i)} \xrightarrow{p} \theta_1,$$
$$\frac{n\sum_{i=1}^n x_i}{(\sum_{i=1}^n \min(x_i,y_i))^2} - \frac{n}{\sum_{i=1}^n \min(x_i,y_i)} \xrightarrow{p} \theta_2$$
$$\frac{3n}{\sum_{i=1}^n \min(x_i,y_i)} - \frac{n\sum_{i=1}^n y_i}{(\sum_{i=1}^n \min(x_i,y_i))^2} - \frac{n\sum_{i=1}^n x_i}{(\sum_{i=1}^n \min(x_i,y_i))^2} \xrightarrow{p} \theta_3$$

The likelihood function of BVE of Proschan-Sullo(2) model is given by

$$L_c(\underline{\lambda}, t_{ix}, t_{iy}, t_{ixy}|x_i, y_i, t_i) = \prod_{j=1}^{6}\prod_{i=1}^{n_j} f_j(x_i, y_i) \quad (25)$$

where

$$\begin{aligned}
f_1(x_i,y_i) &= \theta_1\theta \exp(-\theta y_i) \;\; 0 < x_i < y_i < t_i\\
f_2(x_i,y_i) &= \theta_2\theta \exp(-\theta x_i) \;\; 0 < y_i < x_i < t_i\\
f_3(x_i,y_i) &= \theta_3 \exp(-\theta x_i) \;\; 0 < x_i = y_i < t_i\\
f_4(x_i,y_i) &= \theta_1\theta \exp(-\theta t_{iy}) \;\; 0 < x_i < t_i < t_{iy} = y_i\\
f_5(x_i,y_i) &= \theta_2\theta \exp(-\theta t_{ix}) \;\; 0 < y_i < t_i < t_{ix} = x_i\\
f_6(x_i,y_i) &= f(t_{ixy}) = \exp(-\theta t_{ixy}) \;\; t_i < t_{ixy} = min(x_i,y_i)
\end{aligned}$$

n_1, n_2, n_3, n_4, n_5 and n_6 are number of observations observed to fall in the range corresponding to $f_1(x_i,y_i)$, $f_2(x_i,y_i)$, $f_3(x_i,y_i)$, $f_4(x_i,y_i)$, $f_5(x_i,y_i)$ and $f_6(x_i,y_i)$ respectively. $f_1(x_i,y_i)$, $f_2(x_i,y_i)$, $f_4(x_i,y_i)$ and f_{5i} are the pdf with respect to Lebesque measure in R^2, $f_3(x_i,y_i)$ and $f_6(x_i,y_i)$ are the pdf with respect to Lebesque measure in R^1. t_{ix}, t_{iy} and t_{ixy} are the missing data or unobserved data defined earlier.

Let the range space corresponding to $f_j(x_i,y_i)$ be A_j, $j = 1,2,...,5$. $A_1,...,A_5$ are disjoint sets or regions and letting $B = A_1 + A_2 + A_3$. The posterior pdf $p(\underline{\lambda}, t_{ix}, t_{iy}, t_{ixy}|x_i, y_i, t_i)$ is

$$p(\underline{\lambda}, t_{ix}, t_{iy}, t_{ixy}|x_i, y_i, t_i) = \theta_1^{n_1+n_4} \times \theta^{n_1+n_2+n_4+n_5} \times \theta_2^{n_2+n_5} \times \theta_3^{n_3}\theta^{n_6} \exp[-\theta(\sum_{i\in B}\max(x_i,y_i)$$
$$+ \sum_{i\in A_4} t_{iy} + \sum_{i\in A_5} t_{ix} + \sum_{i\in A_6} t_{ixy})] \times \prod_{i=1}^{3}\theta_i^{\alpha_i-1}\exp(-\beta_i\theta_i) \quad (26)$$

Table 1. Bayesian EM and ML estimation of parameters in BVE of M-O Model based on Censored Sampling

Parameters	λ_1 .08	λ_2 .1	λ_3 .03	λ_1 .1	λ_2 .12	λ_3 .04
			$n = 50$			
Est_{bys}	.08	.104	.032	.099	.12	.04
SE_{bys}	.0084	.0098	.008	.013	.014	.009
Est_{mle}	.078	.098	.028	.096	.117	.037
SE_{mle}	.015	.018	.0102	.019	.023	.0125
			$n = 75$			
Est_{bys}	.08	.104	.032	.098	.121	.042
SE_{bys}	.007	.008	.006	.012	.01	.007
Est_{mle}	.078	.097	.028	.096	.117	.037
SE_{mle}	.012	.014	.008	.015	.019	.0104
			$n = 100$			
Est_{bys}	.08	.101	.031	.096	.120	.043
SE_{bys}	.0057	.007	.006	.009	.0096	.0067
Est_{mle}	.077	.097	.028	.097	.117	.037
SE_{mle}	.0103	.0123	.007	.013	.016	.0088

E-step:
The unobserved random variables, $T_x|T_x > Y$, $T_y|T_y > X$, and T_{xy} follow exponential distribution with failure rates, θ, θ, and θ respectively as stated in equation (1.10). By loss of memory property(LMP) of exponential distribution, the conditional distributions of $T_x|T_x > t > Y$, $T_y|T_y > t > X$, and $T_{xy}|T_{xy} > t$ still follow exponential distribution with failure rates, θ, θ, and θ respectively and the corresponding p.d.f. of these unobserved random variables are

$$\begin{aligned} f(t_x|T_x > t > Y) &= \theta e^{-\theta(t_x - t)} \\ f(t_y|T_y > t > X) &= \theta e^{-\theta(t_y - t)} \\ f(t_{xy}|T_{xy} > t) &= \theta e^{-\theta(t_{xy} - t)} \end{aligned} \tag{27}$$

The expected values of the conditional unobserved random variables, $T_x|T_x > t > Y$, $T_y|T_y > t > X$, and $T_{xy}|T_{xy} > t$ are as follows.

$$E(T_x|T_x > t > Y) = E(T_y|T_y > t > X) = E(T_{xy}|T_{xy} > t) = t + \frac{1}{\theta} \tag{28}$$

Now $Q(\underline{\lambda}|\underline{\lambda}^{(k)})$ [the equation (4.7) with expectations given in (4.9)] can be computed as

Table 2. Bayesian EM and ML estimation of parameters in BVE of B-B Model based on Censored Sampling

Parameters	λ_1	λ_2	λ_3	λ_1	λ_2	λ_3
	.08	.1	.03	.1	.12	.04
			$n = 50$			
Est_{bys}	.083	.1	.033	.1	.118	.045
SE_{bys-em}	.0107	.0109	.0103	.014	.013	.014
Est_{mle}	.085	.15	.025	.103	.126	.03
SE_{mle}	.025	.028	.033	.03	.034	.038
			$n = 75$			
Est_{bys}	.083	.1	.032	.099	.117	.048
SE_{bys}	.0078	.0082	.0075	.01	.0106	.0106
Est_{mle}	.083	.103	.021	.105	.125	.026
SE_{mle}	.021	.023	.026	.024	.027	.03
			$n = 100$			
Est_{bys}	.083	.1	.031	.099	.117	.049
SE_{bys}	.007	.007	.007	.0089	.009	.0088
Est_{mle}	.084	.104	.018	.107	.127	.022
SE_{mle}	.018	.02	.023	.022	.024	.027

follows

$$Q(\underline{\lambda}|\underline{\lambda}^{(k)}) = (n_1 + n_4 + \alpha_1 - 1)log\theta_1 + (n - n_6 - n_3)log\theta + (n_2 + n_5 + \alpha_2 - 1)log\theta_2$$

$$+ (n_3 + \alpha_3 - 1)log\theta_3 + n_6 \log\theta - \theta(\sum_{i\in B} \max(x_i, y_i) + \sum_{i\in B^c} t_i) + (n_4 + n_5 + n_6)/\theta^{(k)} - \sum_{i=1}^{3} \theta_i\beta_i. \tag{29}$$

M-step:
The first and second derivative of $Q(\underline{\lambda}|\underline{\lambda}^{(k)})$ are used to obtain M-step of EM algorithm, Newton-Raphson method is used for the iterative procedure.

When $\theta_3 = 0$, the BVE of Proschan-Sullo model reduces to BVE of Freund model. Substituting $\theta_3 = 0$ in the above expressions, we get corresponding expressions for BVE of Freund model.

5. Simulation Study

We resort to simulation studies as an appropriate way to show our results due to previous parts. In this section two methods of estimation yield from Bayesian EM algorithm and MLE have been compared to verify how the proposed method performs well. Data have been generated from three pdf, comprising M-O, B-B, P-S models in three sample of sizes 50,75 and 100. We carried out the simulation 1000 times for each sample set and for each pdf. To illuminate the right-censoring-structure in our data set, we decided to choose an

Table 3. Bayesian EM and ML estimation of parameters in BVE of P-S(1) Model based on Censored Sampling

Parameters	θ_1	θ_2	θ_3	η_1'	η_2'
	.08	.1	.03	.11	.15
			$n = 50$		
Est_{bys}	.079	.097	.027	.117	.142
SE_{bys}	.009	.01	.006	.019	.026
Est_{mle}	.082	.11	.0303	.1	.121
SE_{mle}	.02	.022	.011	.025	.033
			$n = 75$		
Est_{bys}	.078	.098	.028	.13	.146
SE_{bys}	.007	.008	.005	.016	.02
Est_{mle}	.081	.099	.03	.102	.12
SE_{mle}	.016	.017	.009	.02	.026
			$n = 100$		
Est_{bys}	.079	.098	.029	.13	.148
SE_{bys}	.0066	.0077	.004	.014	.019
Est_{mle}	.081	.099	.03	.101	.12
SE_{mle}	.014	.015	.008	.02	.023

Table 4. Bayesian EM and ML estimation of parameters in BVE of P-S(1) Model based on Censored Sampling

Parameters	θ_1	θ_2	θ_3	η_1'	η_2'
	.1	.12	.04	.13	.16
			$n = 50$		
Est_{bys}	.099	.117	.038	.135	.163
SE_{bys}	.012	.012	.007	.023	.031
Est_{mle}	.11	.129	.044	.122	.149
SE_{mle}	.024	.026	.014	.03	.04
			$n = 75$		
Est_{bys}	.099	.118	.038	.134	.167
SE_{bys}	.009	.01	.006	.02	.025
Est_{mle}	.106	.129	.043	.12	.148
SE_{mle}	.02	.02	.012	.023	.03
			$n = 100$		
Est_{bys}	.099	.118	.039	.132	.169
SE_{bys}	.008	.009	.005	.016	.022
Est_{mle}	.103	.125	.043	.12	.147
SE_{mle}	.016	.017	.01	.02	.027

Table 5. Bayesian EM and ML estimation of parameters in BVE of P-S(2) Model based on Censored Sampling

Parameters	θ_1 .08	θ_2 .1	θ_3 .03	θ_1 .1	θ_2 .12	θ_3 .04
			$n = 50$			
Est$_{bys}$	.07	.093	.026	.086	.11	.034
SE$_{bys}$	.006	.009	.005	.010	.013	.010
Est$_{mle}$	.06	.07	.022	.07	.085	.0285
SE$_{mle}$	.009	.016	.008	.011	.0187	.0109
			$n = 75$			
Est$_{bys}$	.07	.093	.027	.086	.115	.036
SE$_{bys}$	.006	.007	.007	.0099	.01	.0099
Est$_{mle}$	.06	.07	.021	.07	.084	.028
SE$_{mle}$	.007	.012	.007	.0086	.015	.0086
			$n = 100$			
Est$_{bys}$	.07	.094	.028	.086	.118	.037
SE$_{bys}$	.004	.0066	.0042	.004	.009	.0087
Est$_{mle}$	.06	.071	.021	.07	.084	.028
SE$_{mle}$	.0062	.011	.006	.007	.013	.0073

value $t = 20$ as censored value or end of study so each paired observation (x, y) such that $x > t$ or $y > t$ or both happened, was allocated as censoring observation.
Moreover results have been conducted for two different set values as initial values, $\lambda_1 = .08$, $\lambda_2 = .1$ $\lambda_3 = .03$ as the first set and $\lambda_1 = .1$, $\lambda_2 = .12$ $\lambda_3 = .04$ as the second set, for M-O, B-B, P-S(2) and two parameters more for the case of P-S(1) $\eta_1' = .11$ $\eta_2' = .15$ the first set and $\eta_1' = .13$ $\eta_2' = .16$ as the second set. Tables 5.1, 5.2, 5.3, 5.4 and 5.5 shows the results of simulation study.

It is observed from above Tables that the estimated values by Bayesian EM are more efficient than MLEs which means the SE_{bys} (standard error of Bayes EM) were less than SE_{mle} (standard error of MLE) for each sample size and each model. Furthermore the estimation based on Bayes EM's are close to true values as compared to MLEs.

6. Conclusions

In this paper we have used Bayesian EM algorithm for BVE models. We intend to show that the implementation of Bayesian EM algorithm is not difficult for BVE models. We estimate the parameters of prior distribution by moment estimators which is called empirical Bayes and tried to construct an informative prior which has an essential effect on the posterior mode. Here we use two sets of data, the first set to estimate hyper parameters of prior distribution and the second set of data to obtain posterior mode.

In fact EM algorithm is suitable computational method to improve censored part now we suggest to use EM algorithm with Bayesian analysis to obtain estimates through the informative prior.

In survival analysis and clinical trial studies, the right censoring mostly occurs. MLE's are always suggested as an appropriate estimations [see Lawless(1982) lee(2003) and Marubini(2004)]. We suggest Bayes estimators with EM algorithm for censored samples are more efficient estimators as compared to MLEs.

References

[1] Block, H. W. and Basu, A. P. (1974). A continuous bivariate exponential extension. *J.Amer.Statist.Assoc.*, **69**, 1031-37.

[2] Berger, J. O. (1985). Statistical decision theory and bayesian analysis. *Springer-Verlag New York*, Inc.

[3] Carlin, B. P. and Louis, T. A. (1996). Bayes and empirical bayes methods for data analysis. *Chapman & Hall.*

[4] Efron, B. and Tibshirani, R. (1993). An introduction to the bootstrap. London, *Chapman & Hall*, Inc.

[5] Freund, J. E. (1961). A bivariate extension of the exponential distribution. *J.Amer.Statist.Assoc.*, **56**, 971-977.

[6] Gelman, A. and Carlin, J. B. and Stern, H. S. and Rubin, D. B. (2004). Bayesian data analysis, *Chapman & Hall/CRC.*

[7] Hanagal, D. D. (1992a). Some inference results in bivariate exponential distributions based on censored samples. *Commun.Statist.-Theory meth.*, **21(5)**, 1273-1295.

[8] Hanagal, D.D. (1992b). Some inference results in modified Freund's bivariate exponential distribution. *Biometrical Journal*, **34**, 745-56.

[9] Hanagal, D.D. and Ahmadi, K.A. (2008). Estimation of parameters by EM algorithm in bivariate exponential distribution based on censored samples. *Economic Quality Control*, **23**(2), 257-66.

[10] Lawless, J. F. (1982). Statistical Models and Methods for Lifetime Data, *John Wiley & Sons, Inc.*

[11] Lee, E. T. and Wang J. W. (2003). *Statistical Methods for Survival Data Analysis,* John Wiley & Sons, Inc.

[12] Marshall, A. W. and Olkin, I. (1967). A multivariate exponential distribution. *J.Amer.Statist.Assoc.*, **62**, 30-44.

[13] Marubini, Ettore and Valsecchi, M.G. (2004). *Analyzing survival data from clinical trials and observational studies,* John Wiley & Sons Ltd.

[14] McLachlan, G.J., and Krishnan, T. (1997). *The EM algorithm and extensions.* John wiley & Sons,Inc.

[15] Proschan, F. and Sullo, P. (1974). *Estimating the parameters of a bivariate exponential distributions in several sampling situations. Reliability and Biometry,* Edited by F. Proschan and R.J. Surfling, Philadelphia: SIAM; pp. 423-40.

In: New Developments in Applied Statistics
Editor: Mohammad Ahsanullah
ISBN: 978-1-61324-648-1

Chapter 17

AN ALTERNATIVE FAMILY OF ESTIMATORS FOR POPULATION MEAN IN PRESENCE OF NON-RESPONSE WHEN THE POPULATION MEAN OF AUXILIARY CHARACTER IS NOT KNOWN

Housila P. Singh[a] ***and Sunil Kumar***[b]
School of Studies in Statistics, Vikram University Ujjain, M. P., India

Abstract

This paper considers the problem of estimating the population mean $\overline{Y}$ of the study variate y in the situation when information on the study variate y is not available for the initial sample while information on the auxiliary variable x is available for initial sample (i.e. complete sample) but the population mean $\overline{X}$ of the auxiliary character is not known. A family of estimators for $\overline{Y}$ has been suggested and its properties are studied. Usual unbiased estimator, ratio, product and difference type estimators are identified as members of the proposed family. Realistic conditions are obtained in which the proposed family is better than conventional estimators. Optimum sampling and sub sampling aspects are given. For illustration an empirical study is carried out to demonstrate the performance of various estimators.

Keywords: Non-response, Study variate, Auxiliary variate, Bias, Variance, Fixed variance, Minimum cost.

1. Introduction

Consider a finite population $U = \{U_1, U_2, ..., U_N\}$ of size N from which a random sample of size n is drawn without replacement. Let the study variate y take the value y_i on the unit

[a] E-mail address: hpsujn@rediffmail.com (Corresponding author)
[b] E-mail address: sunilbhougal06@gmail.com

$U_i(i=1,2,...,N)$. In surveys on human populations, frequently n_1 units respond in the first attempt while remaining $n_2(=n-n_1)$ units do not provide any response. The initial survey may be conducted through the mail or telephone calls, perhaps computer-aided. When non-response occurs in the initial attempt, Hansen and Hurwitz (1946) introduced a double sampling scheme for estimating population mean $\bar{Y}$ comprising the following steps:

(i) a simple random sample of size n is drawn and the questionnaire is mailed to the sampled units;
(ii) a sub sample of size $r = n_2/k \, (k \geq 1)$ from the n_2 non-responding units in the initial attempt is contacted through personal interviews.

It is to be mentioned that Hansen and Hurwitz (1946) consider mail surveys for the initial attempt and personal interviews at the second stage, their procedure is appreciable to any type of interviewing (e.g. telephone, personal interview), [see Rao (1983)].

In Hansen and Hurwitz (1946) approach, the population of size N is assumed to be composed of two strata of size N_1 and $N_2 = N - N_1$ of "respondents" and "non-respondents". Let $\bar{Y} = \frac{1}{N}\sum_{i=1}^{N} y_i$ and $S_y^2 = \frac{1}{N-1}\sum_{i=1}^{N}(y_i - \bar{Y})^2$ be the population mean and population variance of the study variable y. Let $\bar{Y}_1 = \frac{1}{N_1}\sum_{i=1}^{N_1} y_i$ and $S_{y_1}^2 = \frac{1}{N_1 - 1}\sum_{i=1}^{N_1}(y_i - \bar{Y}_1)^2$ denote the mean and variance of the response group. Similarly, let $\bar{Y}_2 = \frac{1}{N_2}\sum_{i=1}^{N_2} y_i$ and $S_{y2}^2 = \frac{1}{N_2 - 1}\sum_{i=1}^{N_2}(y_i - \bar{Y}_2)^2$ denote the mean and variance of the non-response group. Note that the suffixes 1 and 2 are identified with the response and non-response groups (or strata) respectively. The population mean $\bar{Y}$ can be written as

$$\bar{Y} = W_1\bar{Y}_1 + W_2\bar{Y}_2 \tag{1.1}$$

where $W_1 = (N_1/N)$ and $W_2 = (N_2/N)$

Hansen and Hurwitz (1946) suggested an unbiased estimator for population mean $\bar{Y}$ as

$$\bar{y}^* = w_1\bar{y}_1 + w_2\bar{y}_{2r} \tag{1.2}$$

where $w_1 = n_1/n$ and $w_2 = n_2/n$. The sample mean $\bar{y}_1$ is based on n_1 observations and $\bar{y}_{2r}$ on r units. The variance of $\bar{y}^*$ is given by

$$Var(\bar{y}^*) = \frac{(1-f)}{n}S_y^2 + w_2\frac{(k-1)}{n}S_{y_2}^2 \tag{1.3}$$

[see Cochran(1977,p. 371)]

where $f = n/N$.

It is well known that in estimating population mean, sample survey experts sometimes use auxiliary information to improve precision of the estimates. Let x, denote an auxiliary variate (correlated with study variate y) with population mean $\overline{X} = \sum_{i=1}^{N} x_i / N$. Let $\overline{X}_1$ and $\overline{X}_2$ denote the means of the response and non-response groups (or strata). Let $\overline{x} = \sum_{i=1}^{n} x_i / n$ denote the mean of all the n units. Let $\overline{x}_1 = \sum_{i=1}^{n_1} x_i / n_1$ and $\overline{x}_2 = \sum_{i=1}^{n_2} x_i / n_2$ denote the means of the n_1 responding units and n_2 non-responding units. Further, let $\overline{x}_{2r} = \sum_{i=1}^{r} x_i / r$ denote the mean of the $r(= n_2 / k)$ sub-sampled units. With this background we define an unbiased estimator of population mean $\overline{X}$ as

$$\overline{x}^* = w_1 \overline{x}_1 + w_2 \overline{x}_{2r} \tag{1.4}$$

The variance of $\overline{x}^*$ is given by

$$Var(\overline{x}^*) = \frac{(1-f)}{n} S_x^2 + w_2 \frac{(k-1)}{n} S_{x_2}^2 \tag{1.5}$$

where $S_x^2 = \frac{1}{N-1} \sum_{i=1}^{N} (x_i - \overline{X})^2$ and $S_{x_2}^2 = \frac{1}{N_2 - 1} \sum_{i=1}^{N_2} (x_i - \overline{X}_2)^2$.

The following well-known estimators of $\overline{Y}$ based on auxiliary information are available in the literature:

Case I

Population mean $\overline{X}$ of auxiliary variate x is known, incomplete information on y and x.

Let y and x be positively (high) correlated. In case when the population mean $\overline{X}$ of the auxiliary character x is known and we have incomplete information on y and x characters for the n selected units, the conventional ratio estimator for population mean $\overline{Y}$ proposed by Hansen and Hurwitz (1946) [Cochran (1977)] is given as

$$T_{R1} = \frac{\overline{y}^*}{\overline{x}^*} \overline{X}, \tag{1.6}$$

If the correlation between y and x is negative, then the conventional product estimator for $\overline{Y}$ is defined by

$$T_{P1} = \frac{\overline{y}^*}{\overline{X}}\overline{x}^*, \tag{1.7}$$

The regression estimator $\overline{Y}$ is defined by

$$T_{lr_1} = \overline{y}^* + b^*(\overline{X} - \overline{x}^*) \tag{1.8}$$

where the value of the regression coefficient b^* is obtained from the results of both samples.

Case II

Population mean $\overline{X}$ of the auxiliary variate x is known, incomplete information on y and complete information on x

An alternative ratio estimator, using $\overline{X}$ and the complete information on x and incomplete information on y character for n sample units, suggested by Rao (1986) is given by

$$T_{R2} = \frac{\overline{y}^*}{\overline{x}}\overline{X} \tag{1.9}$$

Similarly an alternative product estimator for $\overline{Y}$ is given by

$$T_{P2} = \frac{\overline{y}^*}{\overline{X}}\overline{x} \tag{1.10}$$

In this situation the regression estimator for $\overline{Y}$ is defined by

$$T_{lr_2} = \overline{y}^* + b(\overline{X} - \overline{x}) \tag{1.11}$$

where $b = \frac{s_{yx}^*}{s_x^2}$; the estimator s_{yx}^* is based on the available data under given sampling design and $s_x^2 = \frac{1}{(n-1)}\sum_{i=1}^{n}(x_i - \overline{x})^2$.

In this paper we have suggested the ratio, product and difference estimator for $\overline{Y}$ in the situation where the information on the study variable y is incomplete while information on the auxiliary variable x is available for all the initial sample units, but the population mean $\overline{X}$ of the auxiliary variable x is not known.

2. The Suggested Estimators

We are considering the situation when information on the study variable y is incomplete while information on the auxiliary variable x is available for the initial sample units, but the population mean $\overline{X}$ of the auxiliary variate x is unknown. In this situation we define the ratio estimator

$$T_{R3} = \bar{y}^* \frac{\bar{x}}{\bar{x}^*}, \tag{2.1}$$

product estimator

$$T_{P3} = \bar{y}^* \frac{\bar{x}^*}{\bar{x}}, \tag{2.2}$$

and the difference estimator

$$T_d = \bar{y}^* + d(\bar{x} - \bar{x}^*) \tag{2.3}$$

where d is a suitably chosen constant.

Further, we suggest a generalized version of $\bar{y}^*, T_{R3}, T_{P3}$ and T_d as

$$T_{(\alpha,d)} = \left[\bar{y}^* + d(\bar{x} - \bar{x}^*\right]\left(\frac{\bar{x}}{\bar{x}^*}\right)^{\alpha} \tag{2.4}$$

where (α, d) are suitably chosen constants. The class of estimators $T_{(\alpha,d)}$ reduces to the following set of estimators:

(i) $T_{(0,0)} = \bar{y}^*$ for $(\alpha, d) = (0,0)$,

(ii) $T_{(1,0)} = \bar{y}^*\left(\frac{\bar{x}}{\bar{x}^*}\right) = T_{R3}$ for $(\alpha, d) = (1,0)$,

(iii) $T_{(-1,0)} = \bar{y}^*\left(\frac{\bar{x}^*}{\bar{x}}\right) = T_{P3}$ for $(\alpha, d) = (-1,0)$, and

(iv) $T_{(0,d)} = \bar{y}^* + d(\bar{x} - \bar{x}^*)$ for $(\alpha, d) = (0, d)$.

To obtain the bias and variance of $T_{(\alpha,d)}$ we write

$$\bar{y}^* = \overline{Y} + \varepsilon_0, \bar{x}^* = \overline{X} + \varepsilon_1, \bar{x} = \overline{X} + \varepsilon_2$$

such that

$$E(\varepsilon_0) = E(\varepsilon_1) = E(\varepsilon_2) = 0 \tag{2.5}$$

and

$$E(\varepsilon_0^2) = Var(\bar{y}^*) = \left[\left(\frac{1-f}{n}\right)S_y^2 + \frac{W_2(k-1)}{n}S_{y_2}^2\right] \tag{2.6}$$

$$E(\varepsilon_1^2) = Var(\bar{x}^*) = \left[\left(\frac{1-f}{n}\right)S_x^2 + \frac{W_2(k-1)}{n}S_{x_2}^2\right] \tag{2.7}$$

$$E(\varepsilon_2^2) = Var(\bar{x}) = \left(\frac{1-f}{n}\right)S_x^2 \tag{2.8}$$

$$E(\varepsilon_0\varepsilon_1) = Cov(\bar{y}^*, \bar{x}^*) = \left(\frac{1-f}{n}\right)S_{yx} + \frac{W_2(k-1)}{n}S_{yx2} \tag{2.9}$$

$$E(\varepsilon_0\varepsilon_2) = Cov(\bar{y}^*, \bar{x}) = \left(\frac{1-f}{n}\right)S_{yx} \tag{2.10}$$

$$E(\varepsilon_1\varepsilon_2) = Cov(\bar{x}^*, \bar{x}) = \left(\frac{1-f}{n}\right)S_x^2 \tag{2.11}$$

Expressing the estimator $T_{(\alpha,d)}$ in terms of $\varepsilon's$ we have

$$T_{(\alpha,d)} = \left[\bar{Y} + \varepsilon_0 + d(\varepsilon_2 - \varepsilon_1)\right]\left(\frac{\bar{X} + \varepsilon_2}{\bar{X} + \varepsilon_1}\right)^{\alpha}$$

$$= \bar{Y}\left[1 + \frac{\varepsilon_0}{\bar{Y}} + \frac{d}{\bar{Y}}(\varepsilon_2 - \varepsilon_1)\right]\left(1 + \frac{\varepsilon_2}{\bar{X}}\right)^{\alpha}\left(1 + \frac{\varepsilon_1}{\bar{X}}\right)^{-\alpha} \tag{2.12}$$

We assume that $\left|\frac{\varepsilon_i}{\bar{X}}\right| < 1, i = 1,2;$ so that terms $\left(1 + \frac{\varepsilon_1}{\bar{X}}\right)^{-\alpha}$ and $\left(1 + \frac{\varepsilon_2}{\bar{X}}\right)^{\alpha}$ are expandable in terms of power series. Expanding the right hand side of (2.12) and neglecting terms of $\varepsilon's$ having power greater than two, we have

$$T_{(\alpha,d)} = \bar{Y}\Bigg[1 + \frac{\varepsilon_0}{\bar{Y}} + \frac{d}{\bar{Y}}(\varepsilon_2 - \varepsilon_1) - \frac{\alpha\varepsilon_1}{\bar{X}} - \frac{\alpha\varepsilon_0\varepsilon_1}{\bar{Y}\bar{X}} + \frac{\alpha(\alpha+1)}{2\bar{X}^2}\varepsilon_1^2 + \frac{\alpha\varepsilon_2}{\bar{X}} - \frac{\alpha d(\varepsilon_1\varepsilon_2 - \varepsilon_1^2)}{\bar{Y}\bar{X}}$$
$$+ \frac{\alpha\varepsilon_0\varepsilon_2}{\bar{Y}\bar{X}} + \frac{\alpha d}{\bar{Y}\bar{X}}(\varepsilon_2^2 - \varepsilon_1\varepsilon_2) - \frac{\alpha^2\varepsilon_1\varepsilon_2}{\bar{X}^2} + \frac{\alpha(\alpha-1)}{2\bar{X}^2}\varepsilon_2^2\Bigg]$$

or

$$(T_{(\alpha,d)} - \bar{Y}) = \Bigg[\varepsilon_0 + \alpha R(\varepsilon_2 - \varepsilon_1) + d(\varepsilon_2 - \varepsilon_1) + (1/\bar{X})\alpha(\varepsilon_0\varepsilon_2 - \varepsilon_0\varepsilon_1) + \left(\frac{\alpha d}{\bar{X}}\right)(\varepsilon_1 - \varepsilon_2)^2$$
$$+ \left(\frac{\alpha R}{2\bar{X}}\right)\left\{\alpha(\varepsilon_1 - \varepsilon_2)^2 + (\varepsilon_1^2 - \varepsilon_2^2)\right\}\Bigg] \tag{2.13}$$

Taking expectations both sides of (2.13) and using the results (2.5)-(2.11) we get the bias of $T_{(\alpha,d)}$ to the first degree of approximation as

$$B\{T_{(\alpha,d)}\} = (1/\overline{X})\left[-\alpha\frac{W_2(k-1)}{n}S_{yx2} + ad\,\frac{W_2(k-1)}{n}S_{x_2}^2 + \left(\frac{\alpha R}{2}\right)\left\{\alpha\frac{W_2(k-1)}{n}S_{x_2}^2 + \frac{W_2(k-1)}{n}S_{x_2}^2\right\}\right]$$

$$= (\alpha/\overline{X})\frac{W_2(k-1)}{n}S_{x_2}^2\left[(R/2)(\alpha+1) + d - \beta_{(2)}\right] \tag{2.14}$$

where $\beta_{(2)} = \dfrac{S_{yx2}}{S_{x_2}^2}$ is the regression coefficient of y on x in the non-response group of the population.

Squaring (2.13) and neglecting terms involving $\varepsilon's$ having power greater than two, we have

$$(T_{(\alpha,d)} - \overline{Y})^2 = [\varepsilon_0 + \alpha R(\varepsilon_2 - \varepsilon_1) + d(\varepsilon_2 - \varepsilon_1)]^2$$

$$= [\varepsilon_0 + (\alpha R + d)(\varepsilon_2 - \varepsilon_1)]^2$$

$$= \left[\varepsilon_0^2 + (\alpha R + d)^2(\varepsilon_2 - \varepsilon_1)^2 + 2(\alpha R + d)(\varepsilon_0\varepsilon_2 - \varepsilon_0\varepsilon_1)\right] \tag{2.15}$$

Taking expectations both sides of (2.15) and using the results in (2.5)-(2.11) we get the variance of $T_{(\alpha,d)}$ to the first degree of approximation as

$$Var\{T_{(\alpha,d)}\} = \left[\left(\frac{1-f}{n}\right)S_y^2 + \frac{W_2(k-1)}{n}S_{y_2}^2 + (\alpha R + d)^2\frac{W_2(k-1)}{n}S_{x_2}^2 - 2(\alpha R + d)\frac{W_2(k-1)}{n}S_{yx_2}\right]$$

$$= \left[\left(\frac{1-f}{n}\right)S_y^2 + \frac{W_2(k-1)}{n}\left\{S_{y_2}^2 + (\alpha R + d)S_{x_2}^2(\alpha R + d - 2\beta_{(2)})\right\}\right] \tag{2.16}$$

which is minimized for

$$\alpha R + d = \frac{S_{yx_2}}{S_{x_2}^2} = \beta_{(2)} \tag{2.17}$$

Substituting (2.17) in (2.16) we get the minimum variance of $T_{(\alpha,d)}$ as

$$\min.Var(T_{(\alpha,d)}) = \left[\left(\frac{1-f}{n}\right)S_y^2 + \frac{W_2(k-1)}{n}S_{y_2}^2(1-\rho_2^2)\right] \tag{2.18}$$

where $\rho_2 = \dfrac{S_{yx_2}}{S_{y_2} S_{x_2}}$ is the correlation coefficient of y on x in the non-response group of the population.

3. Particular Cases

Putting (α, d)= (0, 0), (1, 0), (-1, 0) and $(0, d)$ in (2.14) and (2.16) respectively we get the biases and variances of the estimators $T_{(0,0)}$, $T_{(1,0)}$, $T_{(-1,0)}$, and $T_{(0,d)}$ as

$$B\{T_{(0,0)}\} = B(\bar{y}^*) = 0 \tag{3.1}$$

$$B\{T_{(1,0)}\} = B(T_{R3}) = \frac{W_2(k-1)}{n\bar{X}} S_{x_2}^2 (R - \beta_{(2)}) \tag{3.2}$$

$$B\{T_{(-1,0)}\} = B(T_{P3}) = \frac{W_2(k-1)}{n\bar{X}} \beta_{(2)} S_{x_2}^2 \tag{3.3}$$

$$B\{T_{(0,d)}\} = B(T_d) = 0 \tag{3.4}$$

$$Var\{T_{(0,0)}\} = Var(\bar{y}^*) = \left(\frac{1-f}{n}\right) S_y^2 + \frac{W_2(k-1)}{n} S_{y_2}^2 \tag{3.5}$$

$$Var\{T_{(1,0)}\} = Var(T_{R3}) = \left(\frac{1-f}{n}\right) S_y^2 + \frac{W_2(k-1)}{n} \left\{S_{y_2}^2 + R^2 S_{x_2}^2 - 2RS_{yx2}\right\}$$

$$= \left(\frac{1-f}{n}\right) S_y^2 + \frac{W_2(k-1)}{n} \left\{S_{y_2}^2 + RS_{x_2}^2 (R - 2\beta_{(2)})\right\} \tag{3.6}$$

$$Var\{T_{(-1,0)}\} = Var(T_{P3}) = \left(\frac{1-f}{n}\right) S_y^2 + \frac{W_2(k-1)}{n} \left\{S_{y_2}^2 + R^2 S_{x_2}^2 + 2RS_{yx2}\right\}$$

$$= \left(\frac{1-f}{n}\right) S_y^2 + \frac{W_2(k-1)}{n} \left\{S_{y_2}^2 + RS_{x_2}^2 (R + 2\beta_{(2)})\right\} \tag{3.7}$$

$$Var\{T_{(0,d)}\} = Var(T_d) = \left(\frac{1-f}{n}\right) S_y^2 + \frac{W_2(k-1)}{n} \left\{S_{y_2}^2 + d^2 S_{x_2}^2 - 2dS_{yx2}\right\}$$

$$= \left(\frac{1-f}{n}\right) S_y^2 + \frac{W_2(k-1)}{n} \left\{S_{y_2}^2 + dS_{x_2}^2 (d - 2\beta_{(2)})\right\} \tag{3.8}$$

From (3.5) and (3.6) we note that

$$Var(\bar{y}^*) - Var(T_{R3}) = -\frac{W_2(k-1)}{n} RS_{x_2}^2 (R - 2\beta_{(2)})$$

which is positive if

$$(R - 2\beta_{(2)}) < 0$$

i.e. if
$$\beta_{(2)} > R/2 \tag{3.9}$$

From (3.5) and (3.7) we have

$$Var(\bar{y}^*) - Var(T_{P3}) = -\frac{W_2(k-1)}{n} RS_{x_2}^2 (R + 2\beta_{(2)}) > 0 \ if$$

$$R + 2\beta_{(2)} < 0$$

i.e. if
$$\beta_{(2)} < -R/2 \tag{3.10}$$

From (3.5) and (3.8) we see that

$$Var(\bar{y}^*) - Var(T_d) = -d\frac{W_2(k-1)}{n} S_{x_2}^2 (d - 2\beta_{(2)}) \ > 0 \ if$$

$$\left.\begin{array}{ll} either & 0 < d < 2\beta_{(2)} \\ or & 2\beta_{(2)} < d < 0 \end{array}\right\} \tag{3.11}$$

From (3.6) and (3.8) we note that

$$Var(T_{R3}) - Var(T_d) \ = \ \frac{W_2(k-1)}{n} S_{x_2}^2 \left[R(R - 2\beta_{(2)}) - d(d - 2\beta_{(2)}) \right] > 0 \text{ if}$$

$$\left.\begin{array}{ll} either & 2\beta_{(2)} - R < d < R \\ or & R < d < (2\beta_{(2)} - R) \end{array}\right\} \tag{3.12}$$

Now from (3.7) and (3.8) we have

$$Var(T_{P3}) - Var(T_d) = \frac{W_2(k-1)}{n} S_{x_2}^2 \left[R(R + 2\beta_{(2)}) - d(d - 2\beta_{(2)}) \right] > 0$$

If
$$\left[R(R + 2\beta_{(2)}) - d(d - 2\beta_{(2)}) \right] > 0$$

i.e. if
$$\left.\begin{array}{ll} either & -R < d < R + \beta_{(2)} \\ or & R + 2\beta_{(2)} < d < -R \end{array}\right\} \tag{3.13}$$

The proposed estimator $T_{(\alpha,d)}$ is more efficient than

(i) The usual unbiased estimator $\bar{y}^*$ if

$$\left.\begin{array}{ll} either & 0 < (\alpha R + d) < 2\beta_{(2)} \\ or & 2\beta_{(2)} < (\alpha R + d) < 0 \end{array}\right\} \tag{3.14}$$

(ii) The ratio-type estimator $T_{(1,0)} = T_{R3} = \bar{y}^* \dfrac{\bar{x}}{\bar{x}^*}$ if

$$\left.\begin{array}{ll} either & (2\beta_{(2)} - R) < (\alpha R + d) < R \\ or & R < (\alpha R + d) < (2\beta_{(2)} - R) \end{array}\right\} \tag{3.15}$$

(iii) The product – type estimator $T_{(-1,0)} = T_{P3} = \bar{y}^* \dfrac{\bar{x}^*}{\bar{x}}$ if

$$\left.\begin{array}{ll} either & -R < (\alpha R + d) < (R + 2\beta_{(2)}) \\ or & (R + 2\beta_{(2)}) < (\alpha R + d) < -R \end{array}\right\} \tag{3.16}$$

(iv) The difference estimator T_d if

$$\left.\begin{array}{ll} either & 0 < \alpha < (2/R)(\beta_{(2)} - d) \\ or & (2/R)(\beta_{(2)} - d) < \alpha < 0 \end{array}\right\} \tag{3.17}$$

4. Allowable Departure from Optimum

For sake of simplicity we consider the difference estimator T_d given by (2.4). The variance of T_d is given as

$$Var(T_d) = \left(\frac{1-f}{n}\right) S_y^2 + \frac{W_2(k-1)}{n} \left\{ S_{y_2}^2 + d^2 S_{x_2}^2 - 2d\rho_2 S_{x_2} S_{y_2} \right\}$$

$$= \left(\frac{1-f}{n}\right) S_y^2 + \frac{W_2(k-1)}{n} S_{y_2}^2 \left\{ 1 + d^2 \frac{S_{x_2}^2}{S_{y_2}^2} - 2d\rho_2 \frac{S_{x_2}}{S_{y_2}} \right\}$$

If $\beta_{(2)}$ is the regression coefficient of y on x in the non-response group so that

$$\beta_{(2)} = \frac{S_{yx_2}}{S_{x_2}^2} = \rho_2 \frac{S_{y_2}}{S_{x_2}}$$

The variance of the difference estimator T_d can be written as

$$Var(T_d) = \left(\frac{1-f}{n}\right) S_y^2 + \frac{W_2(k-1)}{n} S_{y_2}^2 (1 - \rho_2^2)(1 + \delta^2) \tag{4.1}$$

where $$\delta^2 = \frac{\rho_2^2}{(1-\rho_2^2)}\left(\frac{\beta_{(2)} - d}{\beta_{(2)}}\right) \quad (4.2)$$

It follows from (3.18) that the variance of T_d at (3.18) is minimum when

$$d = \beta_{(2)} = d_0(say) \quad (4.3)$$

For this optimal choice of d, we get the difference estimator

$$T_{d=\beta_{(2)}} = \bar{y}^* + \beta_{(2)}(\bar{x} - \bar{x}^*) \quad (4.4)$$

which has the variance (or minimum variance of T_d)

$$Var(T_d)_{d=\beta_{(2)}} = \left(\frac{1-f}{n}\right)S_y^2 + \frac{W_2(k-1)}{n} S_{y_2}^2 (1-\rho_2^2) \quad (4.5)$$

The main drawback with the difference estimator T_d is the choice of $\beta_{(2)}$ since it is generally unknown. One possible way out of this difficulty is to use past experience to make a reliable guess of $\beta_{(2)}$ say $\beta_{(2)}^{(0)}$.

This will no doubt result in increased variance [see Sukhatme et al (1984, p.237)].

From (4.1) and (4.5) we have

$$Q = \frac{Var(T_d)_{d=\beta_{(2)}^{(0)}} - Var(T_d)_{d=\beta_{(2)}}}{W_2(k-1)\left(S_{y_2}^2/n\right)(1-\rho_2^2)} = \delta^2 \quad (4.6)$$

which will be less than α % if

$$\delta^2 < \alpha$$

$$\text{i.e. if } \left(\frac{\beta_{(2)} - \beta_{(2)}^{(0)}}{\beta_{(2)}}\right)^2 < \frac{\alpha(1-\rho_2^2)}{\rho_2^2}$$

$$\text{i.e. if } \left|1 - \frac{\beta_{(2)}^{(0)}}{\beta_{(2)}}\right| < \sqrt{\frac{\alpha(1-\rho_2^2)}{\rho_2^2}} \quad (4.7)$$

Thus to ensure only a small increase in Q (i.e. in variance), $\left|1 - \frac{\beta_{(2)}^{(0)}}{\beta_{(2)}}\right|$ must to close to zero if ρ_2 is high but can depart considerably from 0(zero) if ρ_2 is just moderate. If it is not possible to make a reliable guess of $\beta_{(2)}$, the other possibility is to estimate $\beta_{(2)}$ from the sample by

$$\hat{\beta}_{(2)} = \frac{s_{yx(2r)}}{s^2_{x(2r)}} \tag{4.8}$$

yielding the regression estimator

$$\hat{T}_{lr} = \bar{y}^* + \hat{\beta}_2(\bar{x} - \bar{x}^*) \tag{4.9}$$

where $\hat{\beta}_{(2)}$ is the estimate of the regression coefficient $\beta_{(2)}$ of y on x in non-response group of the population and $s_{yx(2r)} = \frac{1}{(r-1)}\sum_{i=1}^{r}(y_i - \bar{y}_{2r})(x_i - \bar{x}_{2r})$ *and* $s^2_{x(2r)} = \frac{1}{(r-1)}\sum_{i=1}^{r}(x_i - \bar{x}_{2r})^2$

It can be shown to the first degree of approximation that

$$Var(\hat{T}_{lr}) = \left(\frac{1-f}{n}\right)S_y^2 + \frac{W_2(k-1)}{n}S^2_{y_2}(1-\rho_2^2)$$

$$= \min.Var(T_d) = Var(T_d)_{d=\beta_{(2)}} \tag{4.10}$$

From (3.5), (3.6), (3.7) and (4.10), we have

$$Var(\bar{y}^*) - Var(\hat{T}_{lr}) = \frac{W_2(k-1)S^2_{y_2}\rho_2^2}{n} \geq 0 \tag{4.11}$$

$$Var(T_{R3}) - Var(\hat{T}_{lr}) = \frac{W_2(k-1)S^2_{x_2}(R-\beta_{(2)})^2}{n} \geq 0 \tag{4.12}$$

$$Var(T_{P3}) - Var(\hat{T}_{lr}) = \frac{W_2(k-1)S^2_{x_2}(R+\beta_{(2)})^2}{n} \geq 0 \tag{4.13}$$

It follows from (4.11), (4.12) and (4.13) that the linear regression estimator $\hat{T}_{lr}$ is better than

(i) usual unbiased estimator $\bar{y}^*$,
(ii) the ratio estimator T_{R3} except when $R = \beta_{(2)}$,in the case where both the estimators T_{R3} and $\hat{T}_{lr}$ are equally efficient,
(iii) the product estimator T_{P3} except when $R = -\beta_{(2)}$, in the case where both the estimators T_{P3} and $\hat{T}_{lr}$ are equally efficient.

5. Optimum Sampling and Subsampling Factors

The cost of sampling is considered to be

$$C = c_0 n + c_1 n_1 + c_2 r \tag{5.1}$$

where c_0 is the initial cost for 'setting up' the survey, c_1 is the cost per unit for obtaining the responses from the n_1 units and processing them, and c_2 is the cost per unit for contacting the sub sampled units and for obtaining and processing responses from them. Generally c_2 is much greater than c_1, since additional effort is needed for contacting the nonrespondents and eliciting responses from them [Rao, 1993]. Since C, as it varies from sample to sample, is a random quantity and to, get over the variable nature of the cost, we work out the expected cost as

$$E(C) = (c_0 + W_1 c_1 + W_2 \frac{c_2}{k})n = C^*(say) \tag{5.2}$$

For obtaining the minimum cost for a given variance or minimum variance for a given cost, is equivalent to minimizing

$$\left(V_0 + \frac{S_y^2}{N}\right)C^* = \left[\frac{1}{n}\left\{S_y^2 + W_2(k-1)A(\alpha,d)\right\} - \frac{1}{N}S_y^2\right]C^* + \frac{S_y^2}{N}C^*$$

$$= \frac{1}{n}\left\{S_y^2 + W_2(k-1)A(\alpha,d)\right\}(c_0 + W_1 c_1 + W_2 \frac{c_2}{k})n$$

$$= W_2(c_0 + W_1 c_1)A(\alpha,d)k + W_2 c_2\{S_y^2 - W_2 A(\alpha,d)\}\frac{1}{k} + \text{terms independent of } n \text{ and } k. \tag{5.3}$$

where $$V_0 = \left[\frac{1}{n}\left\{S_y^2 + W_2(k-1)A(\alpha,d)\right\} - \frac{1}{N}S_y^2\right] \tag{5.4}$$

is the fixed variance of the estimator $T_{(\alpha,d)}$ and

$$A(\alpha,d) = \left[S_{y_2}^2 + (\alpha R + d)S_{x_2}^2(\alpha R + d - 2\beta_{(2)})\right] \tag{5.5}$$

Differentiating (5.3) with respect to k and equating it to zero, we have

$$W_2(c_0 + W_1 c_1)A(\alpha,d) - W_2 c_2\{S_y^2 - W_2 A(\alpha,d)\}\frac{1}{k^2} = 0$$

which yields the optimum value of k as

$$k_{opt} = \sqrt{\frac{c_2}{(c_0 + W_1 c_1)}\left\{\frac{S_y^2}{A(\alpha,d)} - W_2\right\}} \tag{5.6}$$

It follows from (5.6) that the optimal size of the sub sample is large if c_2 is small relative to $(c_0 + c_1 W_1)$ and $A(\alpha, d)$ is large relative to S_y^2. For a specified V, from (2.16) and (5.5), we get the optimum value of n as

$$n_{opt} = \frac{N\left[S_y^2 + (k_{opt} - 1)W_2 A(\alpha, d)\right]}{NV_0 + S_y^2} = n_0\left[1 + \frac{(k_{opt} - 1)W_2 A(\alpha, d)}{S_y^2}\right] \tag{5.7}$$

where $n_0 = \dfrac{NS_y^2}{NV_0 + S_y^2}$ is the sample size required when $W_2 = 0$.

Using optimum values of n and k from (5.6) and (5.7) respectively, we have the minimum cost given by

$$\min.C^* = \frac{A(\alpha, d)}{\left\{V_0 + \dfrac{S_y^2}{N}\right\}}\left[\sqrt{(c_0 + W_1 c_1)}\sqrt{\left(\frac{S_y^2}{A(\alpha, d)} - W_2\right)} + W_2\sqrt{c_2}\right]^2 \tag{5.8}$$

Now for a fixed cost $C^* = C_0^*$, using (5.6) we get the optimum value of n as

$$n_{opt} = \frac{C_0^*}{\left(c_0 + W_1 c_1 + \dfrac{W_2 c_2}{k_{opt}}\right)} \tag{5.9}$$

Using optimum values of k and n from (5.6) and (5.9) respectively in (2.16) we have the minimum variance as

$$\min.Var\left(T_{(\alpha,d)}\right) = \frac{1}{C_0^*}\left[\sqrt{(c_0 + W_1 c_1)}\sqrt{\{S_y^2 - W_2 A(\alpha, d)\}} + W_2\sqrt{c_2 A(\alpha, d)}\right]^2 - \frac{S_y^2}{N} \tag{5.10}$$

Since k has to be greater than 1, there will be no sub sampling if k from (5.8) turns out to be less than 1. It would be apt to mention that, for a given cost C_0^*, when there is no non-response, we obtain the initial sample size

$$n = \frac{C_0^*}{c_0 + c_1} \tag{5.11}$$

However, if there is some non-response, n_{opt} given by (5.9) for a given cost, should not exceed the value of n given by (5.11), i.e.,

$$\frac{C_0^*}{\left(c_0 + W_1 c_1 + \dfrac{W_2 c_2}{k_{opt}}\right)} \le \frac{C_0^*}{c_0 + c_1}$$

which leads to the condition

$$k_{opt} \le \frac{c_2}{c_1} \tag{5.12}$$

Now we give the optimum values of n and k along with minimum cost (for fixed variance) and minimum variance (for given cost C_0^*) for different estimators belonging to the proposed class of estimators $T_{(\alpha,d)}$:

1. For $(\alpha, d) = (0,0), T_{(\alpha,d)} \to \bar{y}^*$ (usual unbiased estimator)

$$A(0,0) = S_{y_2}^2,$$

For fixed variance V_0:

$$k_{opt}^{(0)} = \sqrt{\frac{c_2}{(c_0 + W_1 c_1)}\left\{\frac{S_y^2}{S_{y_2}^2} - W_2\right\}}$$

$$n_{opt}^{(0)} = \frac{1}{\left(V_0 + \dfrac{S_y^2}{N}\right)}\left[S_y^2 + (k_{opt}^{(0)} - 1)W_2 S_{y_2}^2\right]$$

$$\min .C^* = \frac{S_{y_2}^2}{\left(V_0 + \dfrac{S_y^2}{N}\right)}\left[\sqrt{(c_0 + W_1 c_1}\sqrt{\left(\frac{S_y^2}{S_{y_2}^2} - W_2\right)} + W_2\sqrt{c_2}\right]^2$$

For fixed cost C_0^*:

$$k_{opt(0)}^{(0)} = k_{opt}^{(0)} = \sqrt{\frac{c_2}{(c_0 + W_1 c_1)}\left\{\frac{S_y^2}{S_{y_2}^2} - W_2\right\}}$$

$$n_{opt(1)}^{(0)} = \frac{C_0^*}{\left(c_0 + W_1 c_1 + \dfrac{W_2 c_2}{k_{opt}^{(0)}}\right)}$$

$$\min .Var(T_{(0,0)}) = \frac{1}{C_0^*}\left[\sqrt{(c_0 + W_1 c_1)}\sqrt{(S_y^2 - W_2 S_{y_2}^2)} + W_2\sqrt{c_2 S_{y_2}^2}\right]^2 - \frac{S_y^2}{N}$$

$$= \min .Var(\bar{y}^*)$$

2. For $(\alpha, d) = (1,0), T_{(\alpha,d)} \to T_{(1,0)} = T_{R3} = \bar{y}^* \dfrac{\bar{x}}{\bar{x}^*}$;

$$A(1,0) = \left[S_{y_2}^2 + RS_{x_2}^2(R - 2\beta_{(2)})\right]$$

For fixed variance V_0 :

$$k_{opt}^{(1)} = \sqrt{\frac{c_2}{(c_0 + W_1 c_1)}\left\{\frac{S_y^2}{A(1,0)} - W_2\right\}}$$

$$n_{opt}^{(1)} = \frac{1}{\left(V_0 + \frac{S_y^2}{N}\right)}\left[S_y^2 + (k_{opt}^{(1)} - 1)W_2 A(1,0)\right]$$

$$\min .C^* = \frac{A(1,0)}{\left(V_0 + \frac{S_y^2}{N}\right)}\left[\sqrt{(c_0 + W_1 c_1)}\sqrt{\left(\frac{S_y^2}{A(1,0)} - W_2\right)} + W_2\sqrt{c_2}\right]^2$$

For a given cost C_0^* :

$$k_{opt(1)}^{(1)} = k_{opt}^{(1)} = \sqrt{\frac{c_2}{(c_0 + W_1 c_1)}\left\{\frac{S_y^2}{A(1,0)} - W_2\right\}}$$

$$n_{opt(1)}^{(1)} = \frac{C_0^*}{\left(c_0 + W_1 c_1 + \frac{W_2 c_2}{k_{opt}^{(1)}}\right)}$$

$$\min .Var\,(T_{(1,0)}) = \frac{1}{C_0^*}\left[\sqrt{(c_0 + W_1 c_1)}\sqrt{\{S_y^2 - W_2 A(1,0)\}} + W_2\sqrt{c_2 A(1,0)}\right]^2 - \frac{S_y^2}{N}$$

3. For $(\alpha, d) = (-1,0), T_{(\alpha,d)} \rightarrow T_{(-1,0)} = T_{P3} = \bar{y}^* \dfrac{\bar{x}^*}{\bar{x}}$

$$A(-1,0) = \left[S_{y_2}^2 + RS_{x_2}^2 (R + 2\beta_{(2)})\right]$$

For fixed variance V_0 :

$$k_{opt}^{(2)} = \sqrt{\frac{c_2}{(c_0 + W_1 c_1)}\left\{\frac{S_y^2}{A(-1,0)} - W_2\right\}}$$

$$n_{opt}^{(2)} = \frac{1}{\left(V_0 + \frac{S_y^2}{N}\right)}\left[S_y^2 + (k_{opt}^{(2)} - 1)W_2 A(-1,0)\right]$$

$$\min .C^* = \frac{A(-1,0)}{\left(V_0 + \frac{S_y^2}{N}\right)}\left[\sqrt{(c_0 + W_1 c_1)}\sqrt{\left(\frac{S_y^2}{A(-1,0)} - W_2\right)} + W_2\sqrt{c_2}\right]^2$$

For a given cost C_0^* :

$$k^{(2)}_{opt(2)} = k^{(2)}_{opt} = \sqrt{\frac{c_2}{(c_0 + W_1 c_1)}\left\{\frac{S_y^2}{A(-1,0)} - W_2\right\}}$$

$$n^{(2)}_{opt(2)} = \frac{C_0^*}{\left(c_0 + W_1 c_1 + \dfrac{W_2 c_2}{k^{(2)}_{opt}}\right)}$$

$$\min .Var\,(T_{(-1,0)}) = \frac{1}{C_0^*}\left[\sqrt{(c_0 + W_1 c_1)}\sqrt{\{S_y^2 - W_2 A(-1,0)\}} + W_2\sqrt{c_2 A(-1,0)}\right]^2 - \frac{S_y^2}{N}$$

4. For $(\alpha, d) = (0, \beta_{(2)}), T_{(\alpha,d)} \rightarrow T_{(\alpha,\beta)} = \bar{y}^* + \beta_{(2)}(\bar{x} - \bar{x}^*)$

$$A(0, \beta_{(0)}) = \left(S_{y_2}^2 - \beta_{(2)}^2 S_{x_2}^2\right) = S_{y_2}^2(1 - \rho_2^2)$$

For fixed variance V_0 :

$$k^{(3)}_{opt} = \sqrt{\frac{c_2}{(c_0 + W_1 c_1)}\left\{\frac{S_y^2}{A(0, \beta_{(2)})} - W_2\right\}}$$

$$n^{(3)}_{opt} = \frac{1}{\left(V_0 + \dfrac{S_y^2}{N}\right)}\left[S_y^2 + (k^{(3)}_{opt} - 1)W_2 A(0, \beta_{(2)})\right]$$

$$\min .C^* = \frac{A(0, \beta_{(2)})}{\left(V_0 + \dfrac{S_y^2}{N}\right)}\left[\sqrt{(c_0 + W_1 c_1)}\sqrt{\left(\frac{S_y^2}{A(0, \beta_{(2)})} - W_2\right)} + W_2\sqrt{c_2}\right]^2$$

For a given cost C_0^* :

$$k^{(3)}_{opt(3)} = k^{(3)}_{opt} = \sqrt{\frac{c_2}{(c_0 + W_1 c_1)}\left\{\frac{S_y^2}{A(0, \beta_{(2)})} - W_2\right\}}$$

$$n^{(3)}_{opt(3)} = \frac{C_0^*}{\left(c_0 + W_1 c_1 + \dfrac{W_2 c_2}{k^{(3)}_{opt}}\right)}$$

$$\min .Var\,(T_{(0,\beta_{(2)})}) = \frac{1}{C_0^*}\left[\sqrt{(c_0 + W_1 c_1)}\sqrt{\{S_y^2 - W_2 A(0, \beta_{(2)})\}} + W_2\sqrt{c_2 A(0, \beta_{(2)})}\right]^2 - \frac{S_y^2}{N}$$

6. Empirical Study

To illustrate the findings we consider a natural population data earlier considered by Khare and Srivastava (1995).

The population of 100 consecutive trips (after leaving 20 outlier values) measured by two fuel meters for a small family car in normal usage given by Lewisi et al (1991) has been taken into consideration. The measurements of turbine meter (in ml) are considered as main (study) character y and the measurements of displacement meter (in cm^3) are considered as auxiliary character x . 25% last values treated as non-response units.

The values of the parameters are as follows:

$\bar{Y} = 3500.12,\ \bar{X} = 260.84,\ S_y = 2079.30,\ S_x = 156.40,\ \bar{Y}_2 = 3401.08,\ \bar{X}_2 = 259.96,$

$S_{y_2} = 1726.02,\ S_{x_2} = 134.36,\ \rho = 0.985,\ \rho_2 = 0.995,\ N = 100,\ n = 30$. Let $c_0 = Rs.0.10$,

$c_1 = Rs.1.25$ and $c_2 = Rs.2.00$.

Table 6.1. Percent relative efficiency of the estimator $T_{(\alpha,d)}$ with $\alpha = 1$ (i.e. $T_{(1,d)}$) with respect to $\bar{y}^*$.

$d \downarrow$ \ $(1/k) \rightarrow$	(1/5)	(1/4)	(1/3)	(1/2)
-14	99.8255	99.8506	99.8839	99.9305
-13	103.2674	102.7844	102.1491	101.2757
-10	129.4617	124.1991	117.8296	109.9626
-5	176.6417	159.1042	140.5477	120.8805
-2	194.3733	171.1443	147.6748	123.9614
-1.5	195.6461	171.9877	148.1618	124.1665
-1	196.3544	172.4559	148.4314	124.2797
-0.6366	196.5076	172.5571	148.4896	124.3041
-0.5	196.4859	172.5428	148.4813	124.3007
0	196.0383	172.2471	148.3112	124.2293
0.25	195.5994	171.9568	148.1439	124.1589
0.5	195.0194	171.5728	147.9224	124.0657
0.75	194.3009	171.0962	147.6470	123.9496
1	193.4469	170.5286	147.3185	123.8108
3	182.2689	162.9847	142.8814	121.9059
5	165.4562	151.2204	135.6944	118.6944
7	146.1588	137.0599	126.5808	114.3813
9	126.8983	122.1720	116.4065	109.2166
11	109.1791	107.7566	105.9213	103.4631
12	101.1234	100.9603	100.7442	100.4443
13	93.6446	94.5085	95.6822	97.3692

Table 6.1 exhibits that there is enough scope of selecting the scalar d when the scalar α is fixed (i.e. $\alpha = 1$); for obtaining the estimator better than usual unbiased estimator $\bar{y}^*$ and the ratio-type estimator T_{R3}. The largest PRE (196.51%) is observed when d attains its optimum (i.e. d = -0.6366). It is also observed that the ratio-type estimator T_{R3} and the optimum estimator $T_{d=\beta_{(2)}}$ at (4.4) (or the linear regression estimators T_{lr} at (4.9)) are almost equally efficient. It happens due to the high positive correlation between the study variate y and auxiliary variate x in the non response group (i.e. $\rho_2 = 0.995$). Thus in such a case, the ratio-type estimator T_{R3} and regression estimator $\hat{T}_{lr}$ can be used as alternative estimators when the population mean $\bar{X}$ of the auxiliary variate is not known.

Table 6.2. Percent-Relative Efficiency and the expected cost.

Estimator	k_{opt}	$C_0^* = Rs.100.00(fixed)$		$V_0 = 60\times10^3\, ml(fixed)$	
		n_{opt}	$R.E.(.) w.r.t. \bar{Y}^*$	n_{opt}	**Expected cost C(in Rs.)**
$\bar{Y}^*$	1.52	73	100	45	62.35
T_{R3}	14.99	93	135.31	43	46.19
$\hat{T}_{lr}$	16.73	93	136.14	43	45.92

Table 6.2 demonstrates that for the fixed cost the ratio estimator T_{R3} and regression estimator $\hat{T}_{lr}$ are more efficient than the usual unbiased estimator $\bar{y}^*$ .There is marginal gain in efficiency by using the regression estimator $\hat{T}_{lr}$ over the ratio-type estimator T_{R3} due to the high positive correlation coefficient ρ_2 between the study variate y and the auxiliary variate x in the non-response group. Further for the specified variance, the ratio-type estimator T_{R3} and the regression–type estimator $\hat{T}_{lr}$ has fewer expected cost than the usual unbiased estimator $\bar{y}^*$. However the expected cost is minimum for $\hat{T}_{lr}$ than in case of $\bar{y}^*$ and T_{R3}. There is marginal difference in the expected cost of T_{R3} and $\hat{T}_{lr}$, so for such data one can choose T_{R3} or $\hat{T}_{lr}$ in practice.

References

Cochran, W. G. (1977). *Sampling Techniques*, 3rd ed., John Wiley and Sons, New York.

Hansen, M. H., and Hurwitz, W. N. (1946). The problem of non-response in sample surveys. *Journal of American Statistical Association*, **41**, 517-529.

Khare, B. B. and Srivastava, S. (1995). Study of conventional and alternative two phase sampling ratio, product and regression estimators in presence of non response. *Procedings of National Academic Sciences of India*, **65**, A, II, 195-203.

Lewisi, P. A., Joness, P. W., Polak, J. W. and Tillotson, H. T. (1991). The problem of conversion in method comparison studies. *Applied Statistics*, **40**,1, 105-112.

Rao, P. S. R. S. (1983). *Randomization approach. Incomplete data in sample suryey,* Academic Press, Inc., Volume 10, Part III, Chapter 10, 97-105.

Rao, P. S. R. S. (1986). Ratio estimation with sub sampling the non-respondents. *Survey Methodology*, **12**, 2, 217-230.

Sukhatme, P. V., Sukhatme, B. V., Sukhatme, S. and Ashok, C. (1984). *Sampling Theory of Surveys with Applications*, Iowa State University Press, Ames. Iowa, U.S.A., III revised Edition.

In: New Developments in Applied Statistics
Editor: Mohammad Ahsanullah
ISBN: 978-1-61324-648-1

Chapter 18

SHRINKAGE ESTIMATORS FOR CALCULATING RELIABILITY, WEIBULL CASE

Aliakbar Montazer Haghighi* and Mohammed A. Shayib
Department of Mathematics, Prairie View A&M University, Prairie View, Texas 77446 USA

Abstract

The estimation of the reliability $R = P(Y < X)$, assuming X and Y to be independent and identically distributed random variables, has been extensively studied in the literature. Distribution functions such as Normal, Burr Type X, Exponential, Gamma, and Weibull have been considered. In this report, we consider the Weibull type of distributions. Utilizing the Maximum Likelihood Estimation (MLE), Uniformly Minimum Variance Unbiased Estimation (UMVUE), Shrinkage Estimation Procedures, Method of Moments, and other methods of estimation, using simulation, we will estimate the unknown parameters of the Weibull distribution. In particular, we will observe how well the shrinkage estimators will do compared with the MLE based on the mean squared error (MSE). Also, a comparison among the estimators found by the different methods will be done to choose the best results in calculating R. Those results will be made based on the ratio of the parameters and the sample sizes.

Keywords: Reliability, Weibull Distribution, *MLE*, Shrinkage, Populations' Parameters, Sample sizes, Simulation

1. Introduction

The reliability of an item, or a product, is becoming the top priority for the third millennium and technically sophisticated customer. Manufacturers and all other producing entities are sharpening their tools to satisfy that customer. Estimation of that reliability has become a concern for many quality professionals and statisticians. When Y represents the

* E-mail address: amhaghighi@pvamu.edu

random value of a stress that a device will be subjected to in service, and X represents the strength that varies from item to item in the population of devices, then the reliability R is $P(Y < X)$, i.e., the probability that a randomly selected device functions successfully. Different distributions have been assumed for the random variables X and Y.

Explicit consideration of the relevance of $R = P(Y < X)$ seems to have begun with Birnbaum (1956). Subsequently, point estimation of R has been considered for several parametric families of distributions and under different assumptions. A good overview is presented by Johnson (1988) who specifically considered estimation in the Weibull case when the stress and strength distributions have a common shape parameter. Downton (1973) discussed the estimation of R in the normal case. Church and Harris (1970) have discussed the interval estimation of R in the normal case when the distribution of Y is completely specified. Owen et al. (1964) considered inference in the normal case when the parameters are unknown for both distributions but the variances are equal and equal sample sizes were used. Reiser and Gottman (1986) generalized the above result to account for unequal sizes and variances. Gottman et al. (1988) further generalized the normal results to include the presence of covariates for the stress Y. Constantine et al. (1976) treated the gamma case with known shape parameter. Enis and Geisser (1971) considered inference on R when X and Y are exponential using both the Baysian and sampling theory points of view.

The estimation of $R = P(Y < X)$ in the Burr Type X model has been referred to in the literature by Awad and Gharraf (1986), who provided a simulation study that compares three estimates for R. Those estimators are: the minimum variance unbiased, the maximum likelihood, and Bayes estimators. Also, the sensitivity of the Bayes estimator to the prior parameters was considered. Sartawi and Abu-Salih (1991), discussed Bayesian prediction bounds for the Burr Type X model. Ahmad et al. (1997), dealt with the estimation of R in the Burr Type X case, when maximum likelihood, Bayes and empirical techniques were used. Monte-Carlo simulation was carried out to compare the three methods of estimation. In addition to the comparison among the three estimators, some characterization of the Burr Type X distribution was presented. The first characteristic is based on the recurrence relationship between two successively conditional moments of a certain function of the random variable, whereas the second one was given by the conditional variance of that function. Surles and Padgett (1998) addressed the inference on R in Burr Type X case. Some properties of Burr Type X distribution were reviewed, and the existing and new results on estimation of R were discussed. A significance test for R was introduced based on the *MLE* of R. Also, Bayesian inference on R was presented when the parameters are assumed to have independent gamma distributions. An algorithm was given for finding the highest posterior density interval for R.

The value of R has been calculated in the above cited references under the assumed distributions and using different methods of estimations. None of the references has addressed the effect of the sample size and the values of the distribution parameters on the estimation of R. Shayib (2005) considered the effect of parameters ratio and sample sizes on the calculations of R. Haghighi and Shayib (2007) considered the preferred distribution function for estimating $P(Y < X)$ in the Burr Type X, Gamma and Weibull cases. McCool (1991) addressed the inference on $P(Y < X)$ and interval estimation for R, when the underlying distribution was taken to be the two–parameter Weibull distribution. He addressed the case when X and Y have the common, but unknown, shape parameter. In addition to that both samples for X and Y, sharing the same sample size, are Type II censored at the rth $(r \leq n)$ order statistic. Baklizi and Abu Dayyeh (2003) considered shrinkage estimation of R in the exponential case. Also, Baklizi and El-Masri (2004) considered shrinkage estimation of R in the exponential case with common location parameter.

In this paper, and in contrast to the setting in McCool (1991), we consider the case that the distributions of X and Y are independent, sharing the same known shape parameter α, but unknown and different scale parameters, i.e., X and Y are independent random variables with Weibull distributions Weibull(α, β_1) and Weibull(α, β_2), respectively. Moreover, the analysis will be carried without any censoring, i.e., on the complete data in the sample. The Weibull distribution is commonly used for as a model for life lengths and in the study of breaking strengths of materials because of the properties of its failure rate function. Resistors used in the construction of an aircraft guidance system have life time lengths that follow a Weibull distribution with $\alpha = 2$, and $\beta = 10$, (with measurements in thousands of hours). We also check, through simulation, on which method of estimation will give the best value for R. Section 2 will present the Weibull distribution model to be considered as the underlying assumption for the stress Y and strength X simultaneously. The methods of estimation that will be used include: *MLE*, *Method of Moments*, *Shrinkage Estimation Procedures* (see Thompson 1968), and *UMVUE*, are presented in Section 3. Tables I, III, and V correspond to the case when the ratio of the sample sizes $m/n = 1$, and the ratio of the scale parameters is 1, 1/4, and 4. Tables I, III, and V display the values of the estimators, their standard deviations, and the calculated *MSE*, based on 1000 runs. Tables II, IV, and VI show the ranking of the different estimators that were investigated in the report along with the sample sizes used. Conclusions and recommendations are given in the final section.

2. The Weibull Distribution

Why Weibull? The Weibull distribution is a continuous distribution that is used in a variety of situations such as life time. A common application of the Weibull distribution is to model the lifetimes of components such as bearings, ceramics, capacitors, and dielectrics. The Weibull distribution in model fitting is a strong competitor to the gamma distribution. Both the gamma and Weibull distributions are skewed, but they are valuable distributions for model fitting. Weibull distribution is commonly used as a model for life length because of the properties of its failure rate function $h(w) = (\alpha / \beta) w^{\alpha - 1}$. This failure rate function, for $\alpha > 1$, is a monotonically increasing function with no upper bound. This property gives the edge for the Weibull distribution over the gamma distribution, where the failure rate function is always bounded by $1/\beta$, when the probability density function for the gamma distribution is

$$g(y) = \frac{1}{\beta}\left(\frac{y}{\beta}\right)^{\alpha-1} \frac{e^{-y/\beta}}{\Gamma(\alpha)}, \qquad y > 0 \text{, where } \Gamma(\alpha) = \int_0^\infty e^{-u} u^{\alpha-1} du .$$

Another property that gives the edge for the Weibull distribution over the gamma distribution is by varying the values of α and β, a wide variety of curves can be generated. Because of this, the Weibull distribution can be made to fit a wide variety of data sets.

In this paper we will consider the Weibull distribution with two parameters, namely, α and β, where α is the shape parameter while $\beta^{1/\alpha}$ is the scale parameter of the distribution. The probability density function of the Weibull distribution we will use based on this setup is

$$f(w) = \begin{cases} \dfrac{\alpha}{\beta} w^{\alpha-1} e^{-\frac{w^{\alpha}}{\beta}}, & w, \alpha, \beta > 0, \\ 0, & otherwise. \end{cases} \tag{2.1}$$

We will use the notation $W \sim$ Weibull(α , β) for the random variable W having a Weibull probability distribution function with parameters α and β. We will also take the shape parameter as in Hogg and Tanis (1988) to be $1<\alpha < 5$. The expected value and the variance of W in this case are, respectively

$$E(\mathrm{W}) = \beta^{1/\alpha}\Gamma\left(1+\frac{1}{\alpha}\right), \tag{2.2}$$

and

$$Var(W) = \beta^{2/\alpha}\left\{\Gamma\left(1+\frac{2}{\alpha}\right)-\left[\Gamma\left(1+\frac{1}{\alpha}\right)\right]^2\right\}. \tag{2.3}$$

Let X and Y be two independent Weibull random variables with parameters (α, β_1) and (α, β_2), respectively. To conduct our simulation, due to the involved and tedious calculations of the *MLE* of α (see Johnson and Kotz, 1970, p. 255), we choose α = 2, 3, and 4, when the ratio of β_1 / β_2 (the α power of the ratio between the scale parameters of X and Y) is 1, 1/4 and 4. We choose the ratio of the scale parameters to compare the effect measures of the reliability. The sample sizes m for X and n for Y, are taken in the ratio of m/n = 1, 1/2, and 2.

3. Estimation Methods

Three estimation methods will be used; namely, the *MLE*, moments, and shrinkage.

3.1. Maximum Likelihood Estimation (*MLE*) Method

Maximum likelihood (together with some of its variants) is the most widely used method of estimation, and a list of its applications would cover practically the whole field of statistics,

(Lehmann, 1983). Due to the existence of a unique solution of the maximum likelihood equations (Lehmann, 1983, p. 437) and since the shape parameter, α , is given, we will use the *MLE* method to find the estimate of the parameters β_1 and β_2 based on the simulated sample values with an assigned sample sizes. Thus, the *MLE*'s $\hat{\beta}_1$ and $\hat{\beta}_2$ (and at the same time the *MVUE*) of β_1 and β_2 based on a sample sizes *m* and *n*, and the shape parameter α , are given by

$$\hat{\beta}_1 = \frac{1}{m}\sum_1^m x^\alpha \ , \ \hat{\beta}_2 = \frac{1}{n}\sum_1^n y^\alpha \ . \tag{3.1}$$

The *MLE* of *R* in this sequel will be denoted by $\hat{R}$. Johnson (1988) has shown that

$$R = P(Y < X) = \frac{1}{1 + \frac{\beta_2}{\beta_1}} = \frac{1}{1+\rho} \ , \tag{3.2}$$

where

$$\rho = \beta_2 / \beta_1 , \tag{3.3}$$

when the Weibull *pdf* is given in (2.1).

Note that, since we have

$$\hat{\rho} = \frac{\hat{\beta}_2}{\hat{\beta}_1}, \tag{3.4}$$

then

$$\hat{R} = \frac{1}{1+\hat{\rho}} . \tag{3.5}$$

Based on the fact that *R* is monotonic in ρ and so an inference on ρ is equivalent to the inference on *R*. McCool (1991) found confidence interval estimation for ρ .

In this paper we are interested in the selection of the method of estimation that will give the highest value of *R* in the Weibull distribution case, as it appears in (3.2). We will be taking α as given, and it will be assigned the value 2 for the simulation. Simulation results are presented in Tables I, III, and V.

Let $X_1, X_2, ..., X_m$ be a random sample for *X*, and $Y_1, Y_2, ..., Y_n$ be the corresponding random sample for *Y*. Table I has the estimated values of the different estimators, based on 1000 runs, with $m = n$, and m between 5 and 10 (inclusive) with increment of 1 denoted by $m = 5(1)10$. The ratio of β_1 and β_2 are taken to be 1, 1/4, and 4, representing values equal to, less than, and greater than 1. The computed *R* values (from (3.2)) for given values of the parameters are shown in the tables' headings for comparison with simulated values in the tables.

3.2. Shrinkage Estimation Procedures

In some applications, an experimenter often possesses some knowledge of the experimental conditions based on the behavior of the system under consideration, or from past experience, and is, thus, in a position to give an educated guess or an initial estimate of the parameter of interest. Given a prior estimate R_0 of R, we are looking for a so called "shrinkage estimators" that incorporates this information. Following Thompson (1968), we choose a guessed value for R_0, and compute a value of θ, denoted by θ_1, for the estimators of the form

$$\tilde{R} = \theta\hat{R} + (1+\theta)R_0, \qquad 0 \le \theta \le 1. \tag{3.6}$$

Then, using the p-value of the likelihood ratio test (*LRT*), we find three other values for θ denoted by θ_2, θ_3, and θ_4, by different methods. These estimators are given in the next two sections. The values of θ will be used to compute shrinkage estimators of R, denoted by R_1, R_2, R_3, and R_4. Then, we will compare these estimators with each other and with the *MLE* of R.

3.2.a. Shrinking Towards a Pre-specified Value of R, R_0

We choose R_0 such that θ_1 minimizes the mean square error (*MSE*) of the estimator $\tilde{R} = \theta\hat{R} + (1\text{-}\theta)R_0$. To do so, we first find

$$MSE\left(\tilde{R}_1\right) = E(\tilde{R} - R)^2 = E[\theta_1\hat{R} + (1-\theta_1)R_0 - R]^2. \tag{3.7}$$

Then, we will minimize (3.7) to find θ_1 which is

$$\theta_1 = \frac{(R - R_0)\, E(\hat{R} - R_0)}{E(\hat{R}^2) - 2R_0\, E(\hat{R}) + R_0^2}, \tag{3.8}$$

subject to $0 \le \theta \le 1$. Obviously, θ_1 in (3.8) depends on the unknown parameter R. By substituting $\hat{R}$ from (3.5) for R in (3.8), we get

$$\hat{\theta}_1 = \frac{(\hat{R} - R_0)\, E(\hat{R} - R_0)}{E(\hat{R}^2) - 2R_0\, E(\hat{R}) + R_0^2}. \tag{3.9}$$

Now, substituting $\widehat{\theta}_1$ from (3.9) in (3.6), we obtain our first shrinkage estimator as

$$\tilde{R}_1 = \hat{\theta}_1 \hat{R} + (1 - \hat{\theta}_1) R_0 \,. \tag{3.10}$$

The following theorem will yield approximate values of $E(\hat{R})$ and $Var(\hat{R})$, which are needed for (3.9).

Theorem 3.2.1. *Supposed* $X \sim \text{Weibull}(\alpha, \beta_1)$ *and* $Y \sim \text{Weibull}(\alpha, \beta_2)$, *as given in (2.1). Let U and T be independent identically distributed exponential random variables with parameters* β_1 *and* β_2, *and with pdf* $h(u) = (1/\beta_1)e^{-u/\beta_1}$ *and* $h(t) = (1/\beta_2)e^{-t/\beta_2}$, *respectively. Then,* $V = U/T$ *has probability density function*

$$g(v) = \begin{cases} \beta_1 \beta_2 (\beta_2 v + \beta_1)^{-2}, & v > 0 \\ 0, & otherwise. \end{cases} \tag{3.11}$$

Proof: The proof easily follows using the transformations of the variables. □

We consider two special case examples;

(1) Let $m = n$ and $\beta_2 / \beta_1 = 1$. Thus, from (3.4) and (3.5) we have $\hat{\rho} = v$, and $\hat{R} = 1/(1+v)$. Therefore from (3.11), for $\hat{\theta}_1$, we will have E($\hat{R}$) = 1/2, E($\hat{R}^2$) = 1/3, and Var($\hat{R}$) = 1/12.

(2) As in Baklizi and Abu Dayyeh (2003), we choose

$$V = \frac{\beta_1}{\beta_2}\left(\frac{\overline{Y}_n^{\alpha}}{\overline{X}_m^{\alpha}}\right), \tag{3.12}$$

as the ratio in Theorem 3.1, where $\overline{Y}^{\alpha}$ is the sample mean of n value for Y^{α}, and similarly for $\overline{X}^{\alpha}$. In addition, the sample sizes are not necessary equal. Probability distribution function of the random variable defined by (3.12) is $F_{2n,2m}$. From (3.4), (3.5) and Lindley (1969), we get

$$E(\hat{R}) = \left[1 + \frac{\beta_2}{\beta_1} E(V)\right]^{-1} + Var(V)\left[\left(\frac{\beta_2}{\beta_1}\right)^2 \left(1 + \frac{\beta_2}{\beta_1}\right) E(V)\right]^{-3}, \tag{3.13}$$

and

$$Var(\hat{R}) = Var(V)\left(\frac{\beta_2}{\beta_1}\right)^2 \left[1 + \frac{\beta_2}{\beta_1} E(V)\right]^{-2}, \tag{3.14}$$

where $E(V)$ and $Var(V)$ can be determined from the F-distribution. For numerical computation, the *MLE*'s of β_1 and β_2 from (3.1) can be used in (3.13) and (3.14). This will complete discussion on θ_1 given by (3.8).

3.2.b. Shrinking Using the p-Value of the LRT

In section 3.2.a. we found the value for θ, namely θ_1, and that established the first shrinkage estimator. In this subsection, we will pursue another approach to find θ values in order to establish the other shrinkage estimators, namely, R_2, R_3, and R_4. Now, we will find the other three values of θ, which will give us the other estimators by finding the *p*-value of the hypothesis in the Likelihood ratio test (*LRT*) procedure.

For testing

$$H_0: \ R = R_0 \ \text{vs.} \ H_1: \ R \neq R_0, \tag{3.15}$$

the *LRT* is of the form:

$$\text{reject } H_0 \text{ when } \frac{\overline{Y}^\alpha}{\overline{X}^\alpha} < C_1 \text{ or } \frac{\overline{Y}^\alpha}{\overline{X}^\alpha} > C_2, \tag{3.16}$$

where $\overline{Y}^\alpha$ is the sample mean of n value for Y^α, and similarly for $\overline{X}^\alpha$. In addition, the sample sizes are not necessary equal. This follows by noticing that (3.15) is equivalent to

$$H_0: \quad \beta_1 = \left[\frac{R_0}{1-R_0}\right]\beta_2 \quad vs. \quad H_1: \quad \beta_1 \neq \left[\frac{R_0}{1-R_0}\right]\beta_2. \tag{3.17}$$

Theorem 3.2.2. *Let* $X \sim \text{Weibull}(\alpha, \beta_1)$ *and* $Y \sim \text{Weibull}(\alpha, \beta_2)$, *as given in (2.1).*

i) $U = X^\alpha$ has an exponential distribution with parameter β_1 and pdf

$$h(u) = \begin{cases} \dfrac{e^{-u/\beta_1}}{\beta_1}, & u > 0, \\ 0, & otherwise. \end{cases}$$

ii) Letting $\overline{U}$ be the mean of U, $\dfrac{2m\overline{U}}{\beta_1} \sim \chi^2_{2m}$.

iii) $V = Y^\alpha$ has an exponential distribution with parameter β_2 and pdf

$$g(v) = \begin{cases} \dfrac{1}{\beta_2} e^{-v/\beta_2}, & v > 0 \\ 0, & otherwise, \end{cases}$$

iv) Letting $\bar{V}$ be the mean of V, $\frac{2n\bar{V}}{\beta_2} \sim \chi^2_{2n}$,

and

v) $\frac{\beta_1 \bar{V}}{\beta_2 \bar{U}} \sim F_{2n,2m}$.

Proof: Cases (i) and (iii) are clear by using transformation $U = X^{\alpha}$ and $V = Y^{\alpha}$ in distributions of X and Y, respectively. For cases (ii), (iv) and (v) see Baklizi and Abu Dayyeh (2003). □

Now, from (3.17) and Theorem 3.2.2, when $m = n$, we have

$$T \equiv \frac{R_0}{1-R_0} \frac{\bar{V}}{\bar{U}} \sim F_{2m,2m}. \tag{3.18}$$

Because of the skewness of the F-distribution and since we have a two-tailed test, the p-value (3.17) is

$$z = 2\ min\ [P_{H_0}\ (\mathrm{T} > t),\ P_{H_0}\ (T < t)] = 2\ min\ [1\text{-}\ F(t), F(t)], \tag{3.19}$$

where t is the observed value of the test statistic T that has an F distribution function under H_0. The p-value of this test indicates how strongly H_0 is supported by the data. A large p-value indicates that R is close to the prior estimate R_0 (Tse and Tso, 1996). Thus, from (3.6) we can use this p-value to form the shrinkage estimator

$$\tilde{R}_2 = \theta_2 \hat{R} + (1-\theta_2) R_0, \tag{3.20}$$

where θ_2 =1- p-value.

Based on the comments just made, to have the estimator even closer to R_0, we take the square root of the p-value. Thus, a third shrinking estimator is

$$\tilde{R}_3 = \theta_3 \hat{R} + (1-\theta_3) R_0, \tag{3.21}$$

where $\theta_3 = 1\text{-}\sqrt{p\text{-value}}$.

Now, to have the estimator far away from R_0, we take the square of the p-value such that

$$\tilde{R}_4 = \theta_4 \hat{R} + (1-\theta_4) R_0, \tag{3.22}$$

where θ_4 =1- $(p\text{-value})^2$, which is the forth shrinkage estimator.

3.3. Estimation by the Method of Moments

In reference to (2.2) and (2.3), and since α is assumed given, we have two ways of estimating β_1 (or β_2) using the first and the second moments. Using the sample means of X and Y for the first moment (2.2) from (3.4) we have

$$\hat{\rho} = \frac{\overline{Y}^{\alpha}}{\overline{X}^{\alpha}} = \frac{\hat{\beta}_2}{\hat{\beta}_1}, \tag{3.23}$$

Taking sample standard deviations of random variables X and Y (using square root of (2.3)) and again from (3.4) we have

$$\hat{\rho} = \frac{S_2^{\alpha}}{S_1^{\alpha}} = \frac{\hat{\beta}_2}{\hat{\beta}_1}. \tag{3.24}$$

Based on (3.23) and (3.24), we have two more estimators to consider. These are,

$$\hat{R}_5 = \frac{1}{1+\hat{\rho}} = \frac{1}{1+\dfrac{\overline{Y}_n^{\alpha}}{\overline{X}_m^{\alpha}}} \tag{3.25}$$

and

$$\hat{R}_6 = \frac{1}{1+\hat{\rho}} = \frac{1}{1+\dfrac{S_2^{\alpha}}{S_1^{\alpha}}}. \tag{3.26}$$

4. Performance of the Estimators

It is well known that the Mean Square Error (*MSE*) of an estimator with respect to the estimated parameter θ is defined as

$$MSE\left(\hat{\theta}\right) = E\left[\left(\hat{\theta}-\theta\right)^2\right] = Var\left(\hat{\theta}\right) + \left[Bias\left(\hat{\theta},\theta\right)\right]^2, \tag{4.1}$$

where $\hat{\theta}$ is the estimator of θ. In that sense, the *MSE* assesses the quality of the estimator in terms of its variation and unbiasness.

A simulation study is conducted to investigate the performance of the estimators $\tilde{R}_1$, $\tilde{R}_2$, $\tilde{R}_3$, and $\tilde{R}_4$ in relation to the *MLE* and using the θ values found in subsections 3.2.a and 3.2.b. In addition to the 4 shrinkage estimators, R_1, R_2, R_3, and R_4, as expressed in (3.10) and (3.20) – (3.22) and the *MLE* of R, there are two more that are based on the method of moments. These estimators are denoted by R_5 and R_6 expressed in (3.23) and (3.24), respectively.

The simulation was done for 3 cases as follows:

1. The true value of $R = 0.5$, $R_0 = 0.45$, and $\beta_2 / \beta_1 = 1$.
2. The true value of $R = 0.2$, $R_0 = 0.25$, and. $\beta_2 / \beta_1 = 4$.
3. The true value of $R = 0.8$, $R_0 = 0.75$, and $\beta_2 / \beta_1 = 1/4$.

Table I. Weibull Distribution,
The Estimated Average Values of the Estimators, Standard Deviation and MSE
$m = n$, $\beta_1 / \beta_2 = 1$, $\alpha = 2$, $R = 1/2$, $R_0 = .45$

m

Estimator	5	6	7	8	9	10
$R\hat{}$	0.4996	0.4989	0.4973	0.4692	0.5021	0.4992
	0.1515	0.1393	0.1284	0.1249	0.1127	0.1082
	0.0230	0.0194	0.0165	0.0166	0.0127	0.0117
R_1	0.4582	0.4564	0.4607	0.4552	0.4553	0.4552
	0.0271	0.0195	0.0142	0.0148	0.0139	0.0130
	0.0025	0.0023	0.0017	0.0022	0.0022	0.0022
R_2	0.4876	0.4875	0.4715	0.4877	0.4900	0.4924
	0.1280	0.1189	0.0976	0.1030	0.0996	0.0962
	0.0165	0.0143	0.0104	0.0108	0.0100	0.0093
R_3	0.4786	0.4779	0.4576	0.4781	0.4802	0.4628
	0.1010	0.0941	0.0748	0.0818	0.0801	0.0980
	0.0107	0.0093	0.0074	0.0072	0.0068	0.0110
R_4	0.4933	0.4940	0.4860	0.4940	0.4964	0.4986
	0.1437	0.1303	0.1157	0.1148	0.1100	0.1060
	0.0207	0.0177	0.0136	0.0132	0.0121	0.0112
R_5	0.4983	0.4967	0.4944	0.4964	0.4974	0.4975
	0.1565	0.1429	0.1353	0.1271	0.1188	0.1123
	0.0245	0.0204	0.0183	0.0162	0.0141	0.0126
R_6	0.4966	0.4967	0.4938	0.4981	0.4987	0.4995
	0.2230	0.2031	0.1916	0.1792	0.1714	0.1631
	0.0497	0.0412	0.0368	0.0321	0.0294	0.0266

Table II. The ranking of the estimators with $m = n$, $m = 5(1)10$
$m = n$, $\beta_1 / \beta_2 = 1$, $\alpha = 2$, $R = 1/2$, $R_0 = .45$

Samples size	How they rank
5	$R_1, R_3, R_2, R_4, \hat{R}, R_5, R_6$
6	$R_1, R_3, R_2, R_4, \hat{R}, R_5, R_6$
7	$R_1, R_3, R_2, R_4, \hat{R}, R_5, R_6$
8	$R_1, R_3, R_2, R_4, R_5, \hat{R}, R_6$
9	$R_1, R_2, R_3, R_4, R_5, \hat{R}, R_6$
10	$R_1, R_2, R_3, R_4, \hat{R}, R_5, R_6$

Table III. Weibull Distribution, The Estimated Average Values of the Estimators, Standard Deviation and MSE

$m = n$, β_1 / β_2=1/4, α = 2, R= 0.2, R_0=.25

m

Estimator	5	6	7	8	9	10
$\hat{R}$	0.2201	0.2169	0.2105	0.2099	0.2107	0.2111
	0.0125	0.0978	0.0952	0.0881	0.0793	0.0748
	0.0006	0.0099	0.0092	0.0079	0.0064	0.0057
R_1	0.2249	0.2496	0.2489	0.2029	0.2487	0.2486
	0.1031	0.0006	0.0047	0.0187	0.0048	0.0044
	0.0125	0.0025	0.0024	0.0004	0.0024	0.0023
R_2	0.2261	0.2235	0.2177	0.2178	0.2184	0.2185
	0.0892	0.0816	0.0818	0.0783	0.0682	0.0642
	0.0086	0.0072	0.0070	0.0061	0.0050	0.0045
R_3	0.2319	0.2301	0.2252	0.2255	0.2263	0.2261
	0.0693	0.0640	0.0656	0.0621	0.0555	0.0519
	0.0058	0.0050	0.0049	0.0045	0.0038	0.0034
R_4	0.2223	0.2192	0.2129	0.2128	0.2133	0.2137
	0.1013	0.0920	0.0906	0.0842	0.0755	0.0712
	0.0108	0.0088	0.0084	0.0073	0.0059	0.0053
R_5	0.2201	0.2169	0.2107	0.2099	0.2107	0.2166
	0.1079	0.0978	0.0819	0.0881	0.0793	0.0784
	0.0120	0.0099	0.0068	0.0079	0.0064	0.0063
R_6	0.2482	0.2393	0.2210	0.2131	0.2214	0.2245
	0.1829	0.1640	0.0699	0.0696	0.0604	0.1161
	0.0358	0.0284	0.0053	0.0050	0.0041	0.0141

The shrinkage estimators seem to perform better for small sample sizes. This is expected, as the sample size increases, the precision of the *MLE* increases, whereas shrinkage estimators are still affected by the prior guess R_0 (see Baklizi and Abu Dayyeh, 2003).

Table IV. The ranking of the estimators with $m = n$, m = 5(1)10

$m = n$, β_1 / β_2=1/4, α = 2, R= 0.2, R_0= 0.25

Samples size	How they rank
5	$\hat{R}, R_3, R_2, R_4, R_1, R_5, R_6$
6	$R_1, R_3, R_2, R_4, \hat{R}, R_5, R_6$
7	$R_1, R_3, R_6, R_5, R_2, R_4, \hat{R}$
8	$R_1, R_3, R_6, R_2, R_4, \hat{R}, R_5$
9	$R_1, R_3, R_6, R_2, R_4, \hat{R}, R_5$
10	$R_1, R_3, R_2, R_4, \hat{R}, R_5, R_6$

For each combination of $m = n$, R, and R_0 , 1000 samples were generated for X and for Y. The results of the simulation are shown in Tables I, III, and V, with the average values of the estimators, the average standard deviations. *MSE*s of all the estimators are tabulated in Tables I, III, and V. The corresponding ranking of the estimators are given in Tables II, IV and VI.

Table V. Weibull Distribution,
The Estimated Average Values of the Estimators, Standard Deviation and MSE
$m = n$, $\beta_1 / \beta_2 = 4$, $\alpha = 2$, $R = 0.8$, $R_0 = 0.75$

m

Estimator	5	6	7	8	9	10
$\hat{R}$	0.7796	0.7839	0.7843	0.7873	0.7872	0.7883
	0.1129	0.1014	0.0954	0.0873	0.0834	0.0789
	0.0132	0.0105	0.0093	0.0078	0.0071	0.0064
R_1	0.7502	0.7507	0.7513	0.7560	0.7500	0.7705
	0.0027	0.0051	0.0076	0.0249	0.0025	0.0246
	0.0025	0.0026	0.0024	0.0026	0.0025	0.0015
R_2	0.7736	0.7765	0.7775	0.7803	0.7808	0.7821
	0.0950	0.0860	0.0818	0.0749	0.0718	0.0980
	0.0097	0.0079	0.0072	0.0060	0.0055	0.0099
R_3	0.7681	0.7701	0.7710	0.7731	0.7739	0.7751
	0.0750	0.0682	0.0653	0.0602	0.0575	0.0551
	0.0066	0.0055	0.0051	0.0043	0.0040	0.0037
R_4	0.7773	0.7808	0.7817	0.7847	0.7849	0.7862
	0.1066	0.0961	0.0909	0.0832	0.0796	0.0753
	0.0119	0.0096	0.0086	0.0072	0.0066	0.0059
R_5	0.7796	0.7839	0.7843	0.7873	0.7873	0.7883
	0.1129	0.1014	0.0954	0.0873	0.0834	0.0789
	0.0132	0.0105	0.0093	0.0078	0.0071	0.0064
R_6	0.7498	0.7552	0.7604	0.7695	0.7696	0.7711
	0.1907	0.1579	0.1501	0.1325	0.1326	0.1250
	0.0389	0.0289	0.0241	0.0185	0.0184	0.0165

Table VI. The ranking of the estimators with $m = n$, $m = 5(1)10$
$m = n$, $\beta_1 / \beta_2 = 4$, $\alpha = 2$, $R = 0.8$, $R_0 = 0.75$

Samples size	How they rank
5	$R_1, R_3, R_2, \hat{R}, R_5, R_6, R_4$
6	$R_1, R_3, R_2, \hat{R}, R_5, R_6, R_4$
7	$R_1, R_3, R_2, \hat{R}, R_5, R_6, R_4$
8	$R_1, R_3, R_2, \hat{R}, R_5, R_6, R_4$
9	$R_1, R_3, R_2, \hat{R}, R_5, R_6, R_4$
10	$R_1, R_3, \hat{R}, R_5, R_{2,} R_6, R_4$

5. Conclusions

Industry is data rich and yet it is very often that decisions are made based on 2, and, in most cases, on less than 6, readings. Our discussion in this paper was done based on the Weibull distribution. We, in this paper, found different estimators of R, namely $\hat{R}_1, \hat{R}_2, \hat{R}_3, \hat{R}_4, \hat{R}_5, \hat{R}_6$. Using the MSE values as a measure of performance in comparing these estimators, we noted in ranking the estimators, $\hat{R}_1$ stayed at the top followed by MLE of R and the method of moments at the last choice. There was one exception to this conclusion and that was when $m = n = 8$. However, among all 6 shrinkage estimators, R_1 was the best.

References

Ahmad, K. E., Fakhry, M. E. and Jaheen, Z. F. (1997). Empirical Bayes' estimation of $P(Y < X)$ and characterizations of Burr-type X model. *Journal of Statistical Planning and Inference* **64**, 297-308.

Awad, A. M. and Gharraf, M. K. (1986). Estimation of $P(Y < X)$ in the Burr case: a Comparative Study, *Communications in Statistics, Simulation and Computation* **15**: 389-403.

Baklizi, A. and Abu Dayyeh, W. (2003). Shrinkage Estimation of $P(Y < X)$ in the Exponential Case. *Communications in Statistics, Simulation and Computation* **32**(1): 31-42.

Baklizi, A. and El-Masri, A. Q. (2004). Shrinkage Estimation of $P(Y < X)$ in the Exponential Case, with a common location parameter. *Metrika* **99**, 163-171.

Birnbaum, Z. W. (1956). On a use of the Mann-Whitney statistic, Proc. T*hird Berkeley Symp. Math. Statist. Probab,* Vol. 1, Univ. of California Press, 13-17.

Church, J. D. and Harris, B. (1970). The estimation of reliability from stress-strength relationships. *Technometrics* **12**, 49-54.

Cohen, A. C. (1965). *Maximum likelihood estimation in the Weibull distribution based on complete and censored samples*, **7**, 579-588.

Constantine, K., Karson, H., and Tse, R. (1986). "Estimation of $P(Y < X)$ in the gamma case". Communications in Statistics, *Simulation and Computation* **15**: 365-88.

Downton, F. (1973). The estimation of $P (Y < X)$ in the normal case. *Technometrics* **15**, 551-558.

Enis, P. and Geisser, S. (1971). "Estimation of the probability that Y < X". *JASA* **66**, 162-168.

Guttman, I., Johnson, B., Bhattacharyya, G., and Reiser, R. (1988). Confidence limits for stress-Strength models with explanatory variables. *Technometrics* **30**, 361-68.

Haghighi, A. M. and Shayib, M. A. (2007). Preferred distribution function for estimating $P(Y<X)$. *JSM Proceedings*, 1981- 1985.

Hogg, R. V. and Tanis, E. A. (1988). *Probability and Statistical Inference*, 3rd Edition, MacMillan.

Johnson, R. A. (1988). "Stress-Strength models for reliability" in Handbook of Statistics, P. R. Krismaiah and C. R. Rao, eds. *Elsevier*, **27-54.**

Johnson, N. L. and Kotz, S. (1970). *Continuous Univariate Distributions* – 1. John Wiley & Sons, Inc.

Lehmann, E. L. (1983). *Theory of Point Estimation,* John Wiley & Sons, Inc.

Lindley, D. V. (1969). *Introduction to Probability and Statistics from a Bayesian Viewpoint.* Vol. 1, Cambridge University Press.

McCool, J. I. (1991). Inference on P(Y < *X*) in the Weibull case. Communications in Statistics, *Simulation and Computation* **20**(1): 129-148.

Owen, D. B., Craswell, K .J., and Hanson, D. L. (1964). "Nonparametric upper confidence bounds for P(Y < X) when X and Y are normal." *JASA* **50**, 906-924.

Reiser, R., and Guttman, I. (1986). Statistical inference for P(Y < X), the normal case. *Technometrics* **28**, 253-257.

Sartawi, H.A., and Abu-Salih, M.S. (1991). Bayesian prediction bounds for the Burr Type X model. *Communications Statistics-Theory and Methods* **20**, 1307-1330.

Shayib, M. A. (2005). Effects of Parameters and Sample Size on the Estimation of *P*(*Y* < *X*), A Simulation Study, Part I. *JSM Proceedings*, **142 – 144**.

Surles, J.G. and Padgett, W.J. (1998). Inference for *P* (*Y* < *X*) in the Burr type *X* Model. *Journal of Applied Statistical Science* **7**, 225-238.

Thompson, J. (1968). Some shrinkage techniques for estimating the mean. *JASA* **63**,113 – 122

Tse, S. and Tso, G. (1996). Shrinkage estimation of reliability for exponentially distributed lifetimes, Communications in Statistics. *Simulation and Computation* **25**(2): 415-430.

In: New Developments in Applied Statistics
Editor: Mohammad Ahsanullah

ISBN: 978-1-61324-648-1

Chapter 19

Reliability Analysis: A Curve of Concentration Approach

***Rishideep Roy*[1,a] *and Dilip Roy*[2,b]**
[1]Indian Statistical Institute, Calcutta, Kolkata- 700108, WB, India
[2]Department of Business Administration, Burdwan University
Golapbag, Burdwan- 713104, WB, India

Summary

The curve of concentration approach for examining the extent of variation and/ or inequality in a system has been extended for reliability analysis by replacing the incomplete first order moment by the standardized cumulative reliability value. In this process, its connection with the life distribution classes has been explored in general and also with specific reference to the Weibull distribution. Characterization results for the exponential distribution have been reported based on zero concentration and zero reliability index, and those for the Lomax and related distributions have been studied based on area equivalence with the Gini's measure. Further, weighted distribution has been considered for an extended development of the curve of concentration.

Keywords and phrases: Curve of concentration, survival function, reliability, failure rate, characterization, weighted distribution.

1. Introduction

Gini (1912)'s coefficient of concentration is an alternative to the coefficient of variation for measuring the standardized variability of any given distribution. It arises in a natural way (see Stuart and Ord, 1994) when one draws the curve of concentration with cumulative distribution function (cdf) plotted along the abscissa and the incomplete first moment plotted along the ordinate. Then the area between the equiangular line and the curve of concentration

[a] E-mail address: rishideeproy@gmail.com. (Corresponding author)
[b] E-mail address: dr.diliproy@gmail.com

is twice the coefficient of concentration due to Gini. Glasser (1962) obtained variance formulas for the mean difference and the coefficient of concentration.

Gini's coefficient of concentration has found its extensive use in socio-economic studies where level of development is being examined through the degree of inequality in respect of any socio-economic measure of importance. There are, however, some other measures of inequality (see Sen, 1973) including Bonferroni's inequality index which are also being used in practice.

In particular, the Gini's coefficient of concentration can be applied on the income distribution to measure the degree of income inequality. This is a well-known area where Gini's coefficient has its widest application. Of late, Gini's coefficient is being used for examining inequalities in other fields too. Recent trend of its application is in health inequity. Equity in health is one of the basic values that guide the health Organization's technical cooperation with the states. There is a fundamental difference between inequities and inequalities. Inequities represent unjust and avoidable inequalities. As a result, to identify inequities in health the first step is to measure health inequalities. It is essential to carry out these analyses for reducing the inequities in the health profile of a Region. There exists a good number of summary measures for this purpose. But the Gini Coefficient, along with the Concentration Index, has been taken from the field of economics and applied to the study of health inequalities with success (see Wagstaff *et* al, 1991; Brown, 1994).

In the field of plant science, Damgaard, and Weiner (2000) described inequality in plant size or fecundity based on an inequality measure. In the field of demography, Shkolnikov *et al* (2003) have defined Gini coefficient and four other inequality measures on the basis of length-of-life distribution for measuring and analyzing inter-individual inequality in length of life. Thus, Gini's coefficient has been used as a life table function too. In view of its conceptual naturalness and wide applicability in measuring the departure from a defined state of equality, various attempts have been made in the recent past to generalize this concept in the multidimensional framework (see Arnold, 2005).

The current work aims to define the curve of concentration in the reliability context and to propose the corresponding area/ coefficient of concentration and reliability index. Section 2 deals with these new definitions along with their interconnections with the monotonic failure rate based life distribution classes. In section 3 we present characterization results using zero coefficient of concentration and zero reliability index. Also, characterization results based on area equivalence between the reliability-based coefficient of concentration and the Gini's coefficient of concentration have been reported. The curve of concentration concept has been further generalized in terms of the weight function and the weighted distribution.

2. Basic Concepts

Let X be a nonnegative random variable with probability density function (pdf) as f(x), cdf as F(x), and survival function or reliability function as S(x). Then the failure rate, r(x), is given by

$$r(x)=f(x)/S(x),$$

and the incomplete first order moment up to the point x is given by

$$\Phi(x) = \frac{1}{\mu}\int_0^x uf(u)du, \quad (2.1)$$

where μ is the expectation (or, the first order moment) of X.

Following Gini (1912), the curve of concentration is given by (F(x), Φ(x)), $0 \leq F(x) \leq 1$. The corresponding area A_Φ, between the concentration curve and the equiangular line F(x)=Φ(x) works out as

$$A_\Phi = \frac{1}{2}\left[\int_0^1 F d\Phi - \int_0^1 \Phi dF\right], \quad (2.2)$$

which simplifies to

$$A_\Phi = \frac{1}{4\mu}\int_0^\infty \int_0^\infty |x - y| f(x)f(y)dx\,dy. \quad (2.3)$$

Thus, A_Φ is half of the Gini's coefficient of concentration defined by

$$\Delta = \frac{1}{2\mu}\int_0^\infty \int_0^\infty |x - y| f(x)f(y)dx\,dy. \quad (2.4)$$

We would like to extend this concept of curve of concentration and the area of concentration for reliability analysis. For this, let us define the standardized cumulative reliability as

$$\Psi(x) = \frac{1}{\mu}\int_0^x S(u)du. \quad (2.5)$$

It is easy to note that Ψ(x) lies between 0 and 1 and is an increasing function in x. We can therefore define a curve of concentration in terms of (F(x), Ψ(x)), $0 \leq F(x) \leq 1$. To distinguish it from the Gini's curve of concentration, let us refer it by the Reliability Curve of Concentration (RCC). The area between the RCC and the equiangular line F(x)=Ψ(x) will be denoted by

$$A_\Psi = \frac{1}{2}\left[\int_0^1 F d\Psi - \int_0^1 \Psi dF\right]. \quad (2.6)$$

If the RCC falls below the equiangular line the value of A_Ψ will be nonnegative. In case the RCC lies above the equiangular line, the value of A_Ψ will be non-positive. The maximal vertical deviation between the RCC and the equiangular line can be obtained from the equation

$$\frac{d\Psi}{dF} = 1.$$

It is that value of x where failure rate is equal to the reciprocal of μ. Thus, in terms of Reliability Index (RI), R_Ψ , a generalization of Pietra index (1915) can be proposed as

$$R_\Psi = E[1/r(X) - \mu]/2\mu,$$

by defining the failure rate in the stochastic domain. Earlier, failure rate in the stochastic domain has been studied as a transformation by Mukherjee and Roy (1987) for the Weibull distribution and by Roy (1993) for characterizing different life distributions.

For a closer look at the nature of RCC, let us examine this curve for a two-parameter Weibull distribution with pdf given by

$$f(x) = \lambda\alpha x^{\alpha-1}\exp(-\lambda x^{\alpha}),\ 0 < \alpha,\ 0 < \lambda,\ 0 \le x < \infty.$$

The standardized cumulative reliability is

$$\Psi(x) = \frac{1}{\mu}\int_0^x \exp(-\lambda u^{\alpha})du,$$

where $\mu = E(X) = (1/\alpha)(1/\lambda)^{1/\alpha}\Gamma(1/\alpha)$.

Then the area under the RCC of the Weibull distribution works out as

$$A_\Psi = \frac{1}{2\mu}[\int_0^\infty [1-\exp(-\lambda x^{\alpha})]\exp(-\lambda x^{\alpha})dx - \int_0^\infty \int_0^x \exp(-\lambda u^{\alpha})du\,\lambda\alpha x^{\alpha-1}\exp(-\lambda x^{\alpha})dx]$$

$$= \frac{1}{2\mu}[\{\mu - \mu(1/2)^{1/\alpha}\} - \int_0^\infty \{\int_u^\infty \lambda\alpha x^{\alpha-1}\exp(-\lambda x^{\alpha})dx\}\ \exp(-\lambda u^{\alpha})du]$$

$$= \frac{1}{2\mu}[\{\mu - \mu(1/2)^{1/\alpha}\} - \mu(1/2)^{1/\alpha}]$$

$$= \frac{1}{2} - \left(\frac{1}{2}\right)^{1/\alpha}. \tag{2.7}$$

Thus, for the Weibull distribution

$$A_\Psi = 0 \text{ if } \alpha = 1$$
$$> 0 \text{ if } 0 < \alpha < 1$$
$$< 0 \text{ if } \alpha > 1.$$

In other words, Decreasing Failure Rate (DFR) Weibull distribution has positive area of concentration, Increasing Failure Rate (IFR) Weibull distribution has negative area of concentration, and Constant Failure Rate (CFR) Weibull distribution has zero area of concentration. These above observations can be easily generalized to obtain equivalent definitions for the IFR and the DFR classes of life distributions.

RESULT 2.1. RCC is convex (or concave) to F-axis if and only if the corresponding distribution belongs to DFR (or IFR) class of life distribution.
Proof. Note that □

$$\mu\frac{d\Psi}{dF}=\frac{S(x)}{f(x)}=\frac{1}{r(x)} \tag{2.8}$$

and hence from (2.8) we get

$$\mu\frac{d^2\Psi}{dF^2}=-\frac{r'(x)}{r^2(x)}\frac{1}{f(x)}. \tag{2.9}$$

Since in (2.9) both f(x) and r(x) are positive quantities RCC is convex to F-axis if and only if $r'(x)$ is less than zero, i.e., if and only if F belongs to DFR class of life distributions. Similarly, RCC is concave to F-axis if and only if $r'(x)$ is greater than zero, i.e., if and only if F belongs to IFR class of life distributions. ĩ

Remark. It may be noted that the RCC passes through the points (0, 0) and (1, 1) and hence convexity of RCC with respect to F-axis implies positive area of concentration and concavity of RCC with respect to F-axis implies negative area of concentration.

3. Characterization and Other Results

In view of the above discussions on the RCC, let us now examine the situation where the area of concentration under RCC vanishes. Our objective is to determine the distribution for which survival equality is perfect, i.e., survival inequality is nil. The following characterization result is an outcome of the said search. It uniquely determines the exponential distribution under perfect survival equality.

RESULT 3.1. Within the class of monotonic failure rates, the area under the RCC, A_Ψ, vanishes if and only if the underlying distribution is of the exponential form.

Proof. If part of the result follows from (2.7) under the fact that for $\alpha=1$ the Weibull distribution reduces to the exponential form. □

To prove the only if part let us consider $A_\Psi=0$. But from the definition and the measure given at (2.6) we have

$$A_\Psi=\frac{1}{2}\left[\int_0^1 F d\Psi-\int_0^1 \Psi dF\right]$$

$$=\frac{1}{2\mu}[\int_0^\infty S(x)\int_0^x f(u)dudx - \int_0^\infty\int_0^x S(u)dufdx\ ]$$

$$=\frac{1}{2\mu}[\int_0^\infty\int_0^x \{S(x)f(u)\ - S(u)f(x)\}d\ udx\ ] \tag{3.1}$$

Alternatively, A_Ψ can be expressed as

$$A_\Psi=\frac{1}{2\mu}[\int_0^\infty\int_0^x \{r(u)-r(x)\}S(u)S\ (x)dudx\ ]. \tag{3.2}$$

But from the symmetry of the integrand {S(x)f(u)-S(u)f(x)}

$$\int_0^\infty\int_0^\infty \{S(x)f(u)\ - S(u)f(x)\}d\ udx = 0 \tag{3.3}$$

Hence, using (3.3) in (3.2) we get under monotonic failure rate

$$A_\Psi=\frac{1}{4\mu}[\int_0^\infty\int_0^\infty |r(u)\ - r(x)|S(u)S(x)du\ \ dx]. \tag{3.4}$$

Using the given condition that A_Ψ=0 in (3.4) we get

$$\frac{1}{4\mu}[\int_0^\infty\int_0^\infty |r(u)\ - r(x)|S(u)S(x)du\ \ dx]=0.$$

This implies that r(u)=r(x) for all choices of x and u (x, u≥0). Thus, r(x) is necessarily equal to a constant for all x≥0. This in turn implies that the underlying survival function is necessarily of the form

$$S(x)=\exp(-\lambda x),\quad \lambda>0, \text{ for all } x\geq 0.$$

Hence follows the result.

The above result upholds the fact that in survival analysis in the monotonic failure rate classes, the state of perfect survival equality is attained if and only if the life distribution is of the exponential form. This also explains the fact why the exponential distribution is the natural choice in partial ordering of the life distributions (see Shaked and Shanthikumar, 1994) and in defining the NBUE, NWUE, HNBUE, HNWUE and L classes of life distributions (see Barlow. and Proschan, 1975; Kim and Proschan, 1988).

It is easy to note the following characterization of the exponential distribution in terms of the Reliability Index.

RESULT 3.2. The Reliability Index, R_Ψ, is zero if and only if the underlying distribution is exponential.

Proof. follows from the fact that R_Ψ is zero if and only if the failure is a constant and the failure is a constant if and only if the underlying distribution is exponential. □

The following results are of importance when we try to examine the difference between Gini's measure of concentration, A_Φ, and the reliability measure of concentration, A_Ψ. Since we have ensured that for a DFR distribution A_Ψ is positive and for an IFR distribution A_Ψ is negative, one may be interested to know when $A_\Phi = A_\Psi$ or when $A_\Phi = -A_\Psi$. The following two results present sufficient conditions for these likely situations within the DFR and IFR classes.

RESULT 3.3. For a nonnegative distribution, $A_\Phi = A_\Psi$ if it has an inversely linear failure rate and is of the

(i) Lomax form with survival function

$$S(x)=(1+x/k)-1,\ x\geq 0,\ k>0,\ \text{or}, \tag{3.5}$$

(ii) Pareto form with survival function

$$S(x)=c/x,\ x\geq c>0,\ \text{or} \tag{3.6}$$

with a shift in location whenever needed.

Proof. To ensure the result, let us consider a nonnegative distribution with survival function given by S(x). Then, from (3.1) we have □

$$A_\Psi = \frac{1}{2\mu}\left[\int_0^\infty \int_0^x \{S(x)f(u) - S(u)f(x)\}d\ udx\right].$$

$$= \frac{1}{2\mu}\left[\int_0^\infty \int_0^x \{S(x)/f(x) - S(u)/f(u)\}\ f(x)f(u)du\ dx\right] \tag{3.7}$$

We have on the other hand

$$A_\Phi = \frac{1}{2\mu}\left[\int_0^\infty \int_0^x \{x-u\}f(u)f(x)\ \}dudx\right]. \tag{3.8}$$

Thus, by comparing (3.7) with (3.8) we obtain a sufficient condition for $A_\Phi = A_\Psi$ as

$$S(x)/f(x)=k+x$$

Or,

$$r(x)=[k+x]^{-1}, \tag{3.9}$$

in the range {x f(x)>0 }, for some constant k. For k>0, we get the unique determination of the survival function as

$$S(x) = \exp[-\log\{(k+x)/k\}],$$

which simplifies to (3.5).

For k=0, to ensure (3.9) to be defined we need to have $x \geq c > 0$. Then, we get the unique determination the survival function as

$$S(x) = \exp[-\log(x/c)],$$

which simplifies to (3.6). For a choice of $k<0$ the distribution is again of Pareto form except for a shift in location. These prove the result. ◊

Thus, for equality between A_Φ and A_Ψ we may consider some specific choices of distributions and the Lomax distribution is one of them. It may be worth mentioning that the Lomax distribution plays a vital role in many life distributional studies and one reason for the same has been indicated in Roy (2004) in terms of generalized lack of memory property of the Lomax distribution.

RESULT 3.4. For a nonnegative distribution, $A_\Phi = -A_\Psi$ if it has an inversely linear failure rate and is of the

(i) uniform type with survival function

$$S(x) = (c-x)/c, \quad c \geq x \geq 0. \tag{3.10}$$

Proof. To ensure the result, let us consider a nonnegative distribution with survival function given by S(x). Then, from (3.7) we have □

$$A_\Psi = \frac{1}{2\mu}\left[\int_0^\infty \int_0^x \{S(x)/f(x) - S(u)/f(u)\}\ f(x)f(u)du\ dx\right]$$

From (3.8), we have on the other hand

$$A_\Phi = \frac{1}{2\mu}\left[\int_0^\infty \int_0^x \{x - u\} f(u)f(x)\ \} dudx\ \right].$$

Thus, by comparing the above two formulas we obtain a sufficient condition for $A_\Phi + A_\Psi = 0$ as

$$S(x)/f(x) = c-x, \quad c \geq x \geq 0$$

$$r(x) = [c-x]^{-1}, \quad c \geq x \geq 0 \tag{3.11}$$

for some constant $c>0$. Thus, we get the unique determination of the survival function as

$$S(x) = \exp[-\log\{c/(c-x)\}],$$

which simplifies to (3.10). This proves the result. ◊

RESULT 3.5.

(i) Under the monotonicity of $\{1/r(x)+x\}$, $A_\Phi + A_\Psi = 0$ if and only if the underlying distribution is uniform with survival function of the form (3.10).
(ii) Under the monotonicity of $\{1/r(x)-x\}$, $A_\Phi = A_\Psi$ if and only if the underlying distribution is either Lomax with survival function of the form (3.5) or Pareto with survival function of the form (3.6) except for a shift in location.

Proof.

(i) To prove the if part we note from (3.10) that

$$A_\Psi = \frac{1}{2\mu}[\int_0^\infty \int_0^x \{u - x\}\{1/c^2\}dudx]$$

$$= -\frac{1}{2\mu}[\int_0^\infty \int_0^x \{x - u\}f(u)f(x)\ dudx]$$

$$= -A_\Phi.$$

Thus, $A_\Phi + A_\Psi = 0$

To prove the only if part let us consider $A_\Phi + A_\Psi = 0$, which implies

$$\frac{1}{2\mu}[\int_0^\infty \int_0^x \{S(x)f(u) - S(u)f(x)\}d\ udx] + \frac{1}{2\mu}\int_0^\infty \int_0^x \{x - u\}f(x)f(u)du\ dx = 0$$

Or,
$$\int_0^\infty \int_0^x [\{1/r(x) - 1/r(u)\} + \{x - u\}]f(u)f(x)dudx = 0$$

Or,
$$\int_0^\infty \int_0^\infty |\{1/r(x) - 1/r(u)\} + \{x - u\}|f(u)f(x)du\ dx = 0,$$

under the symmetry of the integrand and the monotonicity of $\{1/r(x)+x\}$.

Thus, we have as a necessary condition

$$\{1/r(x) - 1/r(u)\} + \{x - u\} = 0, \text{ for all x and u.}$$

or,
$$r(x) = 1/\{c - x\},$$

where c is a positive constant. Thus, S(x) is necessarily of the form (3.10).

(ii) The if part is easy to verify. To prove the only if part we have from $A_\Phi = A_\Psi$,

$$A_\Phi - A_\Psi = 0$$

Or,
$$\int_0^\infty \int_0^x [\{1/r(x) - 1/r(u)\} - \{x - u\}]f(u)f(x)dudx = 0$$

Or,
$$\int_0^\infty \int_0^\infty |\{1/r(x) - 1/r(u)\} - \{x - u\}|f(u)f(x)du\ dx = 0,$$

under the symmetry of the integrand and the monotonicity of {1/r(x)-x}.

Thus, we have as a necessary condition

$$\{1/r(x) - 1/r(u)\} - \{x-u\} = 0 \text{ , for all x and u.}$$

or, $$r(x) = 1/\{x+k\}, \text{ for some k, a constant.}$$

Rest of the proof follows along the lines of proof of result 3.3.

4. Extension to Weighted Distribution

Let us extend the curve of concentration approach through the weighted class of distributions which was extended in reliability and life testing by Gupta and Kirmani (1990). Let w(x) be a nonnegative weight function associated with the pdf f(x) such that E(w(X)) exists and is a finite quantity denoted by μ_w. Then, we may consider a nondecreasing function $\xi(x)$, where

$$\xi(x) = \frac{1}{\mu_w} \int_0^x w(u) f(u) du \tag{4.1}$$

It is easy to note that $\xi(x)$lies between 0 and 1 and is an increasing function in x. We can therefore define an extended curve of concentration in terms of $(F(x), \xi(x))$, $0 \le F(x) \le 1$. Let us refer it by the Weighted Curve of Concentration (WCC). The area between the WCC and the equiangular line $F(x) = \xi(x)$ will be denoted by

$$A_\xi = \frac{1}{2}\left[\int_0^1 F d\xi - \int_0^1 \xi dF \right]. \tag{4.2}$$

In, particular, if w(x)=x, WCC reduces to Gini's curve of concentration and if w(x)=1/r(x), WCC reduces to Reliability curve of concentration. Thus, WCC is a generalization of both Gini's curve of concentration and reliability curve of concentration. Let us examine some properties of WCC.

It may be noted that if the WCC falls below the equiangular line the value of A_ξ will be nonnegative. In case the WCC lies above the equiangular line the value of A_ξ will be non-positive.

RESULT 4.1. WCC is convex (or concave) to F-axis if and only if the w(x) is an increasing (or, decreasing) function in x.

Proof. Note that □

$$\mu_w \frac{d\xi}{dF} = \frac{w(x) f(x)}{f(x)} = w(x) \tag{4.3}$$

and hence from (4.3) we get

$$\mu_w \frac{d^2\xi}{dF^2} = w'(x)\frac{1}{f(x)}. \tag{4.4}$$

Since in (4.4) f(x) is a positive quantity WCC is convex to F-axis if and only if $w'(x)$ is greater than zero, i.e., w(x) is an increasing function in x. Similarly, WCC is concave to F-axis if and only if $w'(x)$ is less than zero i.e., w(x) is a decreasing function in x..

Let us present some interesting results concerning WCC. Their proofs are in line with the proofs of the similar results of section 3.

RESULT 4.1. Within the class of monotonic weight function, the area under the WCC, A_ξ, vanishes if and only if the weight function is a constant.

RESULT 4.2. (i) Under the monotonicity of {w(x)+x}, $A_\Phi + A_\xi = 0$ if and only if the underlying weight function is linearly decreasing in x.

(ii) Under the monotonicity of {w(x)-x}, $A_\Phi = A_\xi$ if and only if the weight function is linearly increasing in x.

References

Arnold, B. C. (2005). Inequality measure for multivariate distribution. *Metron*, vol. LXIII, 317-327.

Barlow, R.E. and Proschan, F. (1975). *Statistical Theory of Reliability and Life Testing*, Holt, Rinehart and Winston, New York.

Brown M. (1994). Using Gini-style indices to evaluate the spatial patterns of health practitioners: theoretical considerations and an application based on Alberta data. Soc. *Sci.Med.*38(9), 1243-1256.

Damgaard, C. and Weiner, J. (2000). Describing Inequality in Plant Size or Fecundity. *Ecology,* **81**, 1139-1142.

Dixon, P. M., Weiner, J., Mitchell-Olds, T., and Woodley, R. (1987). Bootstrapping the Gini Coefficient of Inequality. *Ecology*. **68**, 1548-1551, Erratum to 'Bootstrapping the Gini Coefficient of Inequality. *Ecology*. **69**, 1307, 1988.

Gini, C. (1912). "Variabilitá e mutabilita." Reprinted in *Memorie di metodologia statistica* (Ed. E. Pizetti and T. Salvemini.) Rome: Libreria Eredi Virgilio Veschi, 1955.

Glasser, G. J. (1962). Variance Formulas for the Mean Difference and Coefficient of Concentration, *Journal of American Statistical Association*. **57**, 648-654.

Gupta, R. C. and Kirmani, S.N.U.A. (1990). The role of weighted distributions in stochastic modeling. Communications in Statistics, *Theory and methods*. **19**, .3147-3162.

Kim, J.S. and Proschan, F. (1988). Laplace ordering and its reliability applications, Technical report. *Department of Statistics*, Florida State University.

Mukherjee, S. P. and Roy, D. (1987). Failure rate transform of the Weibull variate. *Communications in Statistics*, *Theory and methods*. **16**, 281-291.

Piesch, W. (2005). A look at the structure of some extended Ginis, *Metron*. O(2), 263-296.

Pietra, G. (1915). Delle relazioni fra indici di variabilita note Ie II, Atti del Realo Instituto Veneto di Scienze, *Letters ed Arti*. **74**, 775-804.

Roy, D. (1993). Characterizations through failure rate and mean residual life transformations, *Microelectronics and Reliability*. **33**, 141-142.

Roy, D. (2004). A generalization of the lack of memory property and related characterization results, *Communications in Statistics, Theory and Methods*. **33**, 3145- 3158.

Sen, A. (1973). *On Economic Inequality*, Oxford, England: Clarendon Press.

Shaked, M. and Shanthikumar, J. G. (1994). *Stochastic orders and their Applications*, Academic Press, New York.

Shkolnikov, V., Andreev, E. and Begun, A. Z. (2003). Gini coefficient as a life table function Computation from discrete data, decomposition of differences and empirical examples, *Demographic research*, **8**, article 11.

Stuart, A. and Ord, J. K. (1994). Kendall's Advanced Theory of Statistics, vol.I, *Distribution Theory*, Edward Arnold, London.

Yitzhaki, S. and Schechtman, E. (2005). The properties of the extended Gini measures of variability and inequality, *Metron*. O(3), 401-433.

In: New Developments in Applied Statistics
Editor: Mohammad Ahsanullah
ISBN 978-1-61324-648-1

Chapter 20

INFERENCE ON PARETO DISTRIBUTION AS STRESS-STRENGTH MODEL BASED ON GENERALIZED ORDER STATISTICS

K.E. Ahmad, Z.F. Jaheen and Manal M. Yousef*
Assiut University, Egypt
Department of Mathematics, Assiut University, Assiut 71516, Egypt

Abstract

In this paper, the two-parameter Pareto distribution is assumed as a stress-strength model. Estimation of the function $R = Pr(Y < X)$ is obtained based on generalized order statistics. The Bayes method of estimation is used under squared error and linex loss functions. Monte carlo simulation technique is used to compare the Bayes estimates with their corresponding maximum likelihood.

Key Words and Phrases: Pareto distribution; Stress-Strength Model; Generalized Order Statistic; Monte Carlo Simulation.

1. Introduction

The main aim of this paper is to focus on inference on $R = P(Y < X)$, where X and Y are independent Pareto-distributed random variables with the same scale parameter β, but different shape parameters α_1 and α_2, respectively. Now, suppose that X is the strength of a system which is subject to a stress Y, then R is a measure of system performance. The system fails if and only if at any time the applied stress is greater than its strength.

This problem has been studied by several authors. The estimation of reliability from stress-strength relationships considered by Church and Harris (1970). Saleh and Umbach (1985) obtained estimation of Pareto quantiles using selected order statistics. Ahmad et al. (1997) considered the estimation of R when X and Y are Burr type X random variables. A comprehensive treatment of the different stress strength models can be found in the recent monograph of Kotz et al. (2003). Kundu and Gupta (2005) discussed estimation of

*E-mail address: zjaheen@hotmail.com

$R = P(Y < X)$ for the generalized exponential distribution, also in (2006) they obtained estimation of R for Weibull distribution. Baklizi (2008) obtained likelihood and Bayes estimation of the stress-strength reliability based on lower record values from the generalized exponential distribution.

The Pareto distribution is a power law probability distribution that coincides with social, scientific, geophsical and many other types of observable phenomena. Recently it has been used to the study of ozone levels in the upper atmosphere, tinsile strength of nylon carpet fibers. It has played a very important role in the investigation of city population occurrence of natural resources, insurance risk, business failures. The probability density function (*pdf*) and cumulative distribution function (*cdf*) of the Pareto distribution which was denoted by $X \sim P(I)(\beta, \alpha)$ are given, respectively, by

$$f(x;\alpha,\beta) = \alpha\beta^{\alpha}x^{-(\alpha+1)}, \ x \geq \beta, \quad \alpha, \beta > 0, \tag{1.1}$$

$$F(x;\alpha,\beta) = 1 - (\beta/x)^{\alpha}. \tag{1.2}$$

Arnold and Press (1983) obtained Bayes estimation for samples from classical, generalized and multivariate Pareto distributions. Nadarajah and Kotz (2003) discussed several Pareto distributions and derived the corresponding forms for the reliability. Woo (2004) obtained Bayesian analysis of mean parameter in a Pareto distribution.

Kamps (1995) introduced the concept of generalized order statistics (gos) as a unified approach to order statistics, record values and sequential order statistics. The gos are defined using quintile transformation based on the distribution function F. let $X(1,n,m,k),...,X(r,n,m,k)$, $k \geqslant 1$, m is a real number, be gos based on absolutely continuous distribution function F with density function f. The joint density function of the above quantities is given by

$$f^{X(1,n,m.k),\ldots,X(r,n,m,k)}(x_1,\ldots,x_r)$$

$$= C_{r-1}\left(\prod_{i=1}^{r-1}[1-F(x_i)]^m f(x_i)\right)[1-F(x_r)]^{\gamma_r - 1}f(x_r), \tag{3}$$

$$F^{-1}(0+) < x_1 \leq \ldots \leq x_r < F^{-1}(1),$$

where

$$\left.\begin{array}{l} \gamma_r = k + (n-r)(m+1) > 0, \\ C_{r-1} = \prod_{j=1}^{r} \gamma_j, r = 1,2,\ldots,n, \gamma_n = k. \end{array}\right\} \tag{1.4}$$

With $n \in N, k > 0$ and $m \in R$. For more details of gos, see Kamps (1995). In the case of $m = 0$ and $k = 1$, $X(r,n,m,k)$ reduces to the ordinary $r-th$ order statistics and (1.3) is then the joint *pdf* of r order statistics, $X_{1:n} \leq X_{2:n} \leq \ldots \leq X_{r:n}$. For various distributional properties of ordinary order statistics, see David (1981) and Arnold et al. (1992). If $m = -1$ and $k = 1$, then (1.3) being the joint *pdf* of r upper record values. For some distributional properties of record values, see Ahsanullah (1995) and Arnold et al. (1998).

Habibullah and Ahsanullah (2000) obtained estimates for the parameters of the Pareto distribution based on gos. Kamps and Cramer (2001) studied some distribution properties of the *gos* from the Pareto, Power and Weibull distributions. Jaheen (2002) considered the prediction of future *gos* from a general class of continuous distributions which includes the Weibull, compound Weibull, Burr type XII, Pareto, Beta and Gompertz by using Bayesian two-sample prediction technique. Jaheen (2005) discussed Bayes estimation for the parameters of the Burr type XII distribution based on *gos*.

2. Estimation of $R = P(Y < X)$

Suppose that, X and Y follow $P(I)(\beta,\alpha_1)$ and $P(I)(\beta,\alpha_2)$, respectively, and they are independent. Then it can be easily seen that

$$R = P(Y < X) = E[P(Y < X \mid \underline{X})] = \frac{\alpha_2}{\alpha_1+\alpha_2}. \tag{2.1}$$

2.1. Maximum Likelihood Estimation

Suppose that $X(1,n,m,k),\ldots,X(r_1,n,m,k)$, $k > 0$ and $m \in R$ be a generalized ordered random sample of size n_1 drawn from the Pareto distribution whose *pdf* is given by (1.1), the likelihood function corresponding to (1.3) can be written as

$$L(\alpha,\beta \mid \underline{x}) = C_{r_1-1}\alpha_1^{r_1}e^{-T_1}e^{\alpha_1 T_1^*}, \tag{2.2}$$

where $\underline{x} = (x_1,x_2,\ldots,x_{r_1})$ and

$$\left.\begin{aligned} T_1 &= \sum_{i=1}^{r_1}\ln(x_i),\\ T_1^*(\beta;\underline{x}) &= (m+1)\Big(\sum_{i=1}^{r_1-1}\ln(x_i)\Big)+\gamma_{r_1}\ln(x_{r_1})-[(r_1-1)(m+1)+\gamma_{r_1}]\ln(\beta), \end{aligned}\right\} \tag{2.3}$$

γ_{r_1} is as given by (1.4).

Suppose that, the parameter β is known, the Maximum likelihood (ML) estimate of the parameter α_1 can be shown to be of the form

$$\hat{\alpha}_1 = \frac{r_1}{T_1^*(\beta;\underline{x})}, \tag{2.4}$$

where $T_1^*(\beta;\underline{x})$ is as given by (2.3).

Similarly, if $Y(1,n_2,m,k)\ldots,Y(r_2,n_2,m,k)$, $k > 0$ and $m \in R$ is a generalized ordered random sample of size n_2 drawn from the Pareto distribution $P(I)(\beta,\alpha_2)$, then the ML estimate of α_2 will be

$$\hat{\alpha}_2 = \frac{r_2}{T_2^*(\beta;\underline{y})}, \tag{2.5}$$

where $T_2^*(\beta;\underline{y})$ is as given by (2.3) with $\underline{x}$ being replaced by $\underline{y} = (y_1,y_2,\ldots,y_{r_2})$.

Hence, the ML estimate of R, denoted by $\hat{R}_{ML}$, can be obtained from (2.1),(2.4) and (2.5) and takes the following form

$$\hat{R}_{ML} = \frac{r_2 T_1^*(\beta;\underline{x})}{r_1 T_2^*(\beta;\underline{y}) + r_2 T_1^*(\beta;\underline{x})}. \tag{2.6}$$

2.2. Bayesian Inference on R

In this section, we obtain Bayesain estimate of the reliability R under two different loss functions, the squared error and LINEX loss functions based on *gos*.

The gamma distribution was first used by Lwin (1972) as a conjugate prior distribution for the parameter α when β is known. For α_j, $j = 1,2$, this conjugate prior takes the form

$$g(\alpha_j;\mu_j,\nu_j) = \frac{\nu_j^{\mu_j}}{\Gamma(\mu_j)}\alpha_j^{\mu_j-1}e^{-\nu_j\alpha_j}, \quad \alpha_j > 0, \quad \mu_j,\nu_j > 0. \tag{2.7}$$

It follows, from (2.2) and (2.7), that the posterior density of α_1 given $\underset{\sim}{x}$ can be written as

$$\pi_1(\alpha_1 \mid \underset{\sim}{x}) = \frac{(\nu_1+T_1^*)^{\mu_1+r_1}}{\Gamma(\mu_1+r_1)}\alpha_1^{\mu_1+r_1-1}e^{-\alpha_1(\nu_1+T_1^*)}, \quad \alpha_1 > 0. \tag{2.8}$$

Similarly, the posterior density function of α_2 given $\underset{\sim}{y}$ can be written as

$$\pi_2(\alpha_2 \mid \underset{\sim}{y}) = \frac{(\nu_2+T_2^*)^{\mu_2+r_2}}{\Gamma(\mu_2+r_2)}\alpha_2^{\mu_2+r_2-1}e^{-\alpha_2(\nu_2+T_2^*)}, \quad \alpha_2 > 0. \tag{2.9}$$

The joint posterior density function of α_1 and α_2 is

$$\pi(\alpha_1,\alpha_2 \mid \underset{\sim}{x},\underset{\sim}{y}) = A\alpha_1^{r_1+\mu_1-1}\alpha_2^{r_2+\mu_2-1}e^{-[\alpha_1(\nu_1+T_1^*)+\alpha_2(\nu_2+T_2^*)]}, \tag{2.10}$$

where

$$A = \frac{(\nu_1+T_1^*)^{\mu_1+r_1}(\nu_2+T_2^*)^{\mu_2+r_2}}{\Gamma(\mu_1+r_1)\Gamma(\mu_2+r_2)}. \tag{2.11}$$

Using standard procedure of transformations of random variables, we obtain the *pdf* of R as

$$\begin{aligned} f_R(r) &= \frac{(\nu_1+T_1^*)^{\mu_1+r_1}(\nu_2+T_2^*)^{\mu_2+r_2}}{Be(r_1+\mu_1,r_2+\mu_2)} \\ &\times \frac{r^{r_2+\mu_2-1}(1-r)^{r_1+\mu_1}}{[(\nu_2+T_2^*)r+(\nu_1+T_1^*)(1-r)]^{r_1+r_2+\mu_1+\mu_2}}, 0 < r < 1. \end{aligned} \tag{2.12}$$

where $Be(.,.)$ is the standard beta function.

2.2.1. Estimation Based on Squared Error Loss Function

Firstly, under squared error loss function, it is well known that, the Bayes estimate of a function, say R, is the posterior mean of the function. This posterior mean contains an integral which is usually hard to obtain in a simple closed form. We will use Lindley's (1980) approximation form to compute the Bayes estimate of R.

3. Lindley's Approximation

Lindley (1980) developed approximation procedures for the evaluation of the ratio of two integrals in the form

$$\int_\theta w(\lambda)e^{L(\lambda)}d\lambda \Big/ \int_\theta g(\lambda)e^{L(\lambda)}d\lambda, \tag{2.13}$$

where $\lambda \equiv (\lambda_1, ..., \lambda_N)$, $L(\lambda)$ is the logarithm of the likelihood function, $w(\lambda)$ and $g(\lambda)$ are arbitrary functions of λ. Suppose that $g(\lambda)$ is the prior density of λ and $w(\lambda) = U(\lambda)g(\lambda)$. From (2.13), the posterior expectation of the function $U(\lambda)$ given the data $\underline{x} \equiv (x_1, ..., x_n)$ is given by

$$E(U(\lambda) \mid \underline{x}) = \frac{\int (U(\lambda) e^{L(\lambda)+\rho(\lambda)} d\lambda}{\int e^{L(\lambda)+\rho(\lambda)} d\lambda}, \\ = \frac{\int U(\lambda) e^{Q(\lambda)} d\lambda}{\int e^{Q(\lambda)} d\lambda}. \tag{2.14}$$

Which is the Bayes estimator of $U(\lambda)$ under squared error loss function where $\rho(\lambda) = \log_e g(\lambda)$ and $Q(\lambda) = L(\lambda) + \rho(\lambda)$ is the logarithm of the posterior distribution of λ except for normalizing constant. Expanding $\rho(\lambda)$ and $L(\lambda)$ in (2.14) into a Taylor series expansion about the ML estimate of λ, or $Q(\lambda)$ about the posterior mode of λ, Lindley obtained the required expression for $E[U(\lambda) \mid \underline{x}]$. For more details, see Lindley (1980).

In this paper, we consider Lindley's approximation from expanding about the posterior mode. For the following two parameter case $\lambda \equiv (\lambda_1, \lambda_2)$, lindley's approximtion leads to

$$\hat{U}_B = E[U(\lambda) \mid \underline{x}] + \frac{1}{2}[B + Q_{30}B_{12} + Q_{21}C_{12} + Q_{12}C_{21} + Q_{03}B_{21}], \tag{2.15}$$

where

$$B = \sum_{i=1}^{2}\sum_{j=1}^{2}(U_{ij}\tau_{ij}), \qquad Q_{\eta\zeta} = \partial^{\eta+\zeta}Q/\partial^{\eta}\lambda_1\partial^{\zeta}\lambda_2,$$

where $\eta, \zeta = 0, 1, 2, 3$ and $\eta + \zeta = 3$. For $i, j = 1, 2$

$$U_i = \partial U/\partial\lambda_i, \qquad U_{ij} = \partial^2 U/\partial\lambda_i\partial\lambda_j,$$

For $i \neq j$,

$$B_{ij} = (U_i\tau_{ii} + U_j\tau_{ij})\tau_{ii}, \qquad C_{ij} = 3U_i\tau_{ii}\tau_{ij} + U_j(\tau_{ii}\tau_{jj} + 2\tau_{ij}^2),$$

τ_{ij} is the (i, j) th element in the inverse of the matrix $Q^* = (-Q^*_{ij})$; $i, j = 1, 2$ such that $Q^* = \partial^2 Q/\partial\lambda_1\partial\lambda_2$. Expansion (2.14) is to be evaluated at $(\tilde{\lambda_1}, \tilde{\lambda_2})$, the mode of the posterior density.

In our case, $(\lambda_1, \lambda_2) \equiv (\alpha_1, \alpha_2)$ and Q is given by

$$Q = \ln \pi(\alpha_1, \alpha_2, \mid \underline{x}, \underline{y}) \\ \propto (r_1 + \mu_1 - 1)\ln\alpha_1 + (r_2 + \mu_2 - 1)\ln\alpha_2 - \alpha_1(\nu_1 + T_1^*) - \alpha_2(\nu_2 + T_2^*). \tag{2.16}$$

The joint posterior mode is obtained from (2.16) and is given by

$$\tilde{\alpha_1} = \frac{r_1 + \mu_1 - 1}{\nu_1 + T_1^*} \quad \text{and} \quad \tilde{\alpha_2} = \frac{r_2 + \mu_2 - 1}{\nu_2 + T_2^*}. \tag{2.17}$$

The elements of the inverse of the matrix $Q^* = (-Q^*_{ij})$, $i, j = 1, 2$ are given by

$$\tau_{11} = \frac{\alpha_1^2}{(r_1 + \mu_1 - 1)}, \qquad \tau_{22} = \frac{\alpha_2^2}{(r_2 + \mu_2 - 1)}, \qquad \tau_{12} = \tau_{21} = 0,$$

and

$$Q_{30} = \frac{\partial^3 Q}{\partial\alpha_1^3} = \frac{2(r_1 + \mu_1 - 1)}{\alpha_1^3}, \quad Q_{03} = \frac{\partial^3 Q}{\partial\alpha_2^3} = \frac{2(r_2 + \mu_2 - 1)}{\alpha_2^3}, \quad Q_{12} = Q_{21} = 0.$$

We obtain the Bayes estimate of a function $U \equiv U(\alpha_1, \alpha_2)$ by substituting the above values in (2.14). It follows that

$$\begin{aligned}\hat{U_B} &= E[U(\alpha_1, \alpha_2) \mid \underline{x}, \underline{y}] \\ &= U + \frac{1}{2V}[\alpha_1 W_1 + \alpha_2 W_2],\end{aligned} \tag{2.18}$$

where

$$V = (r_1 + \mu_1 - 1)(r_2 + \mu_2 - 1),$$

$$W_1 = (r_2 + \mu_2 - 1)(\alpha_1 U_{11} + 2U_1),$$

$$W_2 = (r_1 + \mu_1 - 1)(\alpha_2 U_{22} + 2U_2).$$

For the special case, when $U(\alpha_1, \alpha_2) = R$, we have

$$U_1 = \frac{-\alpha_2}{(\alpha_1 + \alpha_2)^2}, \qquad U_2 = \frac{\alpha_1}{(\alpha_1 + \alpha_2)^2},$$

$$U_{11} = \frac{2\alpha_2}{(\alpha_1 + \alpha_2)^3}, \qquad U_{22} = \frac{-2\alpha_1}{(\alpha_1 + \alpha_2)^3}.$$

Under the squared error loss function, the Bayes estimate of R, denoted by $\hat{R}_{BS}$, can be written as

$$\hat{R}_{BS} = \tilde{R}\left(1 + \frac{\tilde{\alpha}_1 \tilde{R}^2}{\tilde{\alpha}_2^2}\left[\frac{\tilde{\alpha}_1}{(r_2 + \mu_2 - 1)} - \frac{\tilde{\alpha}_2}{(r_1 + \mu_1 - 1)}\right]\right), \tag{2.19}$$

where all functions in (2.19) are evaluated at the posterior mode $(\tilde{\alpha}_1, \tilde{\alpha}_2)$.

Estimation Based on LINEX Loss Function

The asymmetric LINEX loss function was introduced by Varian (1975) as

$$L(\Delta) = e^{a\Delta} - a\Delta - 1, \tag{2.20}$$

where $\Delta = \hat{\phi}(\theta) - \phi(\theta)$, is the estimation error. The sign of a represents the direction and its magnitude represents the degree of symmetry. First, for a close to zero, the LINEX is approximately squared error loss, and therefore, almost symmetric. Second, If $a < 0$, $L(\Delta)$ rises exponentially when $\Delta < 0$ and approximately linearly when $\Delta > 0$.

Under the LINEX loss function, the Bayes estimate of $\phi(\theta)$, denoted by $\hat{\phi}_{BL}$, is given by

$$\hat{\phi}_{BL} = -\frac{1}{a}\ln\left(E_\phi(e^{-a\phi(\theta)})\right), \tag{2.21}$$

where $E_\phi(e^{-a\phi(\theta)})$ exists and is finite, see Calabria and Pulcini (1996).
In case of, $U(\alpha_1, \alpha_2) = e^{-aR}$, where $a > 0$, we have

$$U_1 = \frac{a\alpha_2}{(\alpha_1 + \alpha_2)^2}e^{-aR}, \qquad U_2 = \frac{-a\alpha_1}{(\alpha_1 + \alpha_2)^2}e^{-aR},$$

$$U_{11} = \frac{a^2\alpha_2^2 - 2a\alpha_2(\alpha_1 + \alpha_2)}{(\alpha_1 + \alpha_2)^4}e^{-aR}, \qquad U_{22} = \frac{a^2\alpha_1^2 - 2a\alpha_1(\alpha_1 + \alpha_2)}{(\alpha_1 + \alpha_2)^4}e^{-aR}.$$

So, under LINEX loss function, the Bayes estimate of R, denoted by $\hat{R}_{BL}$, can be written as

$$\hat{R}_{BL} = \frac{-1}{a}\ln\left(e^{-a\tilde{R}}[1+\frac{\tilde{\alpha}_1 S_1+\tilde{\alpha}_2 S_2}{2D}]\right) \quad , \tag{2.22}$$

where

$$S_1 = (r_2+\mu_2-1)(a^2\tilde{\alpha}_2^2\tilde{\alpha}_1 - 2a(\tilde{\alpha}_1+\tilde{\alpha}_2)[\tilde{\alpha}_1\tilde{\alpha}_2+\tilde{\alpha}_2(\tilde{\alpha}_1+\tilde{\alpha}_2)],$$
$$S_2 = (r_1+\mu_1-1)(a^2\tilde{\alpha}_1^2\tilde{\alpha}_2 - 2a(\tilde{\alpha}_1+\tilde{\alpha}_2)[\tilde{\alpha}_1\tilde{\alpha}_2+\tilde{\alpha}_1(\tilde{\alpha}_1+\tilde{\alpha}_2)],$$
$$D = (r_1+\mu_1-1)(r_2+\mu_2-1)(\tilde{\alpha}_1+\tilde{\alpha}_2)^4.$$

Now, we will discuss two special cases of generalized order statistics.

3.1. Upper Record Sample

As we mentioned, in this case $m=-1$ and $k=1$, we have the Bayes estimate of R under squared error and LINEX loss function is

$$\left.\begin{aligned} \hat{R}_{ML} &= \frac{\hat{\alpha}_{2mr}}{\hat{\alpha}_{1mr}+\hat{\alpha}_{2mr}},\\ \hat{R}_{BS} &= \tilde{R}(1+\tilde{\alpha}_1\tilde{R^2}W^*),\\ \hat{R}_{BL} &= (-1/a)\ln(E^* e^{-a\tilde{R}}), \end{aligned}\right\} \tag{2.23}$$

where

$$\left.\begin{aligned} W^* &= (\tilde{\alpha}_1/(\tilde{\alpha}_2(n_2+\mu_2-1)) - 1/(n_1+\mu_1-1)),\\ E^* &= 1+(\tilde{\alpha}_1(n_2+\mu_2-1)Q+\tilde{\alpha}_2(n_1+\mu_1-1)P)/G,\\ G &= 2(n_1+\mu_1-1)(n_2+\mu_2-1)(\tilde{\alpha}_1+\tilde{\alpha}_2)^4,\\ Q &= (a\tilde{\alpha}_2)^2\tilde{\alpha}_1 - H_1,\\ P &= (a\tilde{\alpha}_1)^2\tilde{\alpha}_2 + H_2,\\ H_1 &= 2a\tilde{\alpha}_2(\tilde{\alpha}_1^2+\tilde{\alpha}_1\tilde{\alpha}_2-(\tilde{\alpha}_1+\tilde{\alpha}_2)^2),\\ H_2 &= 2a\tilde{\alpha}_1(\tilde{\alpha}_2^2+\tilde{\alpha}_1\tilde{\alpha}_2-(\tilde{\alpha}_1+\tilde{\alpha}_2)^2). \end{aligned}\right\} \tag{2.24}$$

with

$$\hat{\alpha}_{1mr} = \frac{r_1}{T_{1r}^*}, \qquad \hat{\alpha}_{2mr} = \frac{r_2}{T_{2r}^*},$$
$$\tilde{\alpha}_1 = \frac{r_1+\mu_1-1}{\nu_1+T_{1r}^*}, \qquad \tilde{\alpha}_2 = \frac{r_2+\mu_2-1}{\nu_2+T_{2r}^*},$$

where

$$T_{1r}^* = \ln(x_{r_1}/\beta), \qquad T_{2r}^* = \ln(y_{r_2}/\beta).$$

3.2. Type II Censored Sample

In this case, $m=0$ and $k=1$, we have the Bayes estimate of R under squared error and LINEX loss function is

$$\left.\begin{aligned} \hat{R}_{ML} &= \frac{\hat{\alpha}_{2mc}}{\hat{\alpha}_{1mc}+\hat{\alpha}_{2mc}},\\ \hat{R}_{BS} &= \tilde{R}(1+\tilde{\alpha}_1\tilde{R^2}W^*),\\ \hat{R}_{BL} &= (-1/a)\ln(E^* e^{-a\tilde{R}}), \end{aligned}\right\} \tag{2.25}$$

where E^* and W^* are given by (2.24) with n_i, $i = 1,2$, replaced by r_i,$i = 1,2$ and $\tilde{\alpha}_{jr}$, $j = 1,2$, being replaced by

$$\tilde{\alpha_1} = \frac{r_1 + \mu_1 - 1}{\nu_1 + T^*_{1c}}, \qquad \tilde{\alpha_2} = \frac{r_2 + \mu_2 - 1}{\nu_2 + T^*_{2c}},$$

with

$$\hat{\alpha}_{1mc} = \frac{r_1}{T^*_{1c}}, \qquad \hat{\alpha}_{2mc} = \frac{r_2}{T^*_{2c}},$$

where

$$T^*_{1c} = \sum_{j=1}^{r_1-1} \ln x + (n - r_1 + 1) \ln x_{r_1} - n \ln \beta,$$

$$T^*_{2c} = \sum_{j=1}^{r_2-1} \ln y + (m - r_2 + 1) \ln y_{r_2} - m \ln \beta.$$

Table 1. Estimated Risk (ER) of the Estimates of R in case of Upper Record for Different Values of n_1, n_2, $\beta = 2$ and 1000 repetitions. ($\mu_1 = 6$, $\nu_1 = 4$,$\mu_2 = 5$,$\nu_2 = 2$)

n_1	n_2	**ER**$(\hat{R}_{ML})$	**ER**$(\hat{R}_{BS})$	**ER**$(\hat{R}_{BL})$		
				a=-2	**a=1**	**a=2**
7	5	0.00413	0.00079	0.00035	0.00086	0.00114
10	8	0.00200	0.00049	0.00026	0.00045	0.00058
12	10	0.00141	0.00038	0.00023	0.00034	0.00043

Table 2. Estimated Risk (ER) of the Estimates of R in case of Upper Record for Different Values of n_1, n_2, $\beta = 3$ and 1000 repetitions. ($\mu_1 = 6$, $\nu_1 = 5$,$\mu_2 = 3$,$\nu_2 = 1$)

n_1	n_2	**ER**$(\hat{R}_{ML})$	**ER**$(\hat{R}_{BS})$	**ER**$(\hat{R}_{BL})$		
				a=-2	**a=1**	**a=2**
7	5	0.00243	0.00136	0.00030	0.00069	0.00088
10	8	0.00121	0.00085	0.00023	0.00032	.00039
12	10	0.00085	0.00066	0.00023	0.00026	0.00029

Table 3. Estimated Risk (ER) of the Estimates of R in case of TypeII Censord samples for Different sample sizes, $\beta = 2$ and 1000 repetitions. ($\mu_1 = 6, \nu_1 = 4, \mu_2 = 5, \nu_2 = 2$)

n_1	r_1	n_2	r_2	**ER**$(\hat{R}_{ML})$	**ER**$(\hat{R}_{BS})$	**ER**$(\hat{R}_{BL})$		
						a=-2	**a=1**	**a=2**
35	30	30	25	0.00399	0.00203	0.00202	0.00218	0.00224
60	55	45	40	0.00236	0.00156	0.00158	0.00166	0.00169
80	75	70	65	0.00162	0.00119	0.00121	0.00124	0.00125

Table 4. Estimated Risk (ER) of the Estimates of R in case of TypeII Censord samples for Different sample sizes, $\beta = 3$ and 1000 repetitions. ($\mu_1 = 6, \nu_1 = 5, \mu_2 = 3, \nu_2 = 1$)

n_1	r_1	n_2	r_2	**ER**$(\hat{R}_{ML})$	**ER**$(\hat{R}_{BS})$	**ER**$(\hat{R}_{BL})$		
						a=-2	**a=1**	**a=2**
25	20	20	15	0.00447	0.00220	0.00166	0.00184	0.00209
50	45	40	35	0.00203	0.00139	0.00126	0.00132	0.00138
75	70	65	60	0.00113	0.00088	0.00083	0.00085	0.00087

4. Monte Carlo Simulation

The ML and Bayes estimate of the reliability R are compared based on Monte Carlo simulation as follows:

1. For given values of the prior parameters μ_1, ν_1, μ_2 and ν_2 we generate α_1, α_2 from gamma distribution.
2. We generate upper record and type II censored samples of different sizes from Pareto distribution with *pdf* (1.1).
3. The ML estimates of α_1 and α_2 are computed from (2.4) and (2.5), then we obtained the ML estimate of R from (2.6).
4. The Bayes estimates of R are computed from (2.19) and (2.22), based on squared error and LINEX loss functions, respectively.
5. The squared deviations $(R^* - R)^2$ are computed where R^* stands for an estimate (ML, Bayes) of R.

6. Steps 2-6 were repeated at least 1000 times for given prior parameters and for different sample sizes n and m.

The Bayes risks of the three estimates were estimated by

$$ER(R^*) = \sum (R^* - R)^2/1000$$

The computational results are displayed in the following Tables.

5. Concluding Remarks

It may be observed, from Tables (1),(2),(3) and (4) that for different values of the prior parameters μ_i, ν_i, $i = 1, 2$, the estimated risks of the estimates (ML, BS, BL) decreases as the sample sizes n_1 and n_2 increase. On the other hand, the estimated risk of the Bayes estimate of R based on squared error loss function is less than the estimated risk of the maximum likelihood estimate. For different values of a the estimated risk of the Bayes estimate based on LINEX loss function decreases as the sample sizes n_1 and n_2 increase.

References

[23] Ahmad, K. E., Fakhry, M. E. and Jaheen, Z. F. (1997). Empirical Bayes estimation of $P(X < Y)$, and characterization of the Burr-type X model. *J. Statis. Plann. Infer.*, **64**, 297-308

[23] Ahsanullah, M. (1995). *Record Statistics.* Nova Science Publishers, Inc. Commack, New York

[23] Arnold, B. C. and Press, S. (1983). Bayesian inference for Pareto population. *Journal of Econometrice*, **21**, 287-306

[23] Arnold, B. C., Balakrishnan, N. and Nagaraja, H. N. (1992). *A First Course In Order Statistics. New York:John Wiley and Sons, Inc*

[23] Arnold, B. C., Balakrishnan, N. and Nagaraja, H. N. (1998). *Records.*, John Wiley and Sons, Inc., New York

[23] Baklizi, A. (2008). Likelihood and Bayesian estimation of $P(X < Y)$ using lower record values from the generalized exponential distribution. *Comput. Statist. and Data Analy.*, **52**, 3468-3473

[23] Calabria, R. and Pulcini, G. (1996). Point estimation under asymmetric loss functions for left truncated exponential samples. *Commun. Statist. Theor. Meth.* **25**, 285-600

[23] Church, J. D. and Harris, B. (1970). The estimation of reliability from stress-strength relationships. *Technometrices.* **12**, 49-54

[23] David, H. A. (1981). *Order Statistic*, 2nd ed.New York:John Wiley and Sons, Inc

[23] Habibullah, M., Ahsanullah, M. (2000). Estimation of parameters of a Pareto distribution by generalized order statistics. *Commun. Statist. Theor. Meth.* **29 (7)**, 1597-1609

[23] Jaheen, Z. F. (2002). On Bayesain prediction of generalized order statistics. *J. Statis. Theor. Applicat.* **1(3)**, 191-204

[23] Jaheen, Z. F. (2005). Estimation Based on generalized order statistics from the Burr Model. *Commun. Statist. Theor. Meth.* **34**, 785-794

[23] Kamps, U. (1995). *A Concept of Generalized Order Statistics.* B.G. Teubner Stuttgart

[23] Kamps, U. and Cramer, E. (2001). On distributions of generalized order statistics. *Statistics* **35**, 269-280

[23] Kotz, S., Lumelskii, Y. and Pensky, M. (2003). *The Stress- Strength Model and Its Generalizations.* Singapore: World Scientific Press

[23] Kundu, D. and Gupta, R. D. (2005). Estimation of $P(Y < X)$ for the generalized exponential distribution. *Metrika.* **61(3)**, 291- 308

[23] Kundu, D. and Gupta, R. D. (2006). Estimation of $P(Y < X)$ for Weibull. *IEEE Trans. Reliab.* **55(2)**, 270-280

[23] Lwin, T. (1972). Estimation of the tail of the Paretian law. *Skandinavisk Aktuarietidskrift.* **55**, 170-178

[23] Lindley, D.V. (1980). Approximate Bayesian methods. *J. Trabajos de Estadistica.* **31**, 223-237

[23] Nadarajah, S. and Kotz, S. (2003). Reliability for Pareto models. *Metron*, vol. LXI, 191-204

[23] Saleh, E. and Umbach, D. (1985). Large sample estimation of Pareto quantiles using selected order statistics. *Metrika*, **32**, 49-56

[23] Varian, H. R. (1975). A Bayesian approach to real estate assessment. In *Studies in Bayesian Econometrics and Statistics in honor of Leonard J. Savage*, eds. Stephan E. Fienberg and Arnold Zellner, Amesterdam: North-Holland, 195-208

[23] Woo, J. (2004). Bayesian analysis of mean parameter in a Pareto distribution. *Calcutta Statistical Association Bulletin.* **55**, 129-137

In: New Developments in Applied Statistics
Editor: Mohammad Ahsanullah
ISBN: 978-1-61324-648-1

Chapter 21

ANALYZING INCOMPLETE CATEGORICAL DATA: A LOGISTIC REGRESSION APPROACH

***Hoo Ling Ping*[a] *and M. Ataharul Islam*[b]**
School of Mathematical Sciences, Universiti Sains Malaysia, 11800 Penang, Malaysia

Summary

During the process of collecting data, we might not able to get the fully observed data. This will cause us to have the incomplete data. An inappropriate conclusion and confusion occur when the researchers ignore, truncate, censored or collapse those data as it might contain important information. Maximum likelihood estimate and EM algorithm are the most common ways for researchers to solve for the missing data. Here is an alternative method is proposed. In this paper, we consider a logistic regression with the conditional probabilities to obtain the cell probabilities instead of MLE and EM algorithm. We consider a set of data from a clinical investigation of 1502 infants. In this trial, subjects were given either Palivizumab or a placebo and were followed for five months to see if they were hospitalized for respiratory syncytial virus (RSV). The estimation procedure is illustrated in this paper on the basis of the proposed technique of estimating missing values under various assumptions. We consider logistic regression model to estimate the incomplete categorical data for three way contingency table where we let one of the variable to be the outcome variable and the other two as independent variable. In our work, we demonstrate the use of the logistic regression model to estimate the missing independent variable 1, missing independent variable 2 and missing independent variable 1 and independent variable 2. We compare the expected value of the missing value for the logistic regression model with EM algorithm. An alternative approach is proposed in this paper based on the conditional probabilities as logit link functions for estimating missing cell frequencies. This provides an alternative to the existing techniques based on the MLE and the EM algorithm. The proposed method appears to be more flexible and the prediction because it does not require additional assumption on the underlying probability distributions. In addition, the proposed technique provides estimates similar to that of the estimates based on existing techniques.

Keywords: Incomplete Categorical Data, Logistic Regression Model, EM Algorithm

---

[a] E-mail address: lphoo04@yahoo.com. Corresponding author: ClinResearch Sdn Bhd, 2nd Floor MMA House, 124, Jalan Pahang 53000 Kuala Lumpur, Malaysia.
[b] E-mail address: mataharul@yahoo.com

1. Introduction

During the process of collecting data, we might not able to get the fully observed data. This can be human error, equipment error or even unexpected error. For example, some people refuse to fill in the information on a questionnaire or interview due to embarrassing questions like drug abuse, sexual activities, age or income. The other reason may be in clinical trials, some patients do not follow up their health after certain time. Also some results are missing because of mechanical breakdown in an industrial experiment. All the above examples will give us the incomplete data. Therefore it becomes an obstacle for the researcher to have a proper analysis if these data are ignored, truncated, censored or collapsed.

Vach and Schumacher (1993) compared the approaches among maximum likelihood estimation, pseudo maximum likelihood estimation and probability imputation for logistics regression analysis with incompletely observed categorical covariates. However, Fitzmaurice et al. (1996) considered logistics regression as a likelihood-based regression model to analyze binary data with attrition and at the same year, Fitzmaurice et al. considered the same method to model the association between the binary response in terms of conditional log odds ratios. Ibrahim et al. (1999) proposed a method for estimating parameters in generalized linear models with missing covariates and a non-ignorable missing data mechanism. A multinomial model for the missing data indicators was used and a joint distribution which can be written as a sequence of one-dimensional conditional by consisting of logistic regression was proposed. Ibrahim et al. (1999) have estimated the parameter missing covariate 1 and, or covariate 2 based on assumed values. This model cannot be used for testing the underlying model when the covariates are missing.

Therefore a new estimation procedure is developed in this study for analyzing incomplete categorical data. The conditional probabilities are expressed for the complete and incomplete data and then the logistic regression is expressed as logit link functions for estimating missing cell frequencies. We have demonstrated that estimation procedure of missing values for two or more categorical variables can be performed in a more straightforward manner.

In our study, we consider logistic regression model to estimate the incomplete categorical data for three way contingency table where we let one of the variables to be the outcome variable and the other two as independent variables. In our work, we demonstrate the use of the logistic regression model to estimate the missing independent variable 1, missing independent variable 2 and missing independent variable 1 and independent variable 2. We compare the results of logistics regression model with EM algorithm based on discriminant analysis, sensitivity, specificity, Mcnemar's Test and Mantel-Haenszel Test.

Table 1. Complete data.

		Treatment		
	RSV	**Placebo**	**Palivizumab**	**Total**
BPD	Yes	34	39	73
	No	232	457	689
Premature	Yes	19	9	28
	No	215	497	712
Total		500	1002	1502

This paper is organized as follows: Section 2 provides an example taken from Simonoff (2003). The formulation of the logistics regression and brief formulation of EM algorithm is described in section 3. Section 4 discusses the results. However, test procedures are presented in section 5 and section 6 is conclusions.

2. The Data

We consider a set of data from a clinical investigation of 1502 infants. In this trial, subjects were given either Palivizumab or a placebo and were followed for five months to see if they were hospitalized for respiratory syncytial virus (RSV). This data will be illustrated in Table 1. The row variable denotes whether hospitalized for RSV, the column variable is the treatment and the level variable are the initial condition of the infants. The full name of BPD is bronchopulmonary dysplasia where it is a disease often associated with premature birth.

Table 2. Artificial incomplete data.
a) Complete observed data

		Treatment		
	RSV	**Placebo**	**Palivizumab**	**Total**
BPD	Yes	31	35	66
	No	209	411	620
Premature	Yes	17	8	25
	No	194	447	641
Total		451	901	1352

b) For missing RSV data

		Treatment		
	RSV	**Placebo**	**Palivizumab**	**Total**
BPD	missing	10	20	30
Premature	missing	9	20	29
Total		19	40	59

c) For missing Treatment data

		Treatment
	RSV	**missing**
BPD	Yes	3
	No	28
Premature	Yes	1
	No	28
Total		60

d) For missing RSV and missing Treatment data

		Treatment
	RSV	**missing**
BPD	missing	16
Premature	missing	15
Total		31

Table 2 is created such that some of the observations are missing and hence we get an incomplete categorical data.

3. Logistic Regression and the EM Algorithm

3.1. Logistic Regression Model

From Hosmer and Lemeshow (2000), the specific form of logistic regression model is

$$\pi(x_i) = \frac{e^{\beta_0 + \sum_i^n \beta_i x_i}}{1 + e^{\beta_0 + \sum_i^n \beta_i x_i}} \qquad (3.1)$$

where $\pi(x_i)$= E(Y | x_i), the conditional mean of Y given x_i. The logit transformation is defined as

$$g(x_i) = \ln\left[\frac{\pi(x_i)}{1-\pi(x_i)}\right]$$

$$= \beta_0 + \sum_{i=1}^{n} \beta_i x_i$$

$$= \beta_0 + \beta_1 x_1 + \beta_2 x_2 + \ldots + \beta_n x_n \qquad (3.2)$$

Suppose a sample of k independent observations of the pair (x_{i1}, x_{i2}, …, x_{in}, y_i), i = 1, , …, k, where y_i denotes the value of a dichotomous outcome variable and x_{i1}, …, x_{in} are the first until nth of the independent variables. The outcome variable is normally coded as 0 or 1 to represent the absence or presence of the characteristic respectively. The values of β_0, β_1,…, β_n , the unknown parameters are estimated to fit the logistic regression model in equation (3.1).

For three-way incomplete contingency table, we let one of the variables to be the outcome variable and the other two as independent variable. In our work, we demonstrate the use of the logistic regression model to estimate the missing independent variable 1, missing independent variable 2 and missing independent variable 1 and independent variable 2. We consider the conditional probability base on logistic regression model for the cell probabilities. For complete observed data, the cell probabilities will be

$$p(y, x_1, x_2 \mid \beta) = p(y \mid x_1, x_2, \beta_3)p(x_2 \mid x_1, \beta_2)p(x_1 \mid \beta_1), \qquad (3.3)$$

where

$$p(y \mid x_1, x_2, \beta_3) = \frac{e^{\beta_{30} + \beta_{31} x_1 + \beta_{32} x_2}}{1 + e^{\beta_{30} + \beta_{31} x_1 + \beta_{32} x_2}},$$

$$p(x_2 \mid x_1,\ \beta_2) = \frac{e^{\beta_{20}+\beta_{21}x_1}}{1+e^{\beta_{20}+\beta_{21}x_1}},$$

and

$$p(x_1 \mid \beta_1) \sim B(N,\ \beta_1 = \pi_1)$$

where N is the total observations.

However, if we now consider the complete data where we include the missing data, then r_1 is considered as missing on variable x_1 and r_2 is considered as missing on variable x_2. When x_1 is observed, then $r_1 = 1$, while $r_1 = 0$ when x_1 is missing. Similarly we can define $r_2 = 1$ and $r_2 = 0$. When x_1 is missing, the conditional probability is

$$p(y,\ r_1,\ x_2 \mid \phi) = p(y \mid r_1,\ x_2,\ \phi_3)p(r_1 \mid x_2, y,\ \phi_2)p(x_2 \mid \phi_1),$$

where

$$p(y \mid r_1,\ x_2,\ \phi_3) = \frac{e^{\phi_{30}+\phi_{31}r_1+\phi_{32}x_2}}{1+e^{\phi_{30}+\phi_{31}r_1+\phi_{32}x_2}},$$

$$p(r_1 \mid x_2, y,\ \phi_2) = \frac{e^{\phi_{20}+\phi_{21}x_2+\phi_{22}y}}{1+e^{\phi_{20}+\phi_{21}x_2+\phi_{22}y}},$$

and

$$p(x_2 \mid \phi_1) \sim B(N,\ \phi_1 = \pi_2).$$

Then when x_2 is missing,

$$(y,\ x_1,\ r_2 \mid \alpha) = p(y \mid x_1,\ r_2,\ \alpha_3)p(r_2 \mid x_1, y,\ \alpha_2)p(x_1 \mid \alpha_1),$$

where

$$p(y \mid x_1,\ r_2,\ \alpha_3) = \frac{e^{\alpha_{30}+\alpha_{31}x_1+\alpha_{32}r_2}}{1+e^{\alpha_{30}+\alpha_{31}x_1+\alpha_{32}r_2}},$$

$$p(r_2 \mid x_1, y,\ \alpha_2) = \frac{e^{\alpha_{20}+\alpha_{21}x_1+\alpha_{22}y}}{1+e^{\alpha_{20}+\alpha_{21}x_1+\alpha_{22}y}},$$

and

$$p(x_1 \mid \alpha_1) \sim B(N,\ \alpha_1 = \pi_3).$$

For x_1 and x_2 are missing, we will consider the polytomous logistic regression. This is because we have four possible of missing pattern where we expressed

$$R_1 = (r_1 = 1,\ r_2 = 1) = 0,$$

$$R_2 = (r_1 = 1,\ r_2 = 0) = 1,$$

$$R_3 = (r_1 = 0,\ r_2 = 1) = 2,$$

$$R_4 = (r_1 = 0,\ r_2 = 0) = 3.$$

The conditional for both independent variable x_1 and x_2 are missing will be

$$\mathrm{p}(y,\ r_1,\ r_2 \mid \gamma) = \mathrm{p}(r_2,\ r_1 \mid y, \gamma_2)\ \mathrm{p}(y \mid \gamma_1)$$

where

$$\mathrm{p}(r_2,\ r_1 \mid y, \gamma_2) = \mathrm{p}(R \mid y, \gamma_2).$$

We expressed

$$\mathrm{p}(R_m \mid \mathrm{y}, \gamma_2) = \frac{e^{h_{m-1}(y)}}{1 + e^{h_1(y)} + e^{h_2(y)} + \ldots + e^{h_{m-1}(y)}},\ m = 2, 3, \text{ and } 4,$$

and

$$\frac{p(R_m \mid y)}{p(R_1 \mid y)} = e^{h_{m-1}(y)},\ \text{for } m = 2, 3, \text{ and } 4,$$

$$h_{m-1}(\gamma_2) = \gamma_{1(m-1)0} + \gamma_{1(m-1)1} y,\ \text{for } m = 2, 3, \text{ and } 4.$$

For our case,

$$\mathrm{p}(R_m = 0 \mid y, \gamma_2) = \frac{1}{1 + e^{h_1(y)} + e^{h_2(y)} + e^{h_3(y)}},$$

$$\mathrm{p}(R_m = 1 \mid y, \gamma_2) = \frac{e^{h_1(y)}}{1 + e^{h_1(y)} + e^{h_2(y)} + e^{h_3(y)}},$$

$$\mathrm{p}(R_m = 2 \mid y,\ \gamma_2) = \frac{e^{h_2(y)}}{1 + e^{h_1(y)} + e^{h_2(y)} + e^{h_3(y)}},$$

$$\mathrm{p}(R_n = 3 \mid y,\ \gamma_2) = \frac{e^{h_3(y)}}{1 + e^{h_1(y)} + e^{h_2(y)} + e^{h_3(y)}},$$

where

$$\frac{p(R_2 \mid y)}{p(R_1 \mid y)} = e^{h_1(y)},$$

$$\frac{p(R_3 \mid y)}{p(R_1 \mid y)} = e^{h_2(y)},$$

$$\frac{p(R_4 \mid y)}{p(R_1 \mid y)} = e^{h_3(y)},$$

with

$$h_1(y) = \gamma_{210} + \gamma_{211} y,$$

$$h_2(y) = \gamma_{220} + \gamma_{221} y,$$

$$h_3(y) = \gamma_{230} + \gamma_{231} y,$$

and

$$\text{p}(y \mid \gamma_1) \sim \text{B}(N, \gamma_1 = \pi_4).$$

To obtain the values of $\beta, \alpha, \phi, \text{and}\, \gamma$, maximum likelihood (ML) is considered. Since ML maximizes the probability of obtaining the observed set of data. Likelihood function is applied in order to apply the maximum likelihood. Likelihood function expresses the probability of the observed data as a function of the unknown parameters. The vector or the conditional probability that Y is equal to 1 given x, p(Y = 1|x) = $\pi(x)$ and p(Y = 0|x) = 1 - $\pi(x)$ when Y is coded as 0 or 1. Thus the contribution to the likelihood function is $\pi(x_i)$ where $y_i = 1$ and the contribution to the likelihood function is 1 - $\pi(x_i)$ where $y_i = 0$.

To obtain the complete data log-likelihood for (3.3), we will consider

$$\text{p}(y_i, x_{i1}, x_{i2}, r_{i1}, r_{i2} \mid \beta, \phi, \alpha, \gamma)$$

$$= [\text{p}(y_i, x_{i1}, x_{i2} \mid \beta)]^{\delta_{11i}} [\text{p}(y_i,, r_{i1}, x_{i2} \mid \phi)]^{\delta_{10i}}$$

$$\text{x}\, [\text{p}(y_i, x_{i1}, r_{i2} \mid \alpha)]^{\delta_{01i}} [\text{p}(y_i, r_{i1}, r_{i2} \mid \gamma)]^{\delta_{00i}}. \tag{3.4}$$

and the complete data log-likelihood will be,

$$l(\theta) = l(\theta, y_i, x_{i1}, x_{i2}, r_{i1}, r_{i2})$$

$$= \sum_{i=1}^{k} \{\log\left[p(y_i, x_{i1}, x_{i2} \mid \beta)\right]^{\delta_{11i}} + \log\left[p(y_i, r_{i1}, x_{i2} \mid \phi)\right]^{\delta_{10i}}$$

$$+ \log\left[p(y_i, x_{i1}, r_{i2} \mid \alpha)\right]^{\delta_{01i}} + \log\left[p(y_i, r_{i1}, r_{i2} \mid \gamma)\right]^{\delta_{00i}}\}$$

$$= \sum_{i=1}^{k} \{\delta_{11i} \log\left[p(y_i, x_{i1}, x_{i2} \mid \beta)\right] + \delta_{10i} \log\left[p(y_i, r_{i1}, x_{i2} \mid \phi)\right]$$

$$+ \delta_{01i} \log\left[p(y_i, x_{i1}, r_{i2} \mid \alpha)\right] + \delta_{00i} \log\left[p(y_i, r_{i1}, r_{i2} \mid \gamma)\right]\} \tag{3.5}$$

where $\sum_{p=0}^{1}\sum_{q=0}^{1}\sum_{i=1}^{k}\delta_{pqi} = N$, N is total sample and $\sum_{p=0}^{1}\sum_{q=0}^{1}\delta_{pqi} = 1$, $\delta_{pqi} = 1$ or 0.

Let

$$l_{11}(\theta)=\sum_{i=1}^{k}\delta_{11i}\log\left[p\left(y_i,x_{i1},x_{i2}\mid\beta\right)\right], \tag{3.6}$$

$$l_{10}(\theta)=\sum_{i=1}^{k}\delta_{10i}\log\left[p\left(y_i,r_{i1},x_{i2}\mid\alpha\right)\right], \tag{3.7}$$

$$l_{01}(\theta)=\sum_{i=1}^{k}\delta_{01i}\log\left[p\left(y_i,x_{i1},r_{i2}\mid\phi\right)\right], \tag{3.8}$$

and

$$l_{00}(\theta)=\sum_{i=1}^{k}\delta_{00i}\log\left[p\left(y_i,r_{i1},r_{i2}\mid\gamma\right)\right]. \tag{3.9}$$

When $\delta_{11i}=1$, $\delta_{10i}=1$, $\delta_{01i}=1$, or $\delta_{00i}=1$ then from (3.4), then

$$l(\theta)=l_{11}(\theta)=\sum_{i=1}^{k}\log\left[p\left(y\ ,x_{i1},x_{i2}\mid\beta\right)\right] \tag{3.10}$$

or

$$l(\theta)=l_{10}(\theta)=\sum_{i=1}^{k}\log\left[p\left(y_i,r_{i1},x_{i2}\mid\alpha\right)\right] \tag{3.11}$$

or

$$l(\theta)=l_{01}(\theta)=\sum_{i=1}^{k}\log\left[p\left(y_i,x_{i1},r_{i2}\mid\phi\right)\right] \tag{3.12}$$

and, or

$$l(\theta)=l_{00}(\theta)=\sum_{i=1}^{k}\log\left[p\left(y_i,r_{i1},r_{i2}\mid\gamma\right)\right] \tag{3.13}$$

We consider to differentiate (3.10) with $\beta_{30},\beta_{31},\beta_{32},\beta_{20},\beta_{21}$, and β_1, then we obtain

$$\frac{\partial l_{11}(\theta)}{\partial\beta_{30}}=\sum_{i=1}^{k}\left(1-\frac{e^{\beta_{30}+\beta_{31}x_{i1}+\beta_{32}x_{i2}}}{1+e^{\beta_{30}+\beta_{31}x_{i1}+\beta_{32}x_{i2}}}\right), \tag{3.14}$$

$$\frac{\partial l_{11}(\theta)}{\partial\beta_{31}}=\sum_{i=1}^{k}x_{i1}\left(1-\frac{e^{\beta_{30}+\beta_{31}x_{i1}+\beta_{32}x_{i2}}}{1+e^{\beta_{30}+\beta_{31}x_{i1}+\beta_{32}x_{i2}}}\right), \tag{3.15}$$

$$\frac{\partial l_{11}(\theta)}{\partial \beta_{32}} = \sum_{i=1}^{k} x_{i2}\left(1-\frac{e^{\beta_{30}+\beta_{31}x_{i1}+\beta_{32}x_{i2}}}{1+e^{\beta_{30}+\beta_{31}x_{i1}+\beta_{32}x_{i2}}}\right), \tag{3.16}$$

$$\frac{\partial l_{11}(\theta)}{\partial \beta_{20}} = \sum_{i=1}^{k}\left(1-\frac{e^{\beta_{20}+\beta_{21}x_{i1}}}{1+e^{\beta_{20}+\beta_{21}x_{i1}}}\right), \tag{3.17}$$

$$\frac{\partial l_{11}(\theta)}{\partial \beta_{21}} = \sum_{i=1}^{k} x_{i1}\left(1-\frac{e^{\beta_{20}+\beta_{21}x_{i1}}}{1+e^{\beta_{20}+\beta_{21}x_{i1}}}\right), \tag{3.18}$$

and

$$\frac{\partial l_{11}(\theta)}{\partial \beta_{1}} = \sum_{i=1}^{k} \frac{x_{i1}-\beta_1}{\beta_1(1-\beta_1)}. \tag{3.19}$$

Consider (3.11) with $\phi_{30}, \phi_{31}, \phi_{32}, \phi_{20}, \phi_{21}, \phi_{22}$ and ϕ_1,

$$\frac{\partial l_{10}(\theta)}{\partial \phi_{30}} = \sum_{i=1}^{k}\left(1-\frac{e^{\phi_{30}+\phi_{31}r_{i1}+\phi_{32}x_{i2}}}{1+e^{\phi_{30}+\phi_{31}r_{i1}+\phi_{32}x_{i2}}}\right), \tag{3.20}$$

$$\frac{\partial l_{10}(\theta)}{\partial \phi_{31}} = \sum_{i=1}^{k} r_{i1}\left(1-\frac{e^{\phi_{30}+\phi_{31}r_{i1}+\phi_{32}x_{i2}}}{1+e^{\phi_{30}+\phi_{31}r_{i1}+\phi_{32}x_{i2}}}\right), \tag{3.21}$$

$$\frac{\partial l_{10}(\theta)}{\partial \phi_{32}} = \sum_{i=1}^{k} x_{i2}\left(1-\frac{e^{\phi_{30}+\phi_{31}r_{i1}+\phi_{32}x_{i2}}}{1+e^{\phi_{30}+\phi_{31}r_{i1}+\phi_{32}x_{i2}}}\right), \tag{3.22}$$

$$\frac{\partial l_{10}(\theta)}{\partial \phi_{20}} = \sum_{i=1}^{k}\left(1-\frac{e^{\phi_{20}+\phi_{21}x_{i1}+\phi_{22}y_{i}}}{1+e^{\phi_{20}+\phi_{21}x_{i1}+\phi_{22}y_{i}}}\right), \tag{3.23}$$

$$\frac{\partial l_{10}(\theta)}{\partial \phi_{21}} = \sum_{i=1}^{k} x_{i1}\left(1-\frac{e^{\phi_{20}+\phi_{21}x_{i1}+\phi_{22}y_{i}}}{1+e^{\phi_{20}+\phi_{21}x_{i1}+\phi_{22}y_{i}}}\right), \tag{3.24}$$

$$\frac{\partial l_{10}(\theta)}{\partial \phi_{22}} = \sum_{i=1}^{k} y_{i}\left(1-\frac{e^{\phi_{20}+\phi_{21}x_{i1}+\phi_{22}y_{i}}}{1+e^{\phi_{20}+\phi_{21}x_{i1}+\phi_{22}y_{i}}}\right), \tag{3.25}$$

and,

$$\frac{\partial l_{10}(\theta)}{\partial \phi_1} = \sum_{i=1}^{k} \frac{x_{i2} - \phi_1}{\phi_1(1-\phi_1)}. \tag{3.26}$$

Consider (3.12) with $\alpha_{30}, \alpha_{31}, \alpha_{32}, \alpha_{20}, \alpha_{21}, \alpha_{22}$ and α_1,

$$\frac{\partial l_{01}(\theta)}{\partial \alpha_{30}} = \sum_{i=1}^{k} \left(1 - \frac{e^{\alpha_{30}+\alpha_{31}x_{i1}+\alpha_{32}r_{i2}}}{1+e^{\alpha_{30}+\alpha_{31}x_{i1}+\alpha_{32}r_{i2}}} \right), \tag{3.27}$$

$$\frac{\partial l_{01}(\theta)}{\partial \alpha_{31}} = \sum_{i=1}^{k} x_{i1} \left(1 - \frac{e^{\alpha_{30}+\alpha_{31}x_{i1}+\alpha_{32}r_{i2}}}{1+e^{\alpha_{30}+\alpha_{31}x_{i1}+\alpha_{32}r_{i2}}} \right), \tag{3.28}$$

$$\frac{\partial l_{01}(\theta)}{\partial \alpha_{32}} = \sum_{i=1}^{k} r_{i2} \left(1 - \frac{e^{\alpha_{30}+\alpha_{31}x_{i1}+\alpha_{32}r_{i2}}}{1+e^{\alpha_{30}+\alpha_{31}x_{i1}+\alpha_{32}r_{i2}}} \right), \tag{3.29}$$

$$\frac{\partial l_{01}(\theta)}{\partial \alpha_{20}} = \sum_{i=1}^{k} \left(1 - \frac{e^{\alpha_{20}+\alpha_{21}r_{i1}+\alpha_{22}y_{i}}}{1+e^{\alpha_{20}+\alpha_{21}r_{i1}+\alpha_{22}y_{i}}} \right), \tag{3.30}$$

$$\frac{\partial l_{01}(\theta)}{\partial \alpha_{21}} = \sum_{i=1}^{k} r_{i1} \left(1 - \frac{e^{\alpha_{20}+\alpha_{21}r_{i1}+\alpha_{22}y_{i}}}{1+e^{\alpha_{20}+\alpha_{21}r_{i1}+\alpha_{22}y_{i}}} \right), \tag{3.31}$$

$$\frac{\partial l_{01}(\theta)}{\partial \alpha_{22}} = \sum_{i=1}^{k} y_{i} \left(1 - \frac{e^{\alpha_{20}+\alpha_{21}r_{i1}+\alpha_{22}y_{i}}}{1+e^{\alpha_{20}+\alpha_{21}r_{i1}+\alpha_{22}y_{i}}} \right), \tag{3.32}$$

and,

$$\frac{\partial l_{01}(\theta)}{\partial \alpha_1} = \sum_{i=1}^{k} \frac{x_{i1} - \alpha_1}{\alpha_1(1-\alpha_1)}. \tag{3.33}$$

Consider (3.13) with $\gamma_{2(m-1)0}, \gamma_{2(m-1)1}, \gamma_1$,

$$\frac{\partial l_{00}(\theta)}{\partial \gamma_{2(m-1)0}} = \sum_{i=1}^{k} \left(1 - \frac{e^{\gamma_{2(m-1)0}+\gamma_{2(m-1)1}y_i}}{1+e^{\gamma_{210}+\gamma_{211}y_i}+e^{\gamma_{220}+\gamma_{221}y_i}+e^{\gamma_{230}+\gamma_{231}y_i}} \right), \tag{3.34}$$

$$\frac{\partial l_{00}(\theta)}{\partial \gamma_{2(m-1)1}} = \sum_{i=1}^{k} y_i \left(1 - \frac{e^{\gamma_{2(m-1)0} + \gamma_{2(m-1)1} y_i}}{1 + e^{\gamma_{210} + \gamma_{211} y_i} + e^{\gamma_{220} + \gamma_{221} y_i} + e^{\gamma_{230} + \gamma_{231} y_i}} \right), \tag{3.35}$$

and,

$$\frac{\partial l_{00}(\theta)}{\partial \gamma_1} = \sum_{i=1}^{k} \frac{y_i - \gamma_1}{\gamma_1 (1 - \gamma_1)}. \tag{3.36}$$

It is convenience for us to get the $\hat{\beta}, \hat{\alpha}, \hat{\phi},$ and $\hat{\gamma}$ from logistic regression software since the solutions (3.14)-(3.36) for the iterative method have been programmed. McCullagh and Nelder (1989) showed that (3.14)-(3.36) may be obtained by considering iterative weighted least squares procedure.

3.2. The EM Algorithm

The EM algorithm is a broadly applicable algorithm that provides an iterative procedure for computing MLEs (McLachlan and Krishnan, 1997). There are two steps on each iteration of the EM algorithm, which are called as the *expectation* step or the E-step and *maximization* step or the M-step. Therefore, the algorithm is called as the EM algorithm. This name was given by Dempster, Laird and Rubin (1977) in their fundamental paper.

For Poisson distribution, the complete-data log likelihood will be

$$L_c(\theta_i) = \sum_{i=1}^{n} \left\{ (x_i + z_i) \log(\theta_i) - \theta_i - \log\left[(x_i + z_i)! \right] \right\} \tag{3.37}$$

where x_i is observed value and z_i is missing value.

By differentiating (3.37) with respect to θ_i,

therefore

$$\hat{\theta}_i = x_i + z_i. \tag{3.38}$$

after we solve (3.37).

On the (k+1)th iteration, the E- and M-step will be defined as follow:

E-step:

$$z_{ijk}^{(k+1)} = m_{ijk}^{(k+1)} + c_{ijk}^{(k+1)} + d_{ijk}^{(k+1)},$$

where

$$m_{ijk}^{(k+1)} = m_{+jk} \left(\frac{m_{ijk}^{(k)}}{m_{+jk}^{(k)}} \right),$$

$$c_{ijk}^{(k+1)} = c_{i+k} \left(\frac{m_{ijk}^{(k)}}{m_{i+k}^{(k)}} \right),$$

and

$$d_{ijk}^{(k+1)} = d_{++k}\left(\frac{m_{ijk}^{(k)}}{m_{++k}^{(k)}}\right),$$

m_{+jk} is missing row data, c_{i+k} missing column data and d_{++k} is missing row and column data.

M-step:

$$\hat{\theta}_{ijk}^{(k+1)} = x_{ijk} + z_{ijk}^{(k)}.$$

The E- and M-step will be alternate repeated and until

$$\left|\hat{\theta}_{ijk}^{(k+1)} - \hat{\theta}_{ijk}^{(k)}\right| = \left|\left(x_{ijk} + z_{ijk}^{(k+1)}\right) - \left(x_{ijk} + z_{ijk}^{(k)}\right)\right|$$

$$= \left|z_{ijk}^{(k+1)} - z_{ijk}^{(k)}\right|$$

$$\leq \varepsilon$$

Therefore when $k \rightarrow \infty$

$$\lim_{k\to\infty}\left|\hat{\theta}_{ijk}^{(k+1)} - \hat{\theta}_{ijk}^{(k)}\right| = 0.$$

4. Results

Table 3 displays the parameter estimates we have obtained after considering the model from (3.4). Table 4 is the expected value we obtained by considering the results of Table 3 to estimate the cell probabilities. However, Table 5 shows the results of the expected value from the EM algorithm.

Table 3. Parameter estimations for logistic regression model.

Parameters	parameters estimates	p-value
	Likelihood ratio = < 0.0001	
β_{30}	-0.0641	0.343
β_{31}	0.9846	<0.0001
β_{32}	0.0964	0.41
	Likelihood ratio = < 0.0001	
β_{20}	-0.7557	<0.0001
β_{21}	0.8655	<0.0001
	Likelihood ratio = 0.4295	
ϕ_{30}	-0.0134	0.9594
ϕ_{31}	-0.00601	0.982
ϕ_{32}	0.1468	0.1943

Table 3. Continued

Parameters	parameters estimates	p-value
	Likelihood ratio = 0.9828	
ϕ_{20}	3.1176	<0.0001
ϕ_{21}	0.0526	0.8534
ϕ_{22}	-0.00602	0.982
	Likelihood ratio = < 0.0001	
α_{30}	0.0047	0.9856
α_{31}	1.0079	<0.0001
α_{32}	-0.0382	0.8857
	Likelihood ratio = 0.9895	
α_{20}	3.1331	<0.0001
α_{21}	0.0192	0.9713
α_{22}	-0.0382	0.8857
	Likelihood ratio = NA	
γ_{210}	3.7758	<0.0001
γ_{211}	0.0175	0.0037
γ_{220}	0.6439	0.0028
γ_{221}	0.0153	0.9234
γ_{230}	0.0153	0.945
γ_{231}	-0.00108	0.9961

Table 4. Expected value for logistic regression.

		Treatment		
	RSV	**Placebo**	**Palivizumab**	**Total**
BPD	Yes	40	34	74
	No	227	461	688
Premature	Yes	14	14	28
	No	220	492	712
Total		501	1001	1502

Table 5. Expected value for the EM algorithm.

		Treatment		
	RSV	**Placebo**	**Palivizumab**	**Total**
BPD	Yes	34	39	73
	No	232	458	690
Premature	Yes	19	9	29
	No	215	496	711
Total		500	1002	1502

From Table 3, we observed that the overall model is significant. Therefore, we consider that model to estimate the cell probabilities and cell frequencies. From Table 4 and Table 5, we can see that the marginal column total and marginal row total are almost the same. But there are some obvious difference between the cells (1, 1, 1), (1, 1, 2), (1, 2, 1) and (1, 2, 2) when we consider for the marginal column total probabilities for both Table 4 and Table 5. Therefore, in the next section, we consider apparent error rate (APER), Mcnemar's test, Mantel-Haenszel test, sensitivity and specificity to test our results.

6. Test Procedures

6.1. APER

In this section, we have employed different techniques to examine the misclassification error rate as well as to test for the model based on the logistic regression. The results based on the logistic regression are then compared with the results based on the EM algorithm.

From Johnson and Wichern (1998), APER is calculated as

$$\text{APER} = \frac{n_{1m} + n_{2m}}{n_1 + n_2} \tag{6.1}$$

where n_{1m} is number misclassified for predicted membership on group for variable p_1 as group for variable p_2, n_{2m} is number misclassified for predicted membership on group for variable p_2 as group for variable p_1, n_1 is actual membership on group p_1, and n_2 is actual membership on group p_2. This can be illustrated in Table 6.

From Table 6, n_{1c} is referred to number of p_1 are correctly classified and n_2 is referred to number of p_2 are correctly classified.

Table 6. Illustration for APER.

		Predicted memberhsip		
		p_1	p_2	
Actual	p_1	n_{1c}	$n_{1m}= n_1 - n_{1c}$	n_1
membership	p_2	$n_{2m}= n_2 - n_{2c}$	n_{2c}	n_2

Table 7. APER for 1) logistic regression based estimates with complete data, 2) the EM algorithm with complete data, 3) and logistic regression with the EM algorithm that the

		APER	
Cell	**1**	**2**	**3**
(1,1,1)	0.40%	0%	0.40%
(1,1,2)	0.33%	0%	0.33%
(1,2,1)	0.33%	0%	0.33%
(1,1,2)	0.33%	0%	0.33%
(2,1,1)	0.33%	0%	0.33%
(2,1,2)	0.33%	0%	0.33%
(2,2,1)	0.27%	0.07%	0.2%
(2,2,2)	0.33%	0.07%	0.27%
Overall	0.33%	0.02%	0.32%

We compare the results of Table 4 and Table 5 with Table 1, the complete data, by applying (6.1). The results are shown in Table 7 where the results are compared between the complete data and the method based on logistic regression. The comparison is made for these sets: 1) comparison between logistic regression based estimation with the complete data, 2) the comparison between the EM algorithm with the complete data and 3) comparison between the logistic regression based estimation with the EM algorithm.

From Table 7, we reveal that the values are small for all the comparison that we have considered. It seems that the EM algorithm has the smallest discrimination with the complete data when we consider the overall APER. Therefore, we can say that the EM algorithm can provide slightly better result in predicting complete data. For comparison 2 in Table 7, it appears that some values are 0 indicating no error at all. This implies that the EM algorithm predicts exactly the same value as compared to the complete data. However, although comparison 1 and 3 give higher APER compared to comparison 2, it may be considered as small. It means that the logistic approach can be considered to predict the complete data without much deviation from the reality. In Table 7, each cell gives the small value of APER. Therefore we can conclude that there is no substantial difference between the logistic regression based estimation with complete data and also for the logistic regression based estimation and the EM algorithm estimates.

6.2. McNemar's Test

We further investigate our results with McNemar's Test. As referred by Feinstein (2002), McNemar chi-square test is to test whether cells of (1, 2) and (2, 1) as in Table 6 have the equal value. A goodness-of-fit chi-square test between the observed and expected values is calculated as

$$\chi_M^2 = \frac{(n_{1m} - n_{2m})^2}{n_{1m} + n_{2m}} \tag{6.2}$$

with 1 degree of freedom. Equation (6.2) will become

$$\chi_{M_c}^2 = \frac{(|n_{1m} - n_{2m}| - 1)^2}{n_{1m} + n_{2m}} \tag{6.3}$$

when a continuity correction is considered.

Therefore, we consider McNemar's test with H$_0$: $n_{ijk} = n_{lijk}$, H$_0$: $n_{ijk} = n_{eijk}$, H$_0$: $n_{lijk} = n_{eijk}$, where n_{ijk} is cell frequency of (i, j, k) for complete data, n_{lijk} is expected cell frequency of (i, j, k) for the logistic regression based estimation and n_{eijk} is expected cell frequency of (i, j, k) for the EM algorithm. By applying the formula of (6.2) and (6.3), we obtain the results as shown in Table 8.

Table 8 shows that test results for each cell for comparisons 1, 2, and 3 as described in the previous section. The results reveal that comparison 2 provides the best results. For comparison 2, 6 cells among 8 are predicted matching with observed data where for $\chi_{M_c}^2$ is NA. Result in Table 8 shows that comparison 2 gives the perfect results after we consider continuity correction of McNermar's test. However for comparison 1 and comparison, only one cell (1, 1, 1) shows significant difference between the observed and expected cell frequencies. All other cells indicate match between observed and the estimated cell frequencies based on the logistic regression based estimation with complete data and the logistic regression based estimation with the EM algorithm.

Table 8. Results of Mcnemar's test.

	comparison 1		comparison 2		comparison 3	
Cell	χ^2_M	$\chi^2_{M_c}$	χ^2_M	$\chi^2_{M_c}$	χ^2_M	$\chi^2_{M_c}$
(1,1,1)	6	4.17*	NA	NA	6	4.17*
(1,1,2)	5	3.2	NA	NA	5	3.2
(1,2,1)	5	3.2	NA	NA	5	3.2
(1,2,2)	5	3.2	NA	NA	5	3.2
(2,1,1)	5	3.2	NA	NA	5	3.2
(2,1,2)	5	3.2	NA	NA	5	3.2
(2,2,1)	4	2.25	1	0	3	1.33
(2,2,2)	5	3.2	1	0	4	2.25

* signigicant at 5% level

6.3. Mantel-Haenszel Test

In this section, one of the test procedure known as the Mantel-Haenszel test (Mantel and Haenszel, 1959; Agresti, 1990) is used for examing the comparisons 1, 2 and 3. This test applies conditional independence to K strata of 2x2 tables. This test will be based on the assumption that the odds ratios are constant (Hosmer and Lemeshow, 2000). The test statistics will be noted as

$$\chi^2_{MH} = \sum \left\{ w_i \left[\ln\left(\hat{OR}_i \right) - \ln\left(\hat{OR}_L \right) \right]^2 \right\}, \tag{6.4}$$

with k-1 degrees of freedom, $\hat{OR}_i$ is the odd ratio for each strata,

$$\hat{OR}_L = \exp\left[\sum w_i \ln\left(\hat{OR}_i \right) \Big/ \sum w_i \right],$$

with $w_i = 1 \Big/ \operatorname{var}\left[\ln\left(\hat{OR}_i \right) \right]$, and

$$\operatorname{var}\left[\ln\left(\hat{OR}_i \right) \right] = \frac{1}{n_{11}} + \frac{1}{n_{12}} + \frac{1}{n_{21}} + \frac{1}{n_{22}},$$

as n_{11}, n_{12}, n_{21}, and n_{22} are the cell frequency.

We consider Mantel-Haenszel test by considering the null hypothesis H_0: odds ratios are equal in all the strata. The results are obtained and shown in Table 9 and Table 10. Table 9 shows the odds ratios for each strata we have calculated and Table 10 displays the results of Mantel-Haenszel test.

Table 9. Odds ratio for each strata for the three comparisons we considered.

	Cell	
Odds ratio	**(+, +, 1)**	**(+, +, 2)**
Complete data and logistic regression	39374.72	38383.40
Complete data and EM algorithm	NA	323682.00
Logistic regression and EM algorithm	43807.87	42656.33

Table 10, Mantel-Haenszel test.

	χ^2_{MH}
Complete data and logistic regression	0.0016
Complete data and EM algorithm	NA
Logistic regression and EM algorithm	0.0017

From Table 10, we conclude that we accept H_0 for complete data and logistic regression based estimation (comparison 1) and logistic regression based estimation and the EM algorithm (comparison 3). We expressed that between the strata we consider, which are the cells denoted by (+, +, 1) and (+, +, 2) with complete data and the logistic regression based estimation, both strata show there is common odds ratio. As the test show that each strata have the common odd ratio, and we observed from the table 9 that the odd ratio is very large for all the values we calculated indicating very high association between comparison groups. Test results in Table 10 indicates that there is statistically significant difference in comparison 1 (complete data and logistic regression based estimation) as well as in comparison 3 (logistic regression based estimation and the EM algorithm) are equally efficient in estimates the missing values. However, for the EM algorithm, we cannot calculate the odds ratio for cell (+, +, 1). This is because EM algorithm predicts the exactly same value with the complete data. It shows that EM algorithm is performing well to estimate the complete data. Since odds ratio cannot be obtained in Table 9, it is not possible to compute χ^2_{MH} in Table 10 for comparison 2 (complete data and EM algorithm). However, this is attributable to 100% matching in cell (+, +, 1)

6.4. Sensitivity and Specificity

As referred to Feinsten (2002), the terms *sensitivity* and *specificity* was introduced by Yerushalmy (1969). Definition for sensitivity is the proportion of "true positive" results in diseased cases and specificity is the proportion of "true negative" results in the nondiseased controls. The measures of sensitivity and specificity are used here to find the level of accuracy of the estimates based on the logistic regression as well as the EM algorithm. By referring to Table 6, the calculation of sensitivity and specificity will be as below:

$$\text{Sensitivity} = \frac{n_{1c}}{n_1} \times 100\% \text{ and specificity} = \frac{n_{2c}}{n_2} \times 100\%,$$

and the results are shown in Table 11.

Table 11 reveals that both sensitivity and specificity are nearly 100% for comparisons 1, 2, and 3 for overall and also for almost all the cells. From the sensitivity and specificity, we observe more than 95% for comparison 1, 2, and 3. Therefore we can conclude there is no significance difference for comparison 1 (logistic regression based estimation and complete data), comparison 2 (the EM algorithm and complete data) and comparison 3 (logistic regression based estimation and the EM algorithm.

Table 11. Result for sensitivity and specificity for complete data and logistic regression (1), complete data and the EM algorithm (2), and logistic regression and the EM algorithm (3).

		Sensitivity			Specificity	
Cell	1	2	3	1	2	3
(1,1,1)	100%	100%	85%	99.6%	100%	100%
(1,1,2)	73.7%	100%	100%	100%	100%	99.7%
(1,2,1)	87.2%	100%	100%	100%	100%	99.7%
(1,2,2)	100%	100%	64.3%	99.7%	100%	100%
(2,1,1)	100%	100%	100%	100%	100%	99.7%
(2,1,2)	100%	100%	97.7%	99.7%	100%	100%
(2,2,1)	100%	100%	99.3%	99.6%	99.9%	100%
(2,2,2)	99%	99.8%	100%	100%	100%	99.6%
Overall	98.7%	99.9%	98.7%	99.8%	100%	99.8%

From Table 11, we have 100% of sensitivity for cells (1, 1, 1), (1, 2, 2), (2, 1, 1), (2, 1, 2), and (2, 2, 1) for comparison 1, cells (1, 1, 1), (1, 1, 2), (1, 2, 1), (1, 2, 2), (2, 1, 1), and (2, 1, 2) for comparison 2, and cells (1, 1, 2), (1, 2, 1), (2, 1, 1) and (2, 2, 2) for comparison 3. However 100% of specificity is observed for cells (1, 1, 2), (1, 2, 1), (2, 1, 1) and (2, 2, 2) for comparison 1, all cells except (2, 2, 1) for comparison 2, and cells (1, 1, 1), (1, 2, 1), (1, 2, 2), (2, 1, 1), (2, 2, 1), and (2, 2, 2) for comparison 3 . It shows that those cells were perfectly predicted by logistic regression based estimation, and the EM algorithm. For more than 95% of sensitivity, we have 1 cell in comparison 1, 2 cells in comparison 2, and 2 cells in comparison 3. For more than 95% of specificity, we have 5 cells in comparison 1, 1 cell in comparison 2, and 4 cells in comparison 3. Thus most cells show no substantial difference for logistic regression based estimation and complete data, the EM algorithm and complete data, and also between logistic regression based estimation and EM algorithm. Only 1 cell in comparison 1 and comparison 3 which is cell (1, 1, 2) and (1, 2, 2) respectively is below 80% for the sensitivity.

7. Conclusion

A new method of estimating missing values is proposed in this paper. The proposed method is based on the conditional probabilities for two or more categorical variable. This method has wider scope due to its potential utility in predicting the values based on a logistic regression model. This will enhance the utility of the technique in dealing with missing observations. Several test procedures are employed to compare the efficiency of the proposed method as compared to that of the complete data as well as the traditional EM algorithm approach. It is evident that the proposed method can detect missing values more consistently in many cases.

From the results, we conclude that logistic regression can be an alternative to obtain the expected value compared to the EM algorithm. The proposed method may help us to obtain the expected values for incomplete data without employing traditional algorithm. The proposed method gives us much flexibilition and prediction. We show that logistic regression based method can effectively estimate the value of a categorical variable when we have information on the other categorical variables. Hence, the logistic regression method can be used as a discriminant or classification method as well and this method can be generalized for any number of categorical variables without making the procedure more complex for the general users.

References

Agresti, A. (2002), *Categorical Data Analysis*, 2nd. ed., John Wiley and Sons, New York.

Dempster A. P., Laird N. M. and Rubin D. B. (1977) Maximum likelihood from incomplete data via the EM algorithm (with discussion), *J. Roy. Statist. Soc. B,* **39**, 1-38.

Feinstein, A. R. (2002), *Principles of Medical Statistics*, Chapman and Hall/CRC, Florida.

Fitzmaurice, G. M., Heath, A. F., and Clifford, P. (1996), Logistic Regression Model for Binary Panel Data with Attrition, *Journal of the Royal Statistical Society A*, **159**, 249-263.

Fitzmaurice, G. M., Laird, N. M., and Zahner, E. P., (1996), Multivariate Logistic Models for Incomplete Binary Response, *JASA*, **91**, 99-108

Hosmer, D. W. and Lemeshow, S. (2000), *Applied Logistic Regression*, 2nd. Ed. John Wiley and Sons, New York.

Ibrahim, J. G., Lipsitz, S. R., and Chen, M. H. (1999), Missing Covarites in Generalized Linear Models when the Missing Data Mechanism is Non-ignorable, *Journal of the Royal Statistical Society B*, **61**, 173-190.

Johnson, R. A. and Wichern, D. W. (1998), *Applied Multivariate Statistical Analysis*, Prentice-Hall, Inc., New Jersey.

Mantel, N. and Haenszel, W. (1959), Statistical aspects of the analysis of data from retrospective studies of disease. *J. Natl. Cancer Inst.*, **22**, 719-748.

McCullagh, P. and Nelder, J. A. (1989), *Generalized Linear Models*, 2nd. Ed., Chapman and Hall, London.

McLachlan G. J. and Krishnan T. (1997) *The EM algorithm and extensions*, New York: John wiley.

Simnoff, J. S. (2003), *Analyzing Categorical Data*, Springer-Verlag, New York.

Vach, W. and Schumacher, M. (1993), Logistic regression with incompletely observed Categorical Covariates: A comparison of Three Approaches, *Biometrika*, **80**, 353-362

Yerushalmy, J. (1969) The statistical assessment of the variablility in observer perception and description on roentgenographic pulmonary shadow. *Radiol. Clin. North Am.* **7**, 381-392

In: New Developments in Applied Statistics
Editor: Mohammad Ahsanullah

ISBN 978-1-61324-648-1

Chapter 22

COMPUTING THE MOMENTS OF BIVARIATE AND TRIVARIATE ORDER STATISTICS

H.M. Barakat*
Department of Mathematics, Faculty of Science,
Zagazig University, Zagazig- Egypt

Abstract

The most of the material of this article is an overview of principle results in and related to the computing of the product moments (of any order) of bivariate (and trivariate) order statistics arisen from any arbitrary bivariate (and trivariate) distribution function. The exact explicit expression, as well as formulae to compute easily these moments are presented. Some recent analytic bootstrap-type methods for estimating these moments are exhibited. In addition, universal distribution -free bounds for the differences of any two different product moments and general recurrence relations satisfied by the product moments (of any order) are discussed. Finally, a recent general bivariate and trivariate versions of the basic triangle rule for order statistics arisen from univariate distribution function is presented.

Keywords: Bivariate order statistics; Trivariate order statistics; Moments of order statistics; Recurrence relations.

1. Computing the Moments of Bivariate Order Statistics

If measurements of two characteristics are taken on the same members of the population, then the observed random quantities follow some type of bivariate distribution. Let this distribution be $F(\underline{x}) = F(x_1, x_2) = P(X_1 \leq x_1, X_2 \leq x_2)$. Consider a sequence of n two-dimensional random variables (rv's) $\{\underline{X}_j\} = \{(X_{1j}, X_{2j})\}$, $j = 1, 2, ..., n$, with common distribution function (df) $F(\underline{x})$. The order statistics of the tth $(t = 1, 2\ )$ component (characteristic) are $X_{t,1:n} \leq X_{t,2:n} \leq ... \leq X_{t,n:n}$. For any real numbers M and N, let $\mu^{M,N}_{k_1,k_2:n}$ indicate the product moment $\mathrm{E}(X^M_{1,k_1:n} X^N_{2,k_2:n})$. For convenience the components of the numerical vector $\underline{x} = (x_1, x_2)$ are signified by a subscript. Thus $\underline{x} \leq \underline{y}$ means

*E-mail address: hbarakat2@hotmail.com

$x_t \leq y_t$, $t=1,2$. The survivorship functions of the random vector $\underline{X}=(X_1,X_2)$ and its marginal components X_t, $t=1,2$, are defined respectively by $G=G(\underline{x})=P(\underline{X}>\underline{x})$ and $G_t=G_t(x_t)=1-F_t(x_t)$, $t=1,2$. Moreover, the copula (the dependence function) of the df $F(\underline{x})$ is defined by $C(\underline{u})=C(u_1,u_2)=F(F_1^{-1}(u_1),F_2^{-1}(u_2))$, where $F_t^{-1}(u_t)=\sup\{x_t : F_t(x_t)\leq u_t\}, 0<u_t<1$, is the inverse of the df $F_t(x_t)$. The term dependence function is used by Galambos (1987), Chapter 5. Sklar (1959) introduced the general notion of this function and used the term copula, which seems to be more current in the literature than the first term. This is a function which links the two-dimensional df to its one-dimensional margins and is itself a continuous df (whenever, each margin of F is continuous) on the unit square $[0,1]^2$, with uniform margins. Moreover, if we envision $(U_{1,i},U_{2,i})$, $i=1,2,...,n$, as being identically independent distributed rv's with common df $C(\underline{u})$, and write "$\overset{d}{=}$"to be read as "has the same df as ", then $X_{t,k_t:n}\overset{d}{=}F_t^{-1}(U_{t,k_t:n})$, $t=1,2$, where $U_{t,1:n},U_{t,2:n},\ldots,U_{t,n:n}$ are the order statistics of $U_{t,1},U_{t,2},\ldots,U_{t,n}$, $t=1,2$. Throughout this paper, we use the notation $\nu^{M,N}_{k_1,k_2;n}=\mathrm{E}(U^M_{1,k_1:n}U^N_{2,k_2:n})$, i.e., $\nu^{M,N}_{k_1,k_2:n}$ be the (M,N)th moment of the bivariate order statistics from the bivariate uniform df $C(\underline{u})$ (the copula of $F(\underline{x})$).

By using the elementary formula of multinomial distribution, it is not difficult to obtain the explicit form of the joint df (jdf) $F_{k_1,k_2:n}(\underline{x})=P(X_{1,k_1:n}\leq x_1, X_{2,k_2:n}\leq x_2)$, for any integers values $1\leq k_1,k_1\leq n$. Namely,

$$F_{k_1,k_2:n}(\underline{x})=\sum_{i=k_1}^{n}\sum_{j=k_2}^{n}\sum_{r=0\vee(i+j-n)}^{i\wedge j} C_{i,r,j,n}(F_1(x_1)-F(\underline{x}))^{i-r}F^r(\underline{x})$$

$$(F_2(x_2)-F(\underline{x}))^{j-r}G^{n-i-j+r}(\underline{x}), \tag{1.1}$$

where $C_{i,r,j,n}=\frac{n!}{(i-r)!r!(j-r)!(n-i-j+r)!}$ (see, Exercise 2.2.5, Page 25 in David and Nagaraja, 2003). Also, by using the elementary formula of multinomial distribution, the survivorship function $G_{k_1,k_2:n}(\underline{x})$ of the df $F_{k_1,k_2:n}(\underline{x})$ can be written as

$$G_{k_1,k_2:n}(\underline{x})=\sum_{i=0}^{k_1-1}\sum_{j=0}^{k_2-1}\sum_{r=0\vee(i+j-n)}^{i\wedge j} C_{i,r,j,n}(F_1(x_1)-F(\underline{x}))^{i-r}F^r(\underline{x})$$

$$(F_2(x_2)-F(\underline{x}))^{j-r}(1-F_1(x_1)-F_2(x_2)+F(\underline{x}))^{n-i-j+r}. \tag{1.2}$$

The jdf $F_{k_1,k_2:n}(\underline{x})$ can be easily derived from (1.2). Moreover, (1.2) is more suitable for limiting purposes (see, Barakat 1990, 1998 and 2001).

Maritz (1991) has proposed an exact bootstrap estimator for the product moments $\mu_{k_1,k_2:n}=\mu^{1,1}_{k_1,k_2:n}$, $k_1,k_2=1,2,...,n$, based upon an estimate of the df $F_{k_1,k_2:n}(\underline{x})$ of the bivariate order statistics $(X_{1,k_1:n},X_{2,k_2:n})$, $k_1,k_2=1,2,...,n$. This estimate is given by replacing the empirical estimator $\hat{F}(\underline{x})$, for $F(\underline{x})$ in (1.1). The the empirical estimator $\hat{F}(\underline{x})$, is defined by

$$\hat{F}(\underline{x})=\frac{1}{n}\#\{X_{1i}\leq x_1, X_{2i}\leq x_2\}$$

$$=\begin{cases} 0, & (x_1,x_2)<(X_{1,1:n},X_{[2,1:n]}), \\ \frac{i-1}{n}, & (X_{1,i-1:n},X_{[2,i-1:n]})\leq(x_1,x_2)<(X_{1,i:n},X_{[2,i:n]}), \\ & i=2,\ldots,n, \\ 1, & (x_1,x_2)\geq(X_{1,n:n},X_{[2,n:n]}), \end{cases}$$

where the pair $(X_{1,i:n}, X_{[2,i:n]})$, $i = 1, 2, \ldots, n$, denote the bivariate sample data ordered by X_1, and $X_{[2,i:n]}$ denote the corresponding ith concomitant statistic (see, David 1981, for description of concomitants). The estimated probability mass at (x_1, x_2) is then given by

$$\hat{f}_{k_1,k_2:n}(\underline{x}) = \hat{F}_{k_1,k_2:n}(x_1+\varepsilon, x_2+\varepsilon) - \hat{F}_{k_1,k_2:n}(x_1-\varepsilon, x_2+\varepsilon)$$

$$-\hat{F}_{k_1,k_2:n}(x_1+\varepsilon, x_2-\varepsilon) + \hat{F}_{k_1,k_2:n}(x_1-\varepsilon, x_2-\varepsilon), \ (\varepsilon \text{ is small}),$$

or equivalently

$$\hat{f}_{k_1,k_2:n}(\underline{x}) = \hat{G}_{k_1,k_2:n}(x_1+\varepsilon, x_2+\varepsilon) - \hat{G}_{k_1,k_2:n}(x_1-\varepsilon, x_2+\varepsilon)$$

$$-\hat{G}_{k_1,k_2:n}(x_1+\varepsilon, x_2-\varepsilon) + \hat{G}_{k_1,k_2:n}(x_1-\varepsilon, x_2-\varepsilon), \ (\varepsilon \text{ is small}),$$

obtained by substituting $\hat{F}(\underline{x})$ for $F(\underline{x})$ in (1.1) and (1.2), respectively.

However, if the data are highly correlated such that i equals $[i]$, in the pair $(X_{1,i:n}, X_{[2,i:n]})$, $i = 1, 2, \ldots, n$, it is easy to see that (c.f. Hutson, 2000) $\hat{F}_{k_1,k_2:n}(x_1, x_2) = 0$, for all possible (x_1, x_2) combinations, so the approach of Maritz has some shortcoming.

Relaying on the definition of the classical joint empirical density function estimator

$$\hat{f}_{\underline{X}} \begin{cases} \frac{1}{n}, & (\underline{x}) = (x_1, x_2) = (X_{1,i:n}, X_{[2,i:n]}),\ i = 1, 2, \ldots, n, \\ 0, & \text{otherwise,} \end{cases}$$

and by a simple rank transformation of $\hat{f}(\underline{x})$ to the joint empirical density function of the bivariate order statistics $X_{1,k_1:n}$ and $X_{2,K_2:n}$, Hutson (2000) has shown that the joint empirical density function estimator of the bivariate order statistics $X_{1,k_1:n}$ and $X_{2,k_2:n}$, evaluated at the point $\underline{x} \in \{(x_{1,i-1:n}, x_{1,i:n}] \times (x_{2,j-1:n}, x_{2,j:n}]\}$, $i, j = 1, 2, \ldots, n$, is given by

$$\hat{f}_{k_1,k_2:n}(\underline{x}) = n \sum_{k=1}^{n} (b_{k_1}(i,k) - a_{k_1}(i,k))(b_{k_2}(j,[k]) - a_{k_2}(j,[k])) \mathbf{I}_{\{\gamma_n(i,j,k)>0\}},$$

where

$$\gamma_n(i,j,k) = \min\left\{ \beta_{k_1:n}(\frac{i}{n}) - \frac{k-1}{n}; \frac{k}{n} - \beta_{k_1:n}(\frac{i-1}{n}); \beta_{k_2:n}(\frac{j}{n}) - \frac{[k]-1}{n}; \frac{[k]}{n} - \beta_{k_2:n}(\frac{j-1}{n}) \right\},$$

$a_{k_1}(i,k) = \beta_{k_1:n}(\frac{i-1}{n}) \vee \frac{k-1}{n}$, $b_{k_1}(i,k) = \beta_{k_1:n}(\frac{i}{n}) \wedge \frac{k}{n}$, $\beta_{k_1:n}(.)$ and $\mathbf{I}_{\{\gamma_n(i,j,k)>0\}}$ denote the cumulative df of a beta rv with parameters $k_1 - 1$ and $n - k_1 + 1$ and the indicator function, respectively, $[k]$ denotes the concomitant rank k in the pair $(X_{1,k:n}, X_{2,[k]:n})$, and finally, the two abbreviations $a \wedge b$ and $a \vee b$ denote $\min(a,b)$ and $\max(a,b)$, respectively.

The bootstrap-type estimator of $\mu^{M,N}_{k_1,k_2:n}$ follows straightforward and is given by

$$\hat{\mu}^{M,N}_{k_1,k_2:n} = \sum_{i=1}^{n} \sum_{j=1}^{n} x^{M}_{1,i:n} x^{N}_{2,j:n} \hat{f}_{k_1,k_2:n}(x_{1,i:n}, x_{2,j:n}).$$

One of the important advantage of the above estimator is that it gives the standard bootstrap estimators of the marginal expectations $\mathrm{E}(X^{M}_{1,k_1:n})$ and $\mathrm{E}(X^{N}_{2,k_2:n})$, which are given in Hutson and Ernst (2000).

When $F(\underline{x}) = F(x_1,x_2)$ is implicitly assumed to be of a continuous type and derivable with respect to $\underline{x}$, Barakat (1999) explicitly derived the joint probability density function (jpdf) $f_{k_1,k_2:n}(\underline{x})$ of the bivariate order statistics $(X_{1,k_1:n}, X_{2,k_2:n})$ and from which an exact explicit expression for $\mu^{M,N}_{k_1,k_2:n}$ is derived.

Theorem 1.1 (Barakat, 1999). Let M and N be any two real values. Then, for any continuous bivariate df $F(\underline{x})$, we have

$$\mu^{M,N}_{k_1,k_2:n} = \int_0^1 \int_0^1 (F_1^{-1}(u_1))^M (F_2^{-1}(u_2))^N f_{k_1,k_2:n}(\underline{u}) du_1 du_2, \tag{1.4}$$

where

$$f_{k_1,k_2:n}(\underline{u}) = \sum_{r=0\vee(k_1+k_2-n-1)}^{(k_1-1)\wedge(k_2-1)} \frac{C^{\star}_{k_1,r,k_2,n+1}}{n+1}(u_1 - C(\underline{u}))^{k_1-1-r} C^r(\underline{u})$$

$$\times (u_2 - C(\underline{u}))^{k_2-1-r}(1-u_1-u_2+C(\underline{u}))^{n-k_1-k_2+1+r} C^{1,1}(\underline{u})$$

$$+ \sum_{\varphi=\varphi_*}^{\varphi^*} \sum_{\theta=\theta_*}^{\theta^*} \sum_{r=r_*}^{r^*} C^{\star}_{k_1-\theta,r,k_2-\varphi,n}(u_1 - C(\underline{u}))^{k_1-1-\theta-r} C^r(\underline{u})(u_2 - C(\underline{u}))^{k_2-1-\varphi-r}$$

$$\times (1-u_1-u_2+C(\underline{u}))^{n-k_1-k_2+\theta+\varphi+r}(C^{1,\cdot}(\underline{u}))^{\varphi}(C^{\cdot,1}(\underline{u}))^{\theta}(1-C^{1,\cdot}(\underline{u}))^{1-\varphi}(1-C^{\cdot,1}(\underline{u}))^{1-\theta},$$

$\varphi_* = 0\vee(k_2-n+1)$; $\varphi^* = 1\wedge(k_2-1)$; $\theta_* = 0\vee(k_1-n+1)$; $\theta^* = 1\wedge(k_1-1)$; $r_* = 0\vee(k_1+k_2-n-\theta-\varphi)$; $r^* = (k_1-\theta-1)\wedge(k_2-\varphi-1)$; $C^{\star}_{k_1,r,k_2,n} = \frac{n!}{(k_1-1-r)!r!(k_2-1-r)!(n-k_1-k_2+r)!}$; $C^{1,\cdot}(\underline{u}) = \frac{\partial C(\underline{u})}{\partial u_1}$, $C^{\cdot,1}(\underline{u}) = \frac{\partial C(\underline{u})}{\partial u_2}$ and $C^{1,1}(\underline{u}) = \frac{\partial^2 C(\underline{u})}{\partial u_1 \partial u_2}$.

In particular we get from (1.4),

$$\mu^{M,N}_{1,1:n} = n\int_0^1 \int_0^1 (F_1^{-1}(u_1))^M (F_2^{-1}(u_2))^N (1-u_1-u_2+C)^{n-2}$$

$$[(n-1)(1-C^{1,\cdot})(1-C^{\cdot,1}) + (1-u_1-u_2+C)C^{1,1}]du_1du_2,$$

$$\mu^{M,N}_{n,n:n} = n\int_0^1 \int_0^1 (F_1^{-1}(u_1))^M (F_2^{-1}(u_2))^N C^{n-2}[(n-1)C^{1,\cdot}C^{\cdot,1} + CC^{1,1}]du_1du_2,$$

$$\mu^{M,N}_{1,n:n} = n\int_0^1 \int_0^1 (F_1^{-1}(u_1))^M (F_2^{-1}(u_2))^N (u_2 - C)^{n-2}[(n-1)C^{1,\cdot}(1-C^{\cdot,1}) + (u_2-C)C^{1,1}]du_1du_2$$

and

$$\mu^{M,N}_{n,1:n} = n\int_0^1 \int_0^1 (F_1^{-1}(u_1))^M (F_2^{-1}(u_2))^N (u_1 - C)^{n-2}[(n-1)C^{\cdot,1}(1-C^{1,\cdot}) + (u_1-C)C^{1,1}]du_1du_2$$

When the marginal df's $F_1(x_1)$ and $F_2(x_2)$ are uniform on $(0,1)$, the product moment $\mu^{M,N}_{k_1,k_2:n}$ given in (1.4), becomes $\nu^{M,N}_{k_1,k_2:n}$. Clearly the computation of the latter moment is easier than the computation of the first. By using the well known $\delta-$method (see Arnold et al., 1992, Chap. 5), Barakat (1999) have obtained an approximate formula for computing the product moment $\mu^{1,1}_{k_1,k_2:n}$ for any bivariate df F by using the moment $\nu^{1,1}_{k_1,k_2:n}$. Namely,

$$\mu^{1,1}_{k_1,k_2:n} = F_1^{-1}(\frac{k_1}{n+1})F_2^{-1}(\frac{k_2}{n+1})\left(1 + \frac{k_1+k_2-2k_1k_2}{2(n+1)(n+2)} + \frac{1}{2}\nu^{1,1}_{k_1,k_2:n}\right) + o(n^{-2}).$$

When the dependence structural between X_1 and X_2 is known, but the marginal df's F_1 and F_2 are not, i.e., the copula $C(\underline{u})$ is known while the two marginal df's F_1 and F_2, are not known, by applying the sample quantile function, i.e., $\hat{F}_t^{-1}(u)=X_{t,i:n},\ t=1,2,$ in the region given by $\frac{i-1}{n}\le u\le\frac{i}{n},\ i=1,2,...,n,$ on the relation (1.4), we can easily get an interesting bootstrap-type estimator for $\mu_{k_1,k_2:n}^{M,N}$, which is given by

$$\hat{\mu'}_{k_1,k_2:n}^{M,N}=\sum_{i=1}^{n}\sum_{j=1}^{n}X_{1,i:n}^{M}X_{2,j:n}^{N}W_{i,j;k_1,k_2:n},$$

where

$$W_{i,j;k_1,k_2:n}=\int_{\frac{i-1}{n}}^{\frac{i}{n}}\int_{\frac{j-1}{n}}^{\frac{j}{n}}f_{k_1,k_2:n}(\underline{u})du_1du_2$$

$$=F_{k_1,k_2:n}(\frac{i}{n},\frac{j}{n})-F_{k_1,k_2:n}(\frac{i-1}{n},\frac{j}{n})-F_{k_1,k_2:n}(\frac{i}{n},\frac{j-1}{n})+F_{k_1,k_2:n}(\frac{i-1}{n},\frac{j-1}{n})$$

$$=G_{k_1,k_2:n}(\frac{i}{n},\frac{j}{n})-G_{k_1,k_2:n}(\frac{i-1}{n},\frac{j}{n})-G_{k_1,k_2:n}(\frac{i}{n},\frac{j-1}{n})+G_{k_1,k_2:n}(\frac{i-1}{n},\frac{j-1}{n}).$$

In a technical report in the Southern Methodist university, Wang and Stokes (2005) presented tables containing numerical values of the moments $\mu_{k_1,k_2:n}^{1,1}=\mathrm{E}(X_{1,k_1:n}X_{2,k_2:n})$, $n=2,3,4;\ k_1,k_2\le n,$ where the bivariate order statistics $(X_{1,k_1:n},X_{2,k_2:n})$ are arisen from a standard bivariate df with positive correlation (see, also Wang et al., 2005).

For testing the accuracy of computation of moments, we can use the formula $\sum_{i=1}^{n}\sum_{j=1}^{n}\mu_{i,j:n}^{M,N}=n^2\mathrm{E}(X_1^MX_2^N)$, which follows immediately by using the obvious identity $\sum_{i=1}^{n}\sum_{j=1}^{n}X_{1,i:n}^{M}X_{2,j:n}^{N}=\sum_{i=1}^{n}\sum_{j=1}^{n}X_{1i}^{M}X_{2j}^{N},\quad\forall M,N$.

In addition to computing the moments, (1.4) can be used to obtain a universal distribution-free bounds for the differences of any two different product moments of order statistics, for any arbitrary bivariate uniform df.

Theorem 1.2 (Barakat, 1999). For any bivariate uniform df $C(\underline{u})$ and all $2\le k_1,k_2\le n-1$, we have

$$\frac{-k_1}{(n+1)(n+2)}\le\nu_{k_1,k_2:n}^{1,1}-\nu_{k_1,k_2-1:n}^{1,1}\le\frac{3k_1+1}{(n+1)(n+2)}$$

and

$$\frac{-k_2}{(n+1)(n+2)}\le\nu_{k_1,k_2:n}^{1,1}-\nu_{k_1-1,k_2:n}^{1,1}\le\frac{3k_2+1}{(n+1)(n+2)}.$$

Moreover, for all $n=1,2,...,$ we have $1-\frac{3}{(n+1)}\le\nu_{n,n:n}^{1,1}\le1-\frac{1}{(n+1)(n+2)}$; $\nu_{1,1:n}^{1,1}\le\frac{n+3}{(n+1)(n+2)}$; and $\frac{n-1}{(n+1)(n+2)}\le\nu_{1,n:n}^{1,1},\ \nu_{n,1:n}^{1,1}\le\frac{1}{(n+1)}$.

The following corollary is an easy consequence of Theorem 1.2.

Corollary 1.1. For $2\le k_1,k_2\le n-1$, we have

$$\nu_{k_1,1:n}^{1,1}-\frac{(k_2-1)k_1}{(n+1)(n+2)}\le\nu_{k_1,k_2:n}^{1,1}\le\nu_{k_1,1:n}^{1,1}+\frac{(k_2-1)(3k_1+1)}{(n+1)(n+2)}.$$

Clearly, any discontinuity in the df $F(\underline{x})$, along an arbitrary direction, leads to the nonexistence of the corresponding jpdf $f_{k_1,k_2:n}(\underline{x})$. In this case Theorem 1.2 can not be applied. On the other hand, the product moment $\mu^{M,N}_{k_1,k_2:n}$ is classically defined as the double Stieltjes integral (may be improper) $\mu^{M,N}_{k_1,k_2:n} = \int_a^b \int_c^d x_1^M x_2^N d_{x_1x_2} F_{k_1,k_2:n}(\underline{x})$, where the points (a,b) and (c,d) are the left and the right endpoints of the df $F(\underline{x})$, respectively. On the other hand, if the jpdf $f_{k_1,k_2:n}(\underline{x})$ exists, then $d_{x_1x_2} F_{k_1,k_2:n}(\underline{x}) = \frac{\partial^2 F_{k_1,k_2:n}(\underline{x})}{\partial x_1 \partial x_2} dx_1 dx_2 = f_{k_1,k_2:n}(\underline{x})\, dx_1 dx_2$, and therefore $\int_a^b \int_c^d x_1^M x_2^N d_{x_1x_2} F_{k_1,k_2:n}(\underline{x})$, represents the expected value as a Riemann integral. However, for any finite a,b,c and d, Páez-Borrallo and Zazo (1999) presented a closed formula to compute the bivariate factorized expected value from the knowledge of the jdf. In the particular cases $a=c=0, b=d=0, a=d=0$ and $c=b=0$, it can be shown that this formula gives

$$\mu^{M,N}_{k_1,k_2:n} = MN \int_0^b \int_0^d x_1^{M-1} x_2^{N-1} G_{k_1,k_2:n}(\underline{x}) dx_1 dx_2,$$

$$\mu^{M,N}_{k_1,k_2:n} = MN \int_a^0 \int_c^0 x_1^{M-1} x_2^{N-1} F_{k_1,k_2:n}(\underline{x}) dx_1 dx_2,$$

$$\mu^{M,N}_{k_1,k_2:n} = -MN \int_0^b \int_c^0 x_1^{M-1} x_2^{N-1} P(X_{1,k_1:n} > x_1, X_{2,k_2:n} \le x_2) dx_1 dx_2,$$

and

$$\mu^{M,N}_{k_1,k_2:n} = -MN \int_a^0 \int_0^d x_1^{M-1} x_2^{N-1} P(X_{1,k_1:n} \le x_1, X_{2,k_2:n} > x_2) dx_1 dx_2,$$

respectively. Therefore, (1.1) and (1.2) in these cases are sufficient to determine $\mu^{M,N}_{k_1,k_2:n}$. Barakat and El-Shandidy (2004), extended the cases $a=c=0$ and $b=d=0$ to $(0,0) = (a,c) < (a,b) \le (\infty,\infty)$ and $(-\infty,-\infty) \le (a,c) < (b,d) = (0,0)$, respectively. By applying the same argument of Barakat and El-Shandidy (2004), it can be shown that the other two cases $a=d=0$ and $c=b=0$, can be also extended to $\{0=a<b\le\infty, -\infty\le c<d=0\}$ and $\{-\infty\le a<b=0, 0=c<d\le\infty,\}$ respectively. Moreover, for any bivariate df F, we get

$$\mu^{M,N}_{k_1,k_2:n} = MN \left(\int_{-\infty}^0 \int_{-\infty}^0 x_1^{M-1} x_2^{N-1} F_{k_1,k_2:n}(\underline{x}) dx_1 dx_2 \right.$$

$$- \int_0^\infty \int_{-\infty}^0 x_1^{M-1} x_2^{N-1} P(X_{1,k_1:n} > x_1, X_{2,k_2:n} \le x_2) dx_1 dx_2$$

$$- \int_{-\infty}^0 \int_0^\infty x_1^{M-1} x_2^{N-1} P(X_{1,k_1:n} \le x_1, X_{2,k_2:n} > x_2) dx_1 dx_2$$

$$\left. + \int_0^\infty \int_0^\infty x_1^{M-1} x_2^{N-1} G_{k_1,k_2:n}(\underline{x}) dx_1 dx_2 \right).$$

Therefore, (1.1) and (1.2) in all the above cases are sufficient to determine $\mu^{M,N}_{k_1,k_2:n}$.

2. Recurrence Relations for Moments of Bivariate Order Statistics

The recurrence relations and identities for moments of order statistics have assumed interest in recent years. For example, Balakrishnan and Sultan in (1998) pointed out that there are three main reasons why these recurrence relations and identities for the moments of order statistics are important:

1. they reduce the amount of direct computations,
2. they usefully express the higher order moments of order statistics in terms of the lower order moments and hence make the evaluation of higher order moments easy,
3. they are useful in checking the computation of these moments.

The recurrence relations for order statistics arising from univariate df's, whether from i.i.d. samples or from independent and non-identically distributed samples, is a well documented area. In fact, there is a large literature on this topic, with important contributions by Arnold, et al. (1992), Chap. 5, (and the references therein), Balakrishnan (1994) (and the references therein), David (1995), Thomas and Samuel (1996), Samuel and Thomas (1998) and Balakrishnan and Rao (1998), whereas the recurrence relations for bivariate order statistics have received surprisingly no attention. However, there is a primary question, concerning the area of recurrence relations for moments of bivariate order statistics, which is "*is there a recurrence relation for moments of bivariate order statistics which does not depend on the bivariate df (or on its marginals) from which the bivariate order statistics are arisen?*". In other words, is there a distribution-free recurrence relation for moments of bivariate order statistics?. Such a distribution-free recurrence relation for moments of univariate order statistics exists (e.g., the triangle rule, in Theorem 5.3.1 of Arnold et al., 1992). Recently, by using (1.2) and the formula $\nu_{k_1,k_2:n}^{M,N} = MN\int_0^1\int_0^1 x_1^{M-1}x_2^{N-1}G_{k_1,k_2:n}(\underline{x})dx_1dx_2$, Barakat and El-Shandidy (2005) established general formulae satisfied by $\nu_{k_1,k_2:n}^{M,N}$, $M,N>0$, for any general bivariate uniform distribution.

Theorem 2.1 (Barakat and El-Shandidy, 2005). For any given $1 \leq k_1,k_2 < n-1$ and $M,N>0$, we have

$$\frac{1}{n+1}\left((k_1+1)\nu_{k_1+2,k_2+1:n+1}^{M,N}+(2n-k_1-k_2)\nu_{k_1+1,k_2+1:n+1}^{M,N}+(k_2+1)\nu_{k_1+1,k_2+2:n+1}^{M,N}\right)$$
$$-\frac{1}{n+1}\left(\nu_{k_1+1,2:n+1}^{M,N}+\nu_{2,k_2+1:n+1}^{M,N}\right)-\frac{1}{n+1}\left((k_1+1)\nu_{k_1+2,1:n+1}^{M,N}+(k_2+1)\nu_{1,k_2+2:n+1}^{M,N}\right)$$
$$-\frac{1}{n+1}\left((2n-k_1)\nu_{k_1+1,1:n+1}^{M,N}+(2n-k_2)\nu_{1,k_2+1:n+1}^{M,N}\right)+\frac{1}{n+1}\left(\nu_{2,1:n+1}^{M,N}+2n\nu_{1,1:n+1}^{M,N}+\nu_{1,2:n+1}^{M,N}\right)$$
$$=-\left(\frac{M}{M+1}\nu_{k_1+1,k_2+1:n}^{M+1,N}-2\nu_{k_1+1,k_2+1:n}^{M,N}+\frac{N}{N+1}\nu_{k_1+1,k_2+1:n}^{M,N+1}\right)+\left(\frac{M}{M+1}\nu_{k_1,k_2+1:n}^{M+1,N}+\frac{N}{N+1}\nu_{k_1+1,k_2:n}^{M,N+1}\right)$$
$$+\left[\frac{M}{M+1}\left(\nu_{k+1,1:n}^{M+1,N}+\nu_{1,k_2+1:n}^{M+1,N}\right)-2\left(\nu_{k_1+1,1:n}^{M,N}+\nu_{1,k_2+1:n}^{M,N}\right)+\frac{N}{N+1}\left(\nu_{k_1+1,1:n}^{M,N+1}+\nu_{1,k_2+1:n}^{M,N+1}\right)\right]$$
$$-\left(\frac{M}{M+1}\nu_{k_1,1:n}^{M+1,N}+\frac{N}{N+1}\nu_{1,k_2:n}^{M,N+1}\right)-\left(\frac{M}{M+1}\nu_{1,1:n}^{M+1,N}-2\nu_{1,1:n}^{M,N}+\frac{N}{N+1}\nu_{1,1:n}^{M,N+1}\right).$$

Moreover,

$$n\nu^{M,N}_{1,1:n+1} + \nu^{M,N}_{1,2:n+1} = (n+1)\nu^{M,N}_{1,1:n} - \frac{M(n+1)}{M+1}\nu^{M+1,N}_{1,1:n}$$

and

$$n\nu^{M,N}_{1,1:n+1} + \nu^{M,N}_{2,1:n+1} = (n+1)\nu^{M,N}_{1,1:n} - \frac{N(n+1)}{N+1}\nu^{M,N+1}_{1,1:n}.$$

The following elegant relation is a simple consequence of the last two formulae

$$\frac{1}{n+1}\left(\nu^{M,N}_{1,2:n+1} - \nu^{M,N}_{2,1:n+1}\right) = \frac{N}{N+1}\nu^{M,N+1}_{1,1:n} - \frac{M}{M+1}\nu^{M+1,N}_{1,1:n}.$$

The formulae given in Theorem 2.1 are distribution-free recurrence relations. Moreover, they appear to be the only known general recurrence relations, in the literature, for the moments of the bivariate order statistics. The disadvantage of these formulae is that they contain $\nu^{M,N}_{k_1,k_2:n}$, for different values of k_1, k_2, n and M, N. Thus they can not be regarded as real relations for computation of the moments of bivariate order statistics, but they give a way to test the accuracy of computation of moments if this computation is made by numerical procedures.

David, in an extensive work (1993), used the dropping argument to give a greatly shortened proof to the triangle rule for the single and the product moments of order statistics arisen from a general univariate df. Barakat (2006) developed this method to establish a general recurrence relation (Theorem 2.1) satisfied by the jdf $F_{k_1,k_2:n}(\underline{x})$ which solely depends linearly on $A_{k_1:n}(k_2)$, where $A_{r:n}(s) = P(R_{r:n} = s)$ is the df of the rank $R_{r:n}$ of $X_{[2,r:n]}$, and $X_{[2,r:n]}$, $r = 1,2,\ldots,n$, are the concomitants of order statistics of $X_{1,1:n} \le X_{1,2:n} \le \cdots \le X_{1,n:n}$. On the other hand, the rank $R_{r:n}$ is defined by $R_{r:n} = \sum_{j=1}^{n} I_{(X_{[2,r:n]} - X_{2,j})}$, where I_A denotes the indicator function of the set A. David et al. (1977) obtained the df of $R_{r:n}$ (see, also David and Nagaraja, 2003), which can be explicitly written in the form

$$A_{r:n}(s) = \sum_{i=0\vee(r+s-n-1)}^{r-1\wedge s-1} \frac{n!}{(r-i-1)!i!(s-i-1)!(n-r-s+i+1)!}$$

$$\int_{-\infty}^{\infty}\int_{-\infty}^{\infty} (F_1(x_1) - F(\underline{x}))^{r-i-1}F^i(\underline{x})(F_2(x_2) - F(\underline{x}))^{s-i-1}G^{n-r-s+i+1}(\underline{x})f(\underline{x})dx_1dx_2.$$

Theorem 2.2. (Barakat, 2006) For any given $1 \le k_1, k_2 \le n-1$, we have

$$nF_{k_1,k_2:n-1}(\underline{x}) = k_1F_{k_1+1,k_2:n}(\underline{x}) + (n-k_1-k_2)F_{k_1,k_2:n}(\underline{x}) + k_2F_{k_1,k_2+1:n}(\underline{x})$$

$$\sum_{r=1}^{k_1}\sum_{s=1}^{k_2} A_{r:n}(s)\left(F_{k_1+1,k_2+1:n}(\underline{x}) - F_{k_1+1,k_2:n}(\underline{x}) - F_{k_1,k_2+1:n}(\underline{x}) + F_{k_1,k_2:n}(\underline{x})\right). \quad (2.1)$$

Remark 2.1. Clearly, on letting $x_2 \to \infty$ and $x_1 \to \infty$, we get

$$nF^{(1)}_{k_1:n-1}(x_1) = k_1F^{(1)}_{k_1+1:n}(x_1) + (n-k_1)F^{(1)}_{k_1:n}(x_1)$$

and

$$nF^{(2)}_{k_2:n-1}(x_2) = k_2F^{(2)}_{k_2+1:n}(x_2) + (n-k_2)F^{(2)}_{k_2:n}(x_2),$$

respectively, where $F^{(1)}_{k_1:n}(x_1) = \lim_{x_2\to\infty} F_{k_1,k_2:n}(\underline{x})$ and $F^{(2)}_{k_2:n}(x_2) = \lim_{x_1\to\infty} F_{k_1,k_2:n}(\underline{x})$. The last two relations are the well-known basic recurrence relations (triangle rule) for the marginal univariate order statistics $X_{1,1:n} \le X_{1,2:n} \le ... \le X_{1,n:n}$ and $X_{2,1:n} \le X_{2,2:n} \le ... \le X_{2,n:n}$, respectively.

Remark 2.2. It is worth to mention that the recurrence relation (2.1) is also satisfied by survivorship function $G_{k_1,k_2:n}(\underline{x})$. Moreover, if $F(\underline{x})$ is such that the product moment $\mu^{M,N}_{k_1,k_2:n}$ exists for all $1 \le k_1,k_2 \le n$, then we get, for all $1 \le k_1,k_2 \le n-1$, the following general relation

$$n\mu^{M,N}_{k_1,k_2:n-1} = k_1\mu^{M,N}_{k_1+1,k_2:n} + (n-k_1-k_2)\mu^{M,N}_{k_1,k_2:n} + k_2\mu^{M,N}_{k_1,k_2+1:n}$$

$$+\sum_{r=1}^{k_1}\sum_{s=1}^{k_2} A_{r:n}(s)\left(\mu^{M,N}_{k_1+1,k_2+1:n} - \mu^{M,N}_{k_1+1,k_2:n} - \mu^{M,N}_{k_1,k_2+1:n} + \mu^{M,N}_{k_1,k_2:n}\right). \qquad (2.2)$$

In the special case of bivariate normality, David et al. (1977) gave an illustrative table for the df $A_{r:n}(s)$. Recently, Barakat and El-Shandidy (2004) gave a new representation of $A_{r:n}(s)$ (Theorem 2.1). Namely, for any bivariate df $F(\underline{x})$, with copula $C(\underline{x})$ and for any $2 \le r \le n-1$, $1 \le s \le n-1$, we have

$$A_{r:n}(s) = n\mathrm{E}\left(\int_{U_{2,s-1:n-1}}^{U_{2,s:n-1}}\int_{U_{1,r-1:n-1}}^{U_{1,r:n-1}} C^{1,1}(\underline{u})du_1du_2\right). \qquad (2.3)$$

An advantage of this representation is its ability to extend without any complexity to the multivariate case. Moreover, it gives a new direct approach to compute an approximate formula for the distribution and expected value of the concomitant rank of order statistics. For example, in the special case of bivariate normality with correlation ρ, we get, for any $1 \le s \le n-1$, a primary approximate values of $A_{r:n}(s)$ as

$$\hat{A}^{o}_{r:n}(s) = n\left(\int_{\Phi^{-1}(\frac{s-1}{n})}^{\Phi^{-1}(\frac{s}{n})}\int_{\Phi^{-1}(\frac{r-1}{n})}^{\Phi^{-1}(\frac{r}{n})} f_\rho(\underline{u})du_1du_2\right),\ 2 \le r \le n-1,$$

where $\Phi(.)$ refers to the standard normal df and $f_\rho(\underline{u})$ denotes the bivariate normal density function with correlation ρ (see, Example 2.1 and Table 1 of Barakat and El-Shandidy, 2004). If the values of $A_{r:n}(s)$, for all $1 \le r,s \le n$, are computed, the relation (2.2) enables us to compute the product moment $\mu^{M,N}_{k_1,k_2:n}$, for all $1 \le k_1,k_2 \le n$, of order statistics arisen from any arbitrary bivariate continuous df, in a simple recursive manner. For example, the product moment $\mu^{M,N}_{k_1,k_2:9}$, for all $1 \le k_1,k_2 \le 9$, for bivariate normal distribution, with correlations $\rho = 0.1, 0.2, 0.3, ..., 0.9, 0.95$, can be computed by (2.2) in a simple recursive manner by using Table 1 of $A_{r:n}(s)$, in David et al. (1977). In this case the relation (2.2) reduces the number of independent calculations, which are required for evaluation of the moments $\mu^{M,N}_{k_1,k_2:n}$, $k_1,k_2 = 1,2,...,n$. Moreover, it may also be used as partial checks on direct calculations of such moments. However, $A_{r:n}(s)$ depends only on the dependence structural between the marginals F_1 and F_2 (i.e., depends on C) and not themselves. Therefore, if $A_{r:n}(s)$ for any specific family of bivariate df (such as, Morgenstern family of bivariate df, $F(\underline{x}) = F_1(x_1)F_1(x_1)(1+\alpha G_1(x_1)G_2(x_2)), |\alpha| \leq 1$) is computed, for all $1 \leq r,s \leq n$, (the relation (2.3) may be used in such computation) then the relation (2.2) can be used to compute recursively the product moment $\mu^{M,N}_{k_1,k_2:n}$, for all $1 \leq k_1, k_2 \leq n$, with this family for any arbitrary marginals df's F_1 and F_2.

3. Extention to the Trivariate Case

As we have seen in the preceding section, in view of the result Páez-Borrallo and Zazo (1999), we can compute the multivariate factorized expected value from the knowledge of the jdf. Namely, let $X_{t,1:n} \leq X_{t,2:n} \leq \ldots \leq X_{t,n:n}$, $t = 1,2,\ldots,m$ be the marginal (the tth marginal) order statistics arisen from $m-$dimensional df $F(\underline{x}) = F(x_1,\ldots,x_m)$, which is defined in the region $-\infty < a_t \leq x_t \leq b_t < \infty$, $t = 1,2,\ldots,m$, and let $G(\underline{x}) = G(x_1,\ldots,x_m)$, be the survivorship function of $F(x_1,\ldots,x_m)$. Furthermore, let $\mu^{\underline{\mathrm{M}}}_{\underline{\mathrm{k}}:n} = \mu^{M_1,M_2,\ldots,M_m}_{k_1,k_2,\ldots,k_m:n} = \mathrm{E}\left(\prod_{t=1}^{m} X^{M_t}_{t,k_t:n}\right)$. Finally, let $F_{\underline{\mathrm{k}}:n}(\underline{x})$ and $G_{\underline{\mathrm{k}}:n}(\underline{x})$ be the df and survivorship function of the $m-$dimensional vector of order statistics $(X_{1,k_1:n}, X_{2,k_2:n},\ldots,X_{m,k_m:n})$. Then it can be shown that

$$\mu^{\underline{\mathrm{M}}}_{\underline{\mathrm{k}}:n} = \int_0^{b_1}\cdots\int_0^{b_m}\left(\prod_{t=1}^{m} M_t x_t^{M_t-1}\right) G_{\underline{\mathrm{k}}:n}(\underline{x})dx_1\ldots dx_m,$$

if $a_t = 0$, $t = 1,2,\ldots,m$ and

$$\mu^{\underline{\mathrm{M}}}_{\underline{\mathrm{k}}:n} = (-1)^m\int_{a_1}^{0}\cdots\int_{a_m}^{0}\left(\prod_{t=1}^{m} M_t x_t^{M_t-1}\right) F_{\underline{\mathrm{k}}:n}(\underline{x})dx_1\ldots dx_m,$$

if $b_t = 0$, $t = 1,2,\ldots,m$, respectively. Barakat and El-Shandidy (2004), extended the last two cases to the general cases $0 = a_t < b_t \leq \infty, t = 1,2,\ldots,m$, and $-\infty \leq a_t < b_t = 0, t = 1,2,\ldots,m$, respectively. In particular, when $m = 3$, Barakat (2002) and Barakat et al. (2004) gave an explicit form of $F_{\underline{\mathrm{k}}:n}(\underline{x})$. Namely, let $F_i(x_i)$ and $G_i(x_i), i = 1,2,3$, be the univariate marginals of $F(\underline{x})$ and $G(\underline{x})$ respectively. Furthermore, let $F_{ij}(x_i,x_j)$ and $G_{ij}(x_i,x_j)$, $1 \leq i < j \leq 3$, be the three possible bivariate marginals of $F(\underline{x})$ and $G(\underline{x})$ respectively. Finally, let the abbreviations $\underline{\mathrm{i}} = (i_1,i_2,i_3), \underline{\mathrm{j}} = (j_1,j_2,j_3)$ be adopted . Then

$$F_{n-k_1+1,n-k_2+1,n-k_3+1:n}(\underline{x}) = \sum_{i_1=0}^{k_1-1}\sum_{i_2=0}^{k_2-1}\sum_{i_3=0}^{k_3-1} I_{\underline{\mathrm{i}}:n}(\underline{x}), \tag{3.1}$$

where

$$I_{\underline{\mathrm{i}}:n}(\underline{x}) = \sum_{j_1=0}^{i_2\wedge i_3}\sum_{j_2=0}^{i_3\wedge i_1}\sum_{j_3=0}^{i_1\wedge i_2} J_{\underline{\mathrm{i}};\underline{\mathrm{j}}:n}(\underline{x}),$$

$$J_{\underline{\mathrm{i}};\underline{\mathrm{j}}:n}(\underline{x}) = \sum_{r=\underline{r}_o(\underline{\mathrm{i}},\underline{\mathrm{j}})}^{\bar{r}_o(\underline{\mathrm{i}},\underline{\mathrm{j}},n)} K_{\underline{\mathrm{i}};\underline{\mathrm{j}};r:n}(\underline{x}),$$

$$K_{\underline{\mathrm{i}};\underline{\mathrm{j}};r:n}(\underline{\mathrm{x}})=\frac{n!G^r(\underline{\mathrm{x}})(1-\sum_{i=1}^{3}G_i(x_i)+\sum\sum_{1\le i<j\le 3}G_{ij}(x_i,x_j)-G(\underline{\mathrm{x}}))^{n-i_1-i_2-i_3+j_1+j_2+j_3-r}}{r!(n-i_1-i_2-i_3+j_1+j_2+j_3-r)!}$$

$$\cdot\prod_{\substack{t=1\\1\le t_1<t_2\le 3\\t_1,t_2\ne t}}^{3}\frac{(G_{t_1t_2}(x_{t_1},x_{t_2})-G(\underline{\mathrm{x}}))^{j_t-r}(G_t(x_t)-G_{tt_1}(x_t,x_{t_1})-G_{tt_2}(x_t,x_{t_1})+G(\underline{\mathrm{x}}))^{i_t-j_{t_1}-j_{t_2}+r}}{(j_t-r)!(i_t-j_{t_1}-j_{t_2}+r)!},$$

$\underline{r}_o(\underline{\mathrm{i}},\underline{\mathrm{j}})=\max(0,j_2+j_3-i_1,j_1+j_3-i_2,j_1+j_2-i_3)$, and $\bar{r}_o(\underline{\mathrm{i}},\underline{\mathrm{j}},n)=\min(j_1,j_2,j_3,j_1+j_2+j_3-i_1-i_2-i_3+n)$.

Clearly the main task in obtaining the df $F_{n-k_1+1,n-k_2+1,n-k_3+1:n}(\underline{\mathrm{x}})$ (the df of the vector of upper-upper-upper order statistics) for any (k_1,k_2,k_3) and any sample size n is the huge computation of the required different terms of the sequence $J_{\underline{\mathrm{i}};\underline{\mathrm{j}}:n}(\underline{\mathrm{x}})$. Barakat et al. (2004) established two recurrence relations satisfied by this sequence. These recurrence relations, in addition to providing some checks to test the accuracy of the computation, can considerably reduce the amount of direct computation of the required terms of this sequence. Moreover, Barakat et al. (2004) provided an illustrative example (Example 4.1.), which shows that how these recurrence relations act to reduce these computations.

By replacing the empirical estimator $\hat{F}(\underline{\mathrm{x}})=\hat{F}(x_1,x_2,x_3)$, for $F(\underline{\mathrm{x}})$ in (3.1) we can easily get a corresponding exact bootstrap estimator, to the bootstrap estimator proposed by Maritz (1991) in the two dimensional case, for the product moments $\mu_{k_1,k_2,k_3:n}^{M_1,M_2,M_3}$, $k_1,k_2,k_3=1,2,\dots,n$. Moreover, when the copula $C(\underline{\mathrm{u}})=C(u_1,u_2,u_3)$ is known while the marginal df's F_1,F_2 and F_3, are not known, by applying the sample quantile function, i.e., $\hat{F}_t^{-1}(u)=X_{t,i:n}$, $t=1,2,3$, in the region given by $\frac{i-1}{n}\le u\le\frac{i}{n}$, $i=1,2,\dots,n$, on the relation (3.1), we get the following bootstrap-type estimator for $\mu_{k_1,k_2,k_3:n}^{M_1,M_2,M_3}$

$$\hat{\mu}'^{M_1,M_2,M_3}_{k_1,k_2,k_3:n}=\sum_{i_1=1}^{n}\sum_{i_2=1}^{n}\sum_{i_3=1}^{n}X_{1,i_1:n}^{M_1}X_{2,i_2:n}^{M_2}X_{2,i_3:n}^{M_3}W_{i_1,i_2,i_3;k_1,k_2,k_3:n},$$

where

$$W_{i_1,i_2,i_3;k_1,k_2,k_3:n}=F_{\underline{\mathrm{k}}:n}(\frac{i_1}{n},\frac{i_2}{n},\frac{i_3}{n})-F_{\underline{\mathrm{k}}:n}(\frac{i_1-1}{n},\frac{i_2}{n},\frac{i_3}{n})-F_{\underline{\mathrm{k}}:n}(\frac{i_1}{n},\frac{i_2-1}{n},\frac{i_3}{n})$$
$$-F_{\underline{\mathrm{k}}:n}(\frac{i_1}{n},\frac{i_2}{n},\frac{i_3-1}{n})+F_{\underline{\mathrm{k}}:n}(\frac{i_1-1}{n},\frac{i_2-1}{n},\frac{i_3}{n})+F_{\underline{\mathrm{k}}:n}(\frac{i_1-1}{n},\frac{i_2}{n},\frac{i_3-1}{n})$$
$$+F_{\underline{\mathrm{k}}:n}(\frac{i_1}{n},\frac{i_2-1}{n},\frac{i_3-1}{n})-F_{\underline{\mathrm{k}}:n}(\frac{i_1-1}{n},\frac{i_2-1}{n},\frac{i_3-1}{n}).$$

Finally, the bootstrap estimator of the product moments of the trivariate order statistics, which corresponding the bootstrap estimator proposed by Hutson (2000) in the bivariate case can be written in the form

$$\hat{\mu}_{\underline{\mathrm{k}}:n}^{\underline{\mathrm{M}}}=\sum_{i_1=1}^{n}\sum_{i_2=1}^{n}\sum_{i_3=1}^{n}x_{1,i_1:n}^{M_1}x_{2,i_2:n}^{M_2}x_{3,i_3:n}^{M_3}\hat{f}_{k_1,k_2,k_3:n}(x_{1,i_1:n},x_{2,i_2:n},x_{3,i_3:n}),$$

where $\hat{f}_{k_1,k_2,k_3:n}(\underline{x})$ is the joint empirical density function of the trivariate order statistics $(X_{1,k_1:n}, X_{2,K_2:n}, X_{3,K_3:n})$, which is given by

$$\hat{f}_{k_1,k_2,k_3:n}(\underline{x}) = n\sum_{k=1}^{n}(b_{k_1}(i_1,k)-a_{k_1}(i_1,k))(b_{k_2}(i_2,[k]_2)-a_{k_2}(i_2,[k]_2))$$

$$.(b_{k_3}(i_3,[k]_3)-a_{k_3}(i_3,[k]_3))\mathrm{I}_{\{\gamma_n(i_1,i_2,i_3,k)>0\}},$$

where $\underline{x}=(x_1,x_2,x_3)\in\{(x_{1,i_1-1:n},x_{1,i_1:n}]\times(x_{2,i_2-1:n},x_{2,i_2:n}]\times(x_{3,i_3-1:n},x_{3,i_3:n}]\}$, $i_1,i_2,i_3=1,2,...,n$, and $[k]_2$ and $[k]_3$ denote the concomitant ranks k in the vector

$$(X_{1,k:n},X_{2,[k]_2:n},X_{3,[k]_3:n})=(X_{1,k:n},X_{[2,k:n]},X_{[3,k:n]}),$$

respectively.

To obtain an extension of the recurrence relation (2.1) to the three-dimensional case, let $X_{[2,r:n]}$ and $X_{[3,r:n]}$ denote $X_{2,s:n}$ and $X_{3,t:n}$, respectively, which are paired with $X_{1,r:n}$. Furthermore, Let $A_{r:n}(s,t)=P(R_{1r:n}=s,R_{2r:n}=t)$, where $R_{1r:n}$ and $R_{2r:n}$ are the ranks of concomitants order statistics $X_{[2,r:n]}$ and $X_{[3,r:n]}$, respectively. Finally, let $A_{1r:n}(s)$ and $A_{2r:n}(t)$ be the marginals of $A_{r:n}(s,t)$. Barakat and El-Shandidy (2004) show that

$$A_{r:n}(s,t)=n\mathrm{E}\left(\int_{U_{3,t-1:n-1}}^{U_{3,t:n-1}}\int_{U_{2,s-1:n-1}}^{U_{2,s:n-1}}\int_{U_{1,r-1:n-1}}^{U_{1,r:n-1}}c(u_1,u_2,u_3)du_1du_2du_3\right),$$

for all $2\le r\le n-1$, $1\le s,t\le n-1$, where $c(u_1,u_2,u_3)=\frac{\partial^3C(u_1,u_2,u_3)}{\partial u_1\partial u_2\partial u_3}$, $C(u_1,u_2,u_3)$ is the copula of $F(u_1,u_2,u_3)$, $U_{t,r:n}=F_t(X_{t,r:n})$, $t=1,2,3$, and $U_{t,0,n-1}=0, t=2,3$. The three-dimensional version of the recurrence relation (2.2), is given in the following Theorem.

Theorem 3.1 (Barakat, 2006). For any given $1\le k_t<n-1$, $t=1,2,3$, and $\underline{M}=(M_1,M_2,M_3)$, we have

$$n\mu^{\underline{M}}_{k_1,k_2,k_3:n}=k_1\mu^{\underline{M}}_{k_1+1,k_2,k_3:n}+(n-k_1)\mu^{\underline{M}}_{k_1,k_2,k_3:n}$$

$$+\sum_{r=1}^{k_1}\sum_{s=1}^{k_2}A_{1,r:n}(s)\left(\mu^{\underline{M}}_{k_1+1,k_2+1,k_3:n}-\mu^{\underline{M}}_{k_1+1,k_2,k_3:n}\right)$$

$$+\sum_{r=k_1+1}^{n}\sum_{s=1}^{k_2}A_{1,r:n}(s)\left(\mu^{\underline{M}}_{k_1,k_2+1,k_3:n}-\mu^{\underline{M}}_{k_1,k_2,k_3:n}\right)$$

$$+\sum_{r=1}^{k_1}\sum_{t=1}^{k_3}A_{2,r:n}(t)\left(\mu^{\underline{M}}_{k_1+1,k_2,k_3+1:n}-\mu_{k_1+1,k_2,k_3:n}(\underline{x})\right)$$

$$+\sum_{r=k_1+1}^{n}\sum_{t=1}^{k_3}A_{2,r:n}(t)\left(\mu^{\underline{M}}_{k_1,k_2,k_3+1:n}-\mu_{k_1,k_2,k_3:n}(\underline{x})\right)$$

$$+\sum_{r,s,t=1}^{k_1,k_2,k_3}A_{r:n}(s,t)\left(\mu^{\underline{M}}_{k_1+1,k_2+1,k_3+1:n}-\mu^{\underline{M}}_{k_1+1,k_2,k_3+1:n}-\mu^{\underline{M}}_{k_1+1,k_2+1,k_3:n}+\mu^{\underline{M}}_{k_1+1,k_2,k_3:n}\right)$$

$$+\sum_{r=k_1+1}^{n}\sum_{s,t=1}^{k_2,k_3}A_{r:n}(s,t)\left(\mu^{\underline{M}}_{k_1,k_2+1,k_3+1:n}-\mu^{\underline{M}}_{k_1,k_2+1,k_3:n}-\mu^{\underline{M}}_{k_1,k_2,k_3+1:n}+\mu^{\underline{M}}_{k_1,k_2,k_3:n}\right).$$

Acknowledgement

Most of the material of this paper is presented at the invited session "order statistics "of the Hyderabad international conference on the future of statistical theory, practice and education December 29, 2004 January 1, 2005. The author would like to thank professor H. N. Nagaraja for inviting him to this conference.

References

[1] Arnold, B. C., Balakrishnan, N. and Nagaraja, H. N. (1992). *A first course in order statistics.* John Wiley Sons. Inc.

[2] Balakrishnan, N. (1994). Order statistics from non-identical exponential random variables and some applications. *Comput. Statist. Data. Anal.* **18**, No. 2, 203-225.

[3] Balakrishnan, N. and Sultan, K. S. (1998). Recurrence relations and identities for moments of order statistics. *Handbook of statistics,* Vol. 16 (Order Statistics, Theory & Methods), 149-228.

[4] Balakrishnan, N. and Rao, C. R. (1998). Generalized recurrence relations for moments of order statistics from non-identical Pareto and truncated Pareto random variables with applications to robustness. *Handbook of statistics,* Vol. 16 (Order Statistics, Theory & Methods), 403-438.

[5] Barakat, H. M. (1990). Limit theorems for lower-upper extreme values from two-dimensional distribution function. *J. Statist. Plann. Inference,* **24**, 69-79.

[6] Barakat, H. M. (1998). Asymptotic properties of bivariate random extremes. *J. Statist. Plann. Inference*, **61**, 203-217.

[7] Barakat, H. M. (1999). On moments of bivariate extremes. *Ann. Inst. Statist. Math.*, Vol. 51, No. 2, 351-358.

[8] Barakat, H. M. (2001). The asymptotic distribution theory of bivariate order statistics. *Ann. Inst. Statist. Math.,* Vol. 53, No. 3, 487-497.

[9] Barakat, H. M. (2002). On the confidence region for the bivariate distribution functions. *Journal of Stat. Prob. Lett.* Vol., 56/1, pp 37-43.

[10] Barakat, H. M. (2006). A general recurrence relation for moments of bivariate order statistics with extension. *Journal of Communications in Statistics.* Vol. 35, Issue 11. pp 1985-1992.

[11] Barakat, H. M. and El-Shandidy, M. A. (2005). On the computing of the moments of bivariate order statistics. *J. Statist. Plann. Inference,* **134**, Issue 1, 15-25.

[12] Barakat, H. M. and El-Shandidy, M. A. (2004). Computing the distribution and expected value of the concomitant rank order statistics. *Journal of Communications in Statistics.* Vol. 33, Issue 11. PP 2575-2594.

[13] Barakat, H. M., Nigm, E. M. and Askar, M. M. (2004). On the trivariate extreme value distributions. *J. Statist. Plann. Inference,* **118**, No. 1-2, 19-35.

[14] David, H. A. (1973). Concomitants of order statistics. *Bull. Internat. Statist. Inst.,* **45**, 295-300.

[15] David, H. A., O'Connell, M. J. and Yang, S.S. (1977). Distribution and expected value of the rank of a concomitant od an order statistic. *Ann. Statist.,* **5**, 216-223.

[16] David, H. A. (1993). A note on order statistics for dependent variates. *Amer. Statistician,* **47**, 198-199.

[17] David, H. A. (1995). On recurrence relations of order statistics. *Statist. Probab. Lett.* **24**, 133-138.

[18] David, H. A. and Nagaraja, H. N. (2003). *Order statistics* (Third edition). John Wiley Sons. Inc.

[19] Galambos, J. (1987). The asymptotic theory of extreme order statistics, Krieger, FL (2nd ed.).

[20] Hutson, A. D. (2000). Estimating the covariance of bivariate order statistics with applications. *Statist. Probab. Lett.* **48**, 195-203.

[21] Hutson, A. D. and Ernst, M. D. (2000). The eaxct bootstrap mean and variance of an $L-$ Estimator . *J. Roy. Statist. Soc.* ser. B 62, 1-6.

[22] Maritz, S. (1991). Estimating the covariance matrix of bivariate medians. *Statist. Probab. Lett.* **12**, 305-309.

[23] Samuel, P. and Thomas, P. Y. (1998). Modified expression for a recurrence relation on the product moments of order statistics. *Statist. Probab. Lett.* **37**, 39-59.

[24] Sklar, A. (1959). Fonctions de répartition á n dimensions et leurs marges. *Publ. Inst. statist. Univ. Paris,* **8**, 229-231.

[25] Thomas, P. Y. and Samuel, P. (1996). A note on recurrence relations for the product moments of order statistics. *Statist. Probab. Lett.* **29**, 245-249.

[26] Páez-Borrallo, J. M. and Zazo, S. (1999). A new formula for the computation of multivariate factorized moments by using the joint cumulative distribution function. *Commun. Statist.- Theory Meth.,* **28** (11), 2613-2622.

[27] Wang, X. and Stokes, S. L. (2005). *Moments of bivariate order statistics for the standard normal distribution.* Technical report, Southern Methodist University, Dallas, Texas.

[28] Wang, X. and Stokes, S. L. and Johan, L. (2005). *Concomitant of multivariate order statistics with application to judgment post-stratification.* Technical report, Southern Methodist University, Dallas, Texas.

In: New Developments in Applied Statistics
Editor: Mohammad Ahsanullah
ISBN 978-1-61324-648-1

Chapter 23

LIMIT LAWS FOR RECORD VALUES UNDER POWER NORMALIZATIONS

E.M. Nigm
Department of Mathematics, Faculty of Science
Zagazig University, Zagazig, Egypt

Abstract

This paper deals with, the types of possible nondegenerate limit laws of distributions that could arise as limiting distributions of suitably power normalized record values. Domain of attraction criteria for each of the six types of limit laws can be derived by appealing to the limiting record value distributions to the extreme value distributions under power normalizations.

Key words and phrases :Record Values; Limit Laws; Domain of Attraction; Power Normalization.

1. Introduction

Let $\{X_n, n \geq 1\}$ be a sequence of independent, identically distributed (i.i.d.) random variables (r.v.'s) with common continuous distribution function (cdf) $F(x)$. Let $M_n = \max\{X_1, X_2, ..., X_n\}$ for $n \geq 1$.We say X_j is an upper record value of $\{X_n, n \geq 1\}$ if $M_j > M_{j-1}, \quad j \geq 1$. By definition X_1 is an upper as well as a lower record value. The indices at which the upper record values occur are given by the record times $\{U_{(n)}, n \geq 1\}$ where

$$U(1) = 1,$$

$$U(n) = \min\{j \mid j > U(n-1), \ X_j > X_{U(n-1)}\}, \ n > 1,$$

Let $\{Y_i\}$ denote a sequence of i.i.d. $Exp(1)$ random variables. Exploiting the lack of memory property of the exponential distribution, it follows that the differences between successive upper records will again be i.i.d. standard exponential variables, i.e.,$\{Y_{U(n)} - Y_{U(n-1)}\}$ are i.i.d. $\exp(1)$. Note that if X has cdf F , then

$$R(x) = -\ln(1 - F(x)), -\infty < x < \infty. \tag{1}$$

$R(x)$) has a standard exponential distribution. Consequently, $X \stackrel{d}{\rightarrow} F^{-1}(1-e^{-Y})$, where, as usual, Y is $Exp(1)$. Thus, X is a monotone increasing function of Y and, consequently, the nth record of the X_n sequence, $X_{U(n)}$ is related to the nth record $Y_U(n)$, of the exponential sequence by

$$X_{U(n)} \stackrel{d}{\rightarrow} F^{-1}(1-e^{-Y_{U(n)}}) \tag{2}$$

The distribution function $F_n(x)$ of $X_{U(n)}$ is

$$F(x) = P(X_{U(n)} \leq x) = \int_{-\infty}^{x} \frac{R^{n-1}(u)}{(n-1)!} dF(u)$$

$$= \int_{-\infty}^{x} \frac{R^{n-1}(u)}{(n-1)!} dF(u), \qquad -\infty \leq x \leq \infty. \tag{3}$$

The nth record corresponding to an i.i.d. F sequence is distributed as a smooth function of a sum of (n + 1) i.i.d. standard exponential variables, namely the Y_j. From (2) we may write

$$X_{U(n)} \stackrel{d}{\rightarrow} R^{-1}(\sum_{j=0}^{n} Y_j) \tag{4}$$

where $R^{-1}(x) = F^{-1}(1-e^{-Y})$ represents the inverse of the hazard function R(x) . Note that R^{-1} is non-decreasing with domain $(0,\infty)$. The strong law of large n numbers assures us that $\frac{1}{n+1}\sum_{j=0}^{n} Y_j \stackrel{a.s.}{\rightarrow} 1$ and that $\sum_{j=0}^{n} Y_j \stackrel{a.s.}{\rightarrow} \infty$ as $n \rightarrow \infty$. The central limit theorem also, assures us that as $n \rightarrow \infty$,

$$P(\frac{R(X_{U(n)})-(n+1)}{\sqrt{n+1}} \leq x) \rightarrow \Phi(x), \tag{5}$$

for $-\infty \leq x \leq x\infty$, where $\Phi(x)$ is the standard normal distribution. To every distribution function $F(x)$ corresponds the associated distribution $F_a(x)$ such that

$$R^{\frac{1}{2}} = -\log(1-F_a(x)), \quad -\infty \leq x \leq \infty \tag{6}$$

The limiting distribution of records corresponding to the $cdf\ F$ turns out to be related to the limiting behavior of maxima of samples drawn from the associated $cdf\ F_a$. Recall that there are three possible limit laws for sample maxima M_n (see [(6))]) under linear normalizations α_n, β_n such that as $n \rightarrow \infty$,

$$P(M_n \leq \alpha_n x + \beta_n) \rightarrow G(x)$$

where G(x) , is a nondegenerate d.f. must have one and only one of three types $G_{i;\alpha}(x) = \exp[-\mathcal{U}_{i;\alpha}(x)]$, $i = 1,2,3$, $\alpha > 0$, where

$$\left.\begin{array}{ll}
\text{Type I)}: & \mathcal{U}_{1;\alpha}(x) = \begin{cases} \infty, & x \leq 0, \\ x^{-\alpha}, & x > 0; \end{cases} \\[2ex]
\text{Type II)}: & \mathcal{U}_{2;\alpha}(x) = \begin{cases} (-x)^{\alpha}, & x \leq 0, \\ 0, & x > 0, \end{cases} \\[2ex]
\text{Type III)}: & \mathcal{U}_{3;\alpha}(x) = \mathcal{U}_{3}(x) = e^{-x}, \ \forall\, x.
\end{array}\right\} \tag{7}$$

Resnick (10)) derived a parallel result for record values. He showed that there are only three kinds of distributions that could arise as limiting distributions of suitably normalized record values. The three possibilities are:

$$\left.\begin{array}{ll}
\text{Type I)}: & \Phi_{1;\alpha}(x)=\begin{cases} 0, & x<0,\\ \Phi(\alpha\log(x)), & x\geq 0;\end{cases}\\[2ex]
\text{Type II)}: & \Phi_{2;\alpha}(x)=\begin{cases} \Phi(\log(-x)^{-\alpha}), & x<0,\\ 1, & x\geq 0,\end{cases}\\[2ex]
\text{Type III)}: & \Phi(x)=, \ \forall\, x.
\end{array}\right\} \tag{8}$$

where α is a positive constant. The limit laws are all of the form $\Phi[-\log(-\log G(x))]$ Following Pantcheva (9) and Mohan & Ravi (7) a d.f. F is said to belong to the max domain of attraction of a d.f. $\Psi(x)$ under power normalization if there exist norming constants $a_n>0$ and $b_n>0$, such that, as $n\to\infty$,

$$P\left(\left|\frac{X_{n:n}}{a_n}\right|^{\frac{1}{b_n}}\mathcal{S}(X_{n:n})<x\right)=F^n(a_n|x|^{c_n}\mathcal{S}(x))\xrightarrow{w}\Psi(x), \tag{9}$$

where $\mathcal{S}(x)=-1$ if $x<0$, $=0$ if $x=0$ and $=1$ if $x>0$. In this case we write $F\in D_p(\Psi)$. We call $\Psi(x)$ is a max stable d.f. under power normalization or simply p-max stable d.f. if (9) holds.

We say that two d.f.'s F_1 and F_2 are of the same $p-$type if there exist $A>0$ and $B>0$ such that $F_1(x)=F_2(A|x|^B\mathcal{S}(x))\ \forall x$. Pantcheva (9) showed that any p-max stable d.f. $\Psi(x)$ can be a $p-$type of one of the six d.f.'s $\exp(-\mathcal{V}_{i;\beta}(x))$, $i=1,2,\ldots,6$, $\beta>0$, where

$$\left.\begin{array}{ll}
\mathit{Type\ I)}: & \mathcal{V}_{1;\beta}(x)=\begin{cases} \infty, & x\leq 1,\\ (\log x)^{-\beta}, & x>1;\end{cases}\\[2ex]
\mathit{Type\ II)}: & \mathcal{V}_{2;\beta}(x)=\begin{cases} \infty, & x\leq 0,\\ (-\log x)^{\beta}, & 0<x\leq 1,\\ 0, & x>1;\end{cases}\\[2ex]
\mathit{Type\ III)}: & \mathcal{V}_{3;\beta}(x)=\begin{cases} \infty, & x\leq -1,\\ (-\log(-x))^{-\beta}, & -1<x\leq 0,\\ 0, & x>0;\end{cases}\\[2ex]
\mathit{Type\ IV)}: & \mathcal{V}_{4;\beta}(x)=\begin{cases} (\log(-x))^{\beta}, & x\leq -1,\\ 0, & x>-1;\end{cases}\\[2ex]
\mathit{Type\ V)}: & \mathcal{V}_{5;\beta}(x)=\mathcal{V}_5(x)=\mathcal{U}_{1;1}(x);\\[2ex]
\mathit{Type\ VI)}: & \mathcal{V}_{6;\beta}(x)=\mathcal{V}_6(x)=\mathcal{U}_{2;1}(x).
\end{array}\right\} \tag{10}$$

The necessary and sufficient conditions for a d.f. to belong to $D_p(.)$ for each of the six $p-$max stable laws is obtained by Mohan & Ravi (7) and Subramanya (11). For more details on p - max stable laws see (1), (2), (3),(4), (5), (8), (13). In this paper, we study the asymptotic behavior of the sequence of record values $\{X_{U(n)}, n \geq 1\}$. In section 2, we determine the six types of possible nondegenerate limit laws for the record value sequence by showing that this determination reduces to a problem in extreme value theory under power normalizations. In Section 3, we continue to exploit this duality to derive domain of attraction criteria for each type.

2. The Class of Limit Laws

Tata (1969) (See (12))) showed that a necessary and sufficient condition for the existence of normalizing α_n and $\beta_n > 0$ so that $\frac{(X_{U(n)}-\alpha)}{\beta}$ converges in distribution to a nondegenerate limit law is that there exists a nondecreasing function $g(x)$ such that as $n \longrightarrow \infty$

$$\tfrac{R(\beta_n x+\alpha_n)-1}{\sqrt{n}} \longrightarrow g(x),$$

and in such a case the limiting distribution is given by $\Phi(g(x)$, where (Φ is the standard normal cdf. Now, We rewrite and prove Tata's Theorem 2.1. in the case of a max stable d.f.'s under power normalization

Theorem 1. Let $\{X_n, n \geq 1\}$ be a sequence of i.i.d. r.v.'s with continuous d.f. $F(x)$. In order that there exist normalizing constants $\alpha_n > 0, \beta_n > 0$, $n \geq 1$ such that as $n \longrightarrow \infty$,

$$P\left(|X_{U(n)} \leq a_n |x|^{b_n} \mathcal{S}(x)\right) \rightarrow u(x), \tag{11}$$

where $u(x)$ is a nondegenerate limit law. It is necessary and sufficient that there exist a non-decreasing function $h(x)$ such that as $n \rightarrow \infty$,

$$\frac{R(a_n |x|^{b_n} S(x)) - n}{\sqrt{n}} \longrightarrow h(x) \tag{12}$$

for all continuity points x of $h(x)$. In this case, $u(x)$ has the form

$$u(x) = \Phi(h(x)). \tag{13}$$

Proof Suppose that (11) holds.

$$P(X_{U(n)} \leq a_n |x|^{b_n} \mathcal{S}(x)) = P(R(X_{U(n)}) \leq R(a_n |x|^{b_n} \mathcal{S}(x)))$$

$$= P(\frac{R(X_{U(n)}) - n}{\sqrt{n}} \leq \frac{R(a_n |x|^{b_n} S(x)) - n}{\sqrt{n}}) \rightarrow \Phi(h(x)),$$

after using (5). We see that (11) holds if (12) and (13) are valid.

Conversely, suppose that (12) holds, then we have

$$\frac{R(a_n|x|^{b_n}S(x))-n}{\sqrt{n}} \geq \frac{R(X_{U(n)})-n}{\sqrt{n}} \rightarrow h(x),$$

i.e.

$$R(X_{U(n)}) \leq R(a_n|x|^{b_n}S(x)) \rightarrow h(x).$$

From (5), we get

$$P(R(X_{U(n)}) \leq R(a_n|x|^{b_n}S(x))) \rightarrow \Phi(h(x)).$$

i.e.

$$P(X_{U(n)} \leq a_n|x|^{b_n}S(x)) \rightarrow \Phi(h(x)) = u(x).$$

Hence, the proof of the theorem.

Lemma 1. If (11) or (12) is satisfied with normalizing constants $a_n > 0$ and $b_n > 0$, then

$$(\frac{a_n}{a_{n+1}})^{\frac{1}{b_n}} \longrightarrow 1 \ \ and \ \ \frac{b_n}{b_n+1} \longrightarrow 1 \tag{14}$$

Proof. From (12), we have that

$$\frac{R(a_{n+1}|x||^{b_{n+1}}S(x))-n}{\sqrt{n}} = (\frac{R(a_{n+1}|x|^{b_{n+1}}S(x))-(n+1)}{\sqrt{n+1}})$$

$$(\frac{n+1}{n})^{\frac{1}{2}} + \frac{1}{\sqrt{n}} \rightarrow h(x) \ as \ n \rightarrow \infty$$

Hence

$$P(X_{U(n)} \leq a_{n+1}|x|^{b_{n+1}}S(x)) \rightarrow \Phi(h(x)),$$

and this coupled with (11) give the result by the convergence - to - types Theorem (see Lemma 2.1. in (1)). The following theory describes the class of limit types for record values under power normalizations.

Theorem 2. The class of limit laws for record values is of the form $\Phi[-\log(-\log\Psi(x))]$ where $\Phi(x)$ is an extreme value distribution under power normalization. The limit laws must be the type of one of the following six distributions:

$$\left.\begin{array}{l}
\Phi_{1;\beta}(x)=\begin{cases}0, & x\le 1,\\ \Phi[\beta\log(\log x)], & x>1;\end{cases}\\[2ex]
\Phi_{2;\beta}(x)=\begin{cases}0, & x\le 0,\\ \Phi[\log(-\log x)^{-\beta}], & 0<x\le 1,\\ 1, & x>1;\end{cases}\\[2ex]
\Phi_{3;\beta}(x)=\begin{cases}0, & x\le -1,\\ \Phi[\beta\log(-\log(-x))], & -1<x\le 0,\\ 1, & x>0;\end{cases}\\[2ex]
\Phi_{4;\beta}(x)=\begin{cases}\Phi[\log(\log(-x))^{-\beta}], & x\le -1,\\ 1, & x>-1;\end{cases}\\[2ex]
\Phi_{5;\beta}(x)=\begin{cases}0, & x\le 0,\\ \Phi[\log x], & x>0;\end{cases}\\[2ex]
\Phi_{6;\beta}(x)=\begin{cases}\Phi[-\log(-x)], & x\le 0,\\ 1, & x>0;\end{cases}
\end{array}\right\} \tag{15}$$

where β is a positive constant.

Proof. Once the result (14) is at hand, it is a standard procedure to switch from the discrete parameter n to the continuous parameter s and from (12) we have as $s\longrightarrow\infty$

$$\frac{R(a(s)\,|x|^{b(s)}\,S(x))-s}{\sqrt{s}}\longrightarrow h(x) \tag{16}$$

for all continuity points of $h(.)$. If in (16) we divide by s instead of $\sqrt{s}$, then for all x such that $|h(x)|<\infty$ we have as $s\longrightarrow\infty$,

$$R(a(s)\,|x|^{b(s)}\,S(x))-s \tag{17}$$

Hence as $s\longrightarrow\infty$,

$$\frac{R(a(s)\,|x|^{b(s)}\,S(x))-s}{\sqrt{s}}=R^{\frac{1}{2}}(a(s)\,|x|^{b(s)}\,S(x))-\sqrt{s}$$

$$\frac{R^{\frac{1}{2}}(a(s)\,|x|^{b(s)}\,S(x))+\sqrt{s}}{\sqrt{s}}\longrightarrow h(x).$$

From (17), we have that as $s\longrightarrow\infty$, (16) is equivalent to

$$R^{\frac{1}{2}}(a(s)\,|x|^{b(s)}\,S(x))-\sqrt{s}\longrightarrow\frac{1}{2}h(x) \tag{18}$$

and further that as $s\longrightarrow\infty$,

$$\exp[\sqrt{s}]\exp[-R^{\frac{1}{2}}(a(s)\,|x|^{b(s)}\,S(x))\longrightarrow\exp[-\frac{1}{2}h(x)] \tag{19}$$

Setting $y = \exp[s]$, then (19) becomes

$$y\exp[-R^{\frac{1}{2}}(a(\log y)^2\,|x|^{b(\log y)^2}\,S(x)) \longrightarrow \exp[-\frac{1}{2}h(x)] \tag{20}$$

From (6), then (20) becomes

$$y(1-F_a(a(\log y)^2\,|x|^{b(\log y)^2}\,S(x))) \longrightarrow \exp[-\frac{1}{2}h(x)] \tag{21}$$

This means that $F_a(.)$ is in the domain of attraction of an extreme value distribution under power normalization, and

$$\exp[\frac{1}{2}h(x)] = -\log\Psi(x) \tag{22}$$

where $\Psi(x)$ is an extreme value distribution under power normalization. The factor $\frac{1}{2}$ will appear only as a location or scale parameter and so we have proven the theorem.

3. Domains of Attraction of Record Types

A continuous d.f. $F(.)$ is attracted to the limiting record value distribution $(\Phi(h(x))$ if (11) or (12) holds for a normalizing constants $a_n > 0, b_n > 0$. Alter-natively we write $R(x) \in (\Phi(h(x))$ to indicate attraction to $\Phi(h(x))$. We now derive domain of attraction criteria for each of the six types of limit laws in Theorem 2 through the following theorems.

Theorem 3. Let F be a given $d.f.$ with right-end point $x^0 = \sup\{x : F(x) < 1\}$. Then $R(x) \in \Phi_{i;\beta}(x)$ if and only if $F_a \in D_p(\exp[-v_{i;\beta}/2])$, $i = 1,2,3,4,$

$$\left.\begin{array}{l}
(i) = \begin{cases} x^0, & for\ i=1,\\ 0<x^0<\infty, & for\ i=2\\ x^0=0, & for\ i=3,\\ x^0<0, & for\ i=4.\end{cases}\\[2ex]
\begin{cases} (i)\lim_{i\to\infty}\frac{R(\pm\exp[\pm ty])-R(\pm\exp[\pm y])}{\sqrt{R(\pm\exp[\pm y])}} = \beta\log y, & y>0\\ (ii)\lim_{i\to\infty}\frac{R(x^0\exp[\mp\frac{y}{t}])-R(x^0\exp[\mp\frac{1}{t}])}{\sqrt{R(x^0\exp[\mp\frac{1}{t}])}} = -\beta\log y, & y>0.\end{cases}\\[2ex]
a_n = \begin{cases} 1, & for\ i=1,3,\\ x^0, & for\ i=2\\ -x^0, & for\ i=4.\end{cases}\\[2ex]
b_n = \begin{cases} \log R^{-1}(n), & for\ i=1,\\ \log\frac{x^0}{R^{-1}(n)}, & for\ i=2,\\ -\log(-R^{-1}(n)), & for\ i=3,\\ \log\frac{R^{-1}(n)}{x^0}, & for\ i=4.\end{cases}
\end{array}\right\} \tag{23}$$

Where "+" for $i = 1,4$ and "−" for $i = 2,3$.

Theorem 4. Let F be a given d.f. with right-end point $x^0 = \sup\{x : F(x) < 1\}$. Then $R(x) \in \Phi_5(x)$ if and only if $F_a \in D_p(\exp[-v_5])$,

$$(i) \qquad x^0 > 0$$

and

$$(ii) \lim_{x^0} \frac{R(t\exp[yf(t)]) - R(t)}{\sqrt{R(t)}} = 2y, \; y > 0,$$

for some positive valued function f. If (ii) holds for some f then $\int_a^{x^0} \frac{\exp[-R^{\frac{1}{2}}(x)]}{x} dx < \infty \; for \; 0 < a < x^0$ and (ii) holds with the choice $f(t) = \exp[R^{\frac{1}{2}}(t)] \int_a^{x^0} \frac{\exp[-R^{\frac{1}{2}}(x)]}{x} dx$ The normalizing constants here may be chosen as $a_n = R^{-1}(n)$ and $b_n = f(a_n)$.

Theorem 5. Let F be a given d.f. with right-end point $x^0 = \sup\{x : F(x) < 1\}$. Then $R(x) \in \Phi_6(x)$ if and only if $F_a \in D_p(\exp[-v_6])$,

$$(i) x^0 \leq 0$$

and

$$(ii) \lim_{x^0} \frac{R(t\exp[yf(t)]) - R(t)}{\sqrt{R(t)}} = -2y, \; y > 0$$

for some positive valued function f. If (ii) holds for some f then $\int_a^{x^0} \frac{\exp[-R^{\frac{1}{2}}(x)]}{x} dx < \infty \; for \; a < x^0$ and (ii) holds with the choice $f(t) = -\exp[R^{\frac{1}{2}}(t)] \int_a^{x^0} \frac{\exp[-R^{\frac{1}{2}}(x)]}{x} dx$ The normalizing constants here may be chosen as $a_n = -R^{-1}(n)$ and $b_n = f(-a_n)$.

Proof of Theorem 3. For $i = 1$. We have that $R(x) \in \Phi_{1;\beta}$ if $F_a \in D_p(\exp[-v_{1;\beta/2}])$. Observe that $F_n^a(0) \to 0$ and so $x^0 > 0$. Then we have $F_n^a(a_n|x|_n^b S(x)) \to \Psi_{1;\beta}(x)$ and hence by Theorem 2.1. in [7], $x0 = \infty$, proving (i). From [7] we see that the necessary and sufficient condition for this is that $1 - F_a(x)$ be regularly varying with exponent $\beta/2$. Then

$$\lim_{t\to\infty} \frac{1-F_a(\exp[ty])}{1-F_a(\exp[t]} = y^{\frac{-\beta}{2}}, \; y > o$$

i.e.

$$\lim_{t\to\infty} \frac{\exp[-R^{\frac{1}{2}}(\exp[ty])]}{\exp[-R^{\frac{1}{2}}(\exp[t]]} = y^{\frac{-\beta}{2}}, \; y > o$$

and hence

$$\lim_{t\to\infty}(R^{\frac{1}{2}}(\exp[ty]) - R^{\frac{1}{2}}(\exp[t] = \frac{-\beta}{2} \log y, \; y > o$$

Dividing the above equation by $R^{\frac{1}{2}}(e^t)$ shows that $R^{\frac{1}{2}}(e^t)$ is slowly varying and hence

$$\lim_{t\to\infty} \frac{R(\exp[ty]) - R(\exp[t])}{\sqrt{R(\exp[t]}}$$

$$= \lim_{t\to\infty} \frac{(R^{\frac{1}{2}}(\exp[ty]) - R^{\frac{1}{2}}(\exp[t]))(R^{\frac{1}{2}}(\exp[ty]) + R^{\frac{1}{2}}(\exp[t]))}{\sqrt{R(\exp[t]}} = \beta \log y,$$

giving (ii).
Since $F_a \in D_p(\exp[-v_{1;\beta/2}])$ and from (21), we can set

$$a(\log y)^2 = 1, b(\log y)^2 = \log F_a^1(1-\tfrac{1}{y}).$$

but

$$\log F_a^{-1}(1-\tfrac{1}{y}) = log(R^{1/2})^{-1}(logy) = \log R^{-1}(logy)^2.$$

Hence

$$a_n = a(n) = 1 \text{ , and } b_n = b(n) = \log R^{-1}(n),$$

are acceptable choices of normalizing constants.

Conversely, suppose that (i) and (ii) hold. If $b_n = \log R^{-1}(n) = \log F^{-1}(1-\exp(-n)) = \log F_a^{-1}(1-\frac{1}{n})$, then $n1 - F_a(\exp(b_n x)) \to x^{-\beta/2}, \ x > 0$. Thus $F_a \in D_p(\exp[-v_{1;\beta/2}])$ with the stated norming constants. This completes the proof of Theorem 3 in case $i = 1$.
The proof of Theorems 4 and 5 is similar to the proof of Theorem 3.

References

[1] Barakat, H. M. and Nigm, E. M. (2002). Extreme order statistics under power normalization and random sample size, *Kuwait Journal of Science and Engineering,* Vol. 29, No.1, 27-41.

[2] Barakat, H. M. Nigm, E. M. (2003). Some asymptotic stability properties of extreme order statistics under power normalization. Kuwait *Journal of Science and Engineering*, Vol. 29, No.1, 15-27.

[3] Barakat, H. M., Nigm, E. M. and El-Adll, E. M.(2002). Continuation theorems of the extremes under power normalization. *J. Appl. Math. Computing.* Vol. 10, No. 1-2, pp. 1-15.

[4] Barakat, H. M., Nigm, E. M. and El-Adll, E. M.(2003). Asymptotic properties of random extremes under general normalization from non-identical distributions. *Metrika* To appear.

[5] El-Adll, E.M. (2002). *Asymptotic properties of order statistics under general conditions.* Ph.D. Thesis. Zagazig University, Egypt.

[6] Galambos, J. (1987). *The asymptotic theory of extreme order statistics,* Krieger, FL.

[7] Mohan, N. R. and Ravi, S. (1992). Max domains of attraction of univariate and multivariate p-max stable laws, *Theory Prob. Appl.,* Vol. 37, 632-643.

[8] Nasri-Roudsari, D. (1999). Limit distributions of generalized order statistics under power normalization. *Commun.Statist.-Theory Meth.*, **28**(6),1379-1389

[9] Pantcheva, E. (1985). Limit theorems for extreme order statistics under nonlinear normalization *Lecture Notes in Math.*, No. 1155, 284-309.

[10] Resnick, S., I. (1973). Record laws for record values. *Stochastic Processes and their Applications* **1**, 67-82.

[11] Subramanya, U. R. (1994). On max domains of attraction of univariate $p - max$ stable laws, *Statistics , Probability Letters,* Vol. 19, 271-279.

[12] Tata, M. N. (1969). On outstanding values in a sequence of random variables. *Z. Wahrscheinlichkeits theorie und Verwandte Gebiete* **12**, 9-20.

[13] Weinstein, S. B. (1973). Theory and application of some classical and generalized asymptotic distributions of extreme values, *IEEE Trans. Information Theory IT-19,* No. 2, 148-154.

In: New Developments in Applied Statistics
Editor: Mohammad Ahsanullah
ISBN 978-1-61324-648-1

Chapter 24

IMPROVED ESTIMATION FOR DYNAMIC LINEAR REGRESSION MODEL

Zahirul Hoque
United Arab Emirates University, UAE
RICHARD GERLACH, University of Sydney, Australia

Abstract

This paper studies the preliminary test and shrinkage estimators based on the Kalman filtering procedure applied to a dynamic linear state space regression model. The performance of these estimators, with respect to mean square error, was investigated. It was revealed that under certain conditions both the preliminary test and shrinkage estimators proposed outperform the Kalman filter. This out-performance was not uniform. Further, the shrinkage estimator was found to be superior to the preliminary test estimator over large regions. The results presented in this paper invalidates the global minimum mean square error property of the Kalman filter that is widely used by the engineers for estimation of the parameters of linear state space models.

Keywords and Phrases: Dynamic model, Kalman filter, preliminary test estimator, shrinkage estimator and mean square error.

1. Introduction

Dynamic or time series data is very commonly observed in practice especially in the fields of economics, finance and engineering. For such data the dynamic linear state space model is defined as a system of equations specifying simultaneously (i) how observations of a process are stochastically dependent on the current process parameters (the observation equation) and (ii) how the process parameters might dynamically evolve in time (the state equation); both as a result of the inherent process dynamics and from random noises. To analyse data that follows this dynamic state space model a useful and widely used method of estimation and prediction, known as the Kalman filter, was developed by Kalman (1960) and Kalman and Bucy (1961). As the pioneers were engineers, their works and many of the works thereafter have appeared in engineering journals. For reviews in the statistical literature readers may refer to Carter and Kohn (1996), Gerlach *et al.* (2000), Mainhold and Singpurwalla (1983) and Duncan and Horn (1972).

Estimation and prediction in static models, for example, in simple linear regression model, is straight-forward. For such models the least square estimator is unbiased and has minimum variance among such estimators. However, with respect to the mean square error (MSE) substantial improvement can be achieved via preliminary test estimation (cf. Khan *et al.*, 2002), if relevant prior information is available. In a recent paper Khan *et al.* (2005) investigated the performance of the preliminary test and shrinkage estimators of the intercept parameter of simple linear regression model; the model is static since the regression parameter value is a fixed constant. Their study showed that if the non-sample prior information about the value of the parameter is not too far from the factual value, then compared to the least square estimator both the preliminary test and shrinkage estimators should be considered as improved estimators of the intercept parameter. The superiority of the two estimators is defined in terms of having lower mean square error than the observed data based least squares estimator.

The estimation and prediction of dynamic parameters (i.e. the unknown states) in state space models is naturally not as simple as that for static models. The seminal work of Kalman (1960) and Kalman and Bucy (1961) has generated a large volume of research in this area. Duncan and Horn (1972) investigated the properties of Kalman filter from the viewpoint of the linear dynamic regression model. They showed that the estimator obtained by the Kalman filter holds all interesting properties of the commonly used least squares estimator for static model parameters: e.g. minimum variance, unbiasedness and minimum mean square error properties. Duncan and Horn (1972) also derived the sampling distribution of the estimator obtained by the Kalman filter.

Mainhold and Singpurwalla (1983) illustrated the Bayesian approach for estimation and prediction in state space models. Their study showed that the Bayesian approach yields the same result, as the Kalman filter approach, in an elegant manner. Moreover, the Bayesian approach provides the attractive feature of enabling inference about the parameter through a probability distribution rather than just a point estimate.

In this paper we have investigated the performance of the preliminary test and shrinkage estimators for the simple linear state space regression model via Kalman filtering, under Gaussian errors. Firstly, the recursive estimation equations for the Kalman filter are reviewed. Based on the Kalman filter and some non-sample prior information, about the value of the parameter, the preliminary test estimator (PTE) has been defined. The PTE is biased unless the non-sample prior information is exactly the same as the factual value of the parameter (cf. Khan *et al.*, 2002, 2005). In this paper we compare the MSE of this biased estimator with that of the unbiased Kalman filter. Our result shows that if the non-sample prior information about the value of the parameter is not too far from the factual value, then PTE outperforms the Kalman filter.

As the performance of PTE depends on the level of significance of the preliminary test, we further define the preliminary test based shrinkage estimator (SE) following Khan *et al.* (2005). This estimator is also biased and is a combination of the Kalman filter and non-sample prior information, but does not depend on the level of significance of the preliminary test. We compare the MSE of SE with that of the PTE and Kalman filter. Our investigation reveals that SE outperforms both the PTE and the Kalman filter under certain conditions.

The organization of this paper is as follows. In Section 2 we present the linear state space dynamic regression model and the recursive estimation equations for the Kalman filter. The preliminary test estimator for the model is defined in Section 3.. In the same section the analytical MSE function is presented and numerically compared with that of the Kalman filter. In Section 4. we define the shrinkage estimator and compare its MSE with that of the Kalman filter and the PTE. Finally, some concluding remarks are presented in Section 5..

2. The Model and Estimation Equations

Consider the simple linear dynamic regression model with observation equation:

$$Y_t = X_t\beta_t + e_t, \tag{2.1}$$

and system or state equation:

$$\beta_t = d_t\beta_{t-1} + f_te_t, \tag{2.2}$$

where Y_t is the response; X_t is the known regressor; β_t is the unknown dynamic regression coefficient and e_t is the stochastic disturbance term which is distributed as $N(0, \sigma^2)$. The quantities d_t and f_t are assumed known, as is standard. The system equation (2.2) starts with

$$\beta_1 = \tilde{\beta} + f_1e_1$$

where $\tilde{\beta}$ is the prior mean of β_1; again usually assumed to be known. Our interest is in estimating the unknown state β_t given the measurements $\boldsymbol{Y}$ up to time t. If $\beta_t = \beta_{t-1}$ (i.e. $d_t = 1$), and $f_t = 0$ then the model described above is the usual simple linear (static) regression model.

For model (2.1) – (2.2) the recursive estimation equations are:

$$\begin{aligned} b_{t|t} &= b_{t|t-1} + X_tS_{t|t-1}R_t^{-1}\varepsilon_t & (2.3)\\ S_{t|t} &= S_{t|t-1} - X_t^2S_{t|t-1}^2R_t^{-1}, & (2.4)\end{aligned}$$

where

$$\begin{aligned} b_{t|t-1} &= d_tb_{t-1|t-1} & (2.5)\\ S_{t|t-1} &= d_t^2S_{t-1|t-1} + f_t^2\sigma^2 & (2.6)\\ R_t &= X_t^2S_{t|t-1} + \sigma^2 & (2.7)\\ \varepsilon_t &= Y_t - X_tb_{t|t-1}. & (2.8)\end{aligned}$$

The estimator $b_{t|t}$ is the conditional posterior mean of the unknown state β_t, given the observed data $y_1, \ldots, y_t$ and known prior mean $\tilde{\beta}$ (cf. Duncan and Horn, 1972). The posterior distribution of β_t is normal with variance $S_{t|t}$. R_t is the conditional variance of the observation $y_t|y_1, \ldots, y_{t-1}$. However, not given by Duncan and Horn (1972) was the sampling variance of $b_{t|t}$, which we have derived as:

$$\mathrm{Var}(b_{t|t}) = d_t^2\,\mathrm{Var}(b_{t-1|t-1}) + X_t^2S_{t|t-1}^2R_t^{-1} = \tau \text{ (say)}. \tag{2.9}$$

If σ^2 is known τ does not depend on any observed data. The unknown σ^2 can be estimated using the observed data through maximum likelihood estimation allowing an estimate $\mathrm{Var}(b_{t|t})$ in that case as well. Figure 1 shows a plot of a simulated realisation of

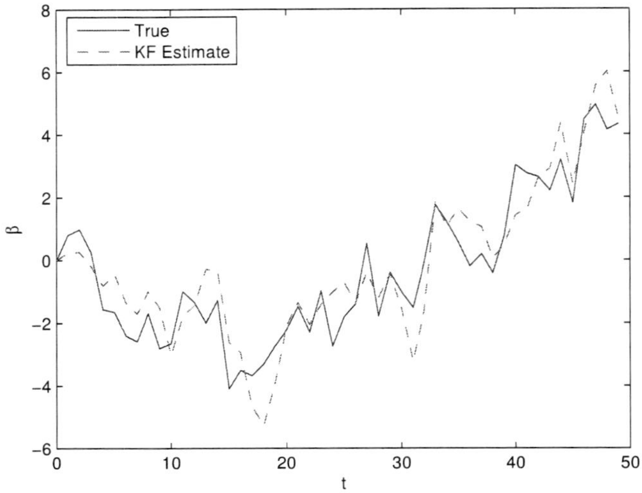

Figure 1. Plot of the simulated values of β_t and their estimated values via Kalman filter against time t.

values of β_t and its estimated values ($b_{t|t}$) against t, where $d_t = f_t = 1$. The plot shows that the Kalman filter estimator is indeed very close to the simulated values of the unknown state. Duncan and Horn (1972) showed that this estimator is unbiased and holds the minimum mean square error property. As the Kalman filter is an unbiased estimator of the dynamic model parameter (state) its MSE is no different from its variance. In this study our interest is in further minimising the achieved MSE of the Kalman filter, by incorporating non-sample prior information about the value of β_t with the Kalman filter, by adapting the methods of Khan *et al.* (2005) to the state space model.

3. Preliminary Test Estimator

According to Khan *et al.* (2005) a better estimator of the parameter of a static regression model can be obtained by incorporating the sample information with non-sample prior information about the value of the parameter. The idea was originally pioneered by Bancroft (1944). In our case we use the Kalman filtering value of β_t as our sample information and incorporate it with non-sample prior information that can be obtained from previous investigations or experts knowledge.

Let the non-sample prior information about the value of β_t be expressed in the form of the null hypothesis

$$H_0 : \beta_t = \beta_{t0}. \tag{3.1}$$

To remove the uncertainty from the non-sample prior information we test the above null hypothesis against the alternative hypothesis

$$H_a : \beta_t \neq \beta_{t0}. \tag{3.2}$$

In this case an appropriate test statistic is

$$Z = \frac{b_{t|t} - \beta_{t0}}{\sqrt{\tau}}. \tag{3.3}$$

Under the null hypothesis the distribution of Z is normal with 0 mean and unit variance. Therefore, under the alternative hypothesis the distribution of Z^2 is non-central χ^2 with 1 d.f. and non-centrality parameter

$$\Delta = \tau^{-1}(\beta_t - \beta_{t0})^2. \tag{3.4}$$

Following Khan *et al.* (2002) the PTE of β_t is defined as

$$\hat{\beta}_t^{\text{PTE}} = b_{t|t}\, I(Z \geq Z_{\alpha/2}) + \hat{\beta}_t\, I(Z < Z_{\alpha/2}) \tag{3.5}$$

where $\hat{\beta}_t = \beta_{t0}$ and $Z_{\alpha/2}$ is the $\alpha/2$-level critical value of the Z-statistic. In the literature $\hat{\beta}_t$ is known as the restricted estimator (RE) of β_t. Under the null hypothesis RE outperforms both the observed data based least square estimator and PTE. An equivalent expression of (3.5) is given by

$$\hat{\beta}_t^{\text{PTE}} = b_{t|t}\, I(\chi^2 \geq \chi^2_\alpha) + \hat{\beta}_t\, I(\chi^2 < \chi^2_\alpha) \tag{3.6}$$

where χ^2_α is the α-level critical value of the chi-square statistic. The MSE function of the PTE of β_t is given by

$$\text{M}(\hat{\beta}_t^{\text{PTE}}) = \tau\left[1 - G_3(\chi^2_\alpha;\Delta) + \Delta\left\{2G_3(\chi^2_\alpha;\Delta) - G_5(\chi^2_\alpha;\Delta)\right\}\right] \tag{3.7}$$

where τ is the MSE of the Kalman filter, $G_j\left(\chi^2_\alpha;\Delta\right)$ is the c.d.f. of non-central χ^2-distribution with j d.f., non-centrality parameter Δ and is evaluated at $\chi^2_{\alpha(1)}$.

3.1. PTE of β_t for Unknown Error Variance

In case of unknown σ^2 we estimate it by using the data and plug into the expression of R_t in (2.6) to get the recursion done. In this case the appropriate test statistic for testing the null hypothesis in (3.1) is the Student's-t statistic with $\nu = (t-1)$ d.f., and the PTE of β_t is defined as

$$\hat{\beta}_t^{\text{PTE}} = b_{t|t}\, I(t \geq t_{\alpha/2}) + \hat{\beta}_t\, I(t < t_{\alpha/2}) \tag{3.8}$$

or equivalently

$$\hat{\beta}_t^{\text{PTE}} = b_{t|t}\, I(F \geq F_\alpha) + \hat{\beta}_t\, I(F < F_\alpha) \tag{3.9}$$

where F_α is the α-level critical value of the F-statistic with $(1, \nu)$ degrees of freedom. Under the alternative hypothesis the distribution of F is the non-central F-distribution with the same degrees of freedom and non-centrality parameter Δ given by (3.4). The MSE function of the PTE of β_t with unknown σ^2 is given by

$$\text{M}(\hat{\beta}_t^{\text{PTE}}) = \tau\left[1 - Q_{3,\nu}(3^{-1}F_\alpha;\Delta) + \Delta\left(2Q_{3,\nu}(3^{-1}F_\alpha;\Delta) - Q_{5,\nu}(5^{-1}F_\alpha;\Delta)\right)\right] \tag{3.10}$$

where $Q_{m,n}(m^{-1}F_\alpha;\Delta)$ is the c.d.f. of non-central F-distribution with $(m,\, n)$ d.f., non-centrality parameter Δ and is evaluated at $(m^{-1}F_\alpha)$.

3.2. Analysis

In this subsection we analyse the MSE of PTE and compare with that of the Kalman filter. The anlysis of the mean square error function of the PTE of β_t with unknown σ^2 is very similar to that of the mean square error function of the PTE of β_t with known σ^2. For the sake of the journal pages we present only the analysis of the mean square error function of the PTE of β_t for known σ^2.

From the expression of the MSE function of PTE in (3.7) it is clear that the performance of this estimator varies with varying Δ, the non-centrality parameter of the non-central χ^2 statistic and α, the level of significance of the preliminary test. Figure 2 shows the plot of the MSEs of Kalman filter and PTE of β_t against Δ, a function of the difference between the true value of the parameter and the value under the non-sample prior information.

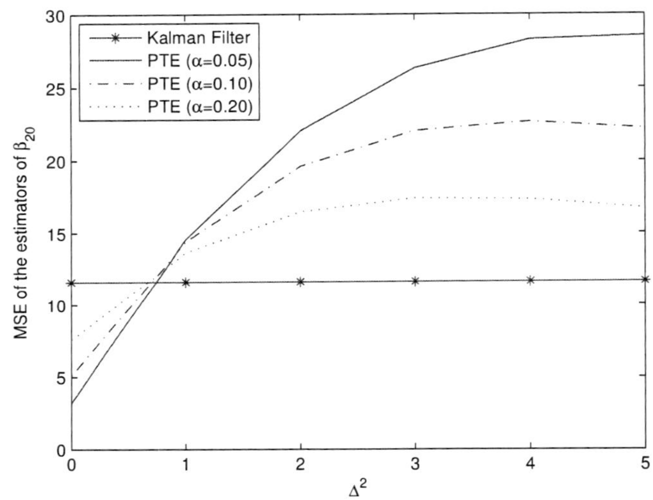

Figure 2. Plot of the MSE of KF and PTE of β_{30} against Δ for unknown error variance.

Under the null hypothesis $\Delta = 0$. Therefore, the MSE of $\hat{\beta}_t^{\text{PTE}}$ is

$$\text{M}(\hat{\beta}_t^{\text{PTE}}) = \tau \left[1 - G_3(\chi^2_\alpha; 0)\right]. \tag{3.11}$$

For any value of α, $G_3(\chi^2_\alpha; 0) \geq 0$. Therefore, for $\Delta = 0$ the performance of the PTE of β_t is at least as good as that of the Kalman filter. As Δ deviates from 0 the MSE of PTE grows larger and crosses the MSE of the Kalman filter at

$$\Delta = G_3(\chi^2_\alpha; \Delta) \left(2G_3(\chi^2_\alpha; \Delta) - G_5(\chi^2_\alpha; \Delta)\right)^{-1}. \tag{3.12}$$

Finally, as $\Delta \to \infty$ both of $G_3(\chi^2_\alpha; \Delta)$ and $G_5(\chi^2_\alpha; \Delta)$ tend to zero, and hence the MSE of the PTE approaches that of the Kalman filter.

As the performance of PTE depends on the level of significance, α, of the preliminary test it is very important to analyse the MSE function of PTE for varying α. If $\alpha = 0$ the value of F_α is ∞, and hence $G_m(\chi^2_\alpha; \Delta) = 1$ regardless of the value of Δ. Therefore, the MSE of PTE becomes the product of that of the Kalman filter and non-centrality parameter Δ. For $\alpha = 1$ the value of F_α is 0, and hence $G_m(\chi^2_\alpha; \Delta) = 0$ for any value of Δ.

Therefore, the MSE of PTE is no different from that of the Kalman filter regardless of the value of Δ.

As a level of significance of any hypothesis test neither 0 nor 1 are practical. On the other hand, for any given value of α between 0 and 1, the PTE does not uniformly outperform a different α based PTE. To find an optimum value of α at which the minimum guaranteed efficiency of PTE relative to the Kalman filter attains we pursue a numerical analysis.

Let the efficiency of PTE relative to Kalman filter be denoted by $\xi(\hat{\beta}_t^{\text{PTE}}; b_{t|t})$ and defined as

$$\begin{aligned} \xi\left(\hat{\beta}_t^{\text{PTE}}; b_{t|t}\right) &= \text{M}(b_{t|t})/\text{M}(\hat{\beta}_t^{\text{PTE}}) \\ &= \left[1 - G_3(\chi_\alpha^2; \Delta) + \Delta\left\{2G_3(\chi_\alpha^2; \Delta) - G_5(\chi_\alpha^2; \Delta)\right\}\right]^{-1}. \end{aligned} \quad (3.13)$$

This efficiency function attains its maximum at $\Delta = 0$. As Δ deviates from the origin, the efficiency decreases and becomes the same as that of the Kalman filter for the value of Δ given by (3.12) and also for very large values of Δ.

To find the optimum level of significance α let us preassign a tolerable relative efficiency, say ξ_0, and consider the set

$$A_\alpha = \left\{\alpha | \xi\left(\hat{\beta}_t^{\text{PTE}}; b_{t|t}\right) \geq \xi_0\right\}. \quad (3.14)$$

An estimator $\hat{\beta}_t^{\text{PTE}}$ is chosen which maximises $\xi\left(\hat{\beta}_t^{\text{PTE}}; b_{t|t}\right)$ over all $\alpha \in A_\alpha$ and Δ. For any given ξ_0, solving the equation

$$\max_\alpha \min_\Delta \xi\left(\hat{\beta}_t^{\text{PTE}}; b_{t|t}\right) = \xi_0 \quad (3.15)$$

for α, the solution α^* provides an optimum level of significance. The procedure is known as *maxmin rule* of the optimum level of significance of the preliminary test. Table 1 provides the maximum and minimum relative efficiencies ξ^* and ξ_0 respectively for a range of values of α. It also provides the value $\Delta = \Delta_0$ at which the minimum efficiency attains. For example, for unknown σ^2 if one wishes to estimate β_{50} with minimum relative efficiency 0.6205, the recommended value of α is 0.15. This minimum guaranteed efficiency attains at $\Delta = 3.4445$. For the same level of significance of the preliminary test the maximum efficiency of PTE is expected to be 1.6475. If σ^2 is known the efficiency of PTE relative to Kalman filter is the same for all values of t. For example, if $\alpha = 0.15$, the maximum and minimum guaranteed efficiencies of the PTE of β_t are 1.7936 and 0.595 respectively, for any value of t. The minimum efficiency attains at $\Delta = 3.58$.

4. Shrinkage Estimator

From the earlier discussion it is seen that the performance of PTE depends on the level of significance of the preliminary test. Moreover, PTE does not provide any smooth transition between the Kalman filter and the restricted estimator. In other words PTE is an extreme choice between the Kalman filter and the prior information based restricted estimator. Therefore, we define preliminary test based shrinkage estimator (SE) that does not

Table 1. Efficiency of PTE relative to Kalman filter for selected values of α and t.

α		σ unknown $t=20$	$t=30$	$t=40$	$t=50$	σ known
0.05	ξ^*	2.5222	2.4002	2.3436	2.3110	3.5929
	ξ_o	0.4494	0.4680	0.4772	0.4827	0.4058
	Δ_o	4.3958	4.2200	4.0990	3.9960	4.6700
0.10	ξ^*	1.9621	1.9043	1.8770	1.8611	2.2764
	ξ_o	0.5373	0.5521	0.5594	0.5637	0.5122
	Δ_o	3.9405	3.7955	3.7165	3.7000	3.9700
0.15	ξ^*	1.7088	1.6739	1.6572	1.6475	1.7936
	ξ_o	0.5989	0.6110	0.6170	0.6205	0.5950
	Δ_o	3.6605	3.5720	3.5550	3.4445	3.5800
0.20	ξ^*	1.5557	1.5323	1.5211	1.5145	1.5389
	ξ_o	0.6479	0.6579	0.6628	0.6657	0.6647
	Δ_o	3.5105	3.4190	3.3805	3.3550	3.3309
0.25	ξ^*	1.4501	1.4336	1.4257	1.4210	1.3820
	ξ_o	0.6892	0.6975	0.7015	0.7039	0.7248
	Δ_o	3.3605	3.2950	3.2600	3.2550	3.1850
0.30	ξ^*	1.3715	1.3595	1.3538	1.3404	1.2766
	ξ_o	0.7251	0.7320	0.7354	0.7374	0.7769
	Δ_o	3.2550	3.1800	3.1655	3.0900	3.0300
0.35	ξ^*	1.3101	1.3013	1.2970	1.2944	1.2022
	ξ_o	0.7570	0.7627	0.7655	0.7671	0.8220
	Δ_o	3.1729	3.1110	3.1050	3.0800	2.9400

depend on the level of significance of the preliminary test but provides a weighted average between the Kalman filter and the restricted estimator. For static model parameter this estimator outperforms PTE (cf. Khan *et al.*, 2005). The SE of β_t is given by

$$\hat{\beta}_t^{\mathrm{SE}} = b_{t|t} - c(b_{t|t} - \beta_{t0})|Z|^{-1} \tag{4.16}$$

where c is the shrinkage constant and Z is from (3.3). In the case of unknown error variance the shrinkage estimator of β_t is obtained by replacing the Z-statistic with Student's-t statistic. The MSE function of the SE of β_t is given by

$$\mathrm{M}(\hat{\beta}_t^{\mathrm{SE}}) = \tau(1 + c^2 - 2c\sqrt{2/\pi}\exp(-\Delta/2)). \tag{4.17}$$

The optimum value of c that minimises the above MSE function is given by

$$c^* = \sqrt{2/\pi}\exp(-\Delta/2). \tag{4.18}$$

To make c^* independent of Δ, we choose $c^* = \sqrt{2/\pi}$. Therefore, the MSE function of the SE of β_t is given by

$$\mathrm{M}(\hat{\beta}_t^{\mathrm{SE}}) = \tau\left(1 - 2/\pi(2\exp(-\Delta/2) - 1)\right). \tag{4.19}$$

We compare the above mean square error function with that of the Kalman filter and preliminary test estimator in the following subsection.

4.1. Analysis

4.1.1. Shrinkage Estimator vs. Kalman Filter

Unlike PTE the performance of the shrinkage estimator does not depend on the level of significance of the preliminary test but it depends on the non-centrality parameter Δ. Under the null hypothesis $\Delta = 0$ and hence the MSE function of the SE of β_t is

$$\mathrm{M}(\hat{\beta}_t^{\mathrm{SE}}) = \tau(1 - 2/\pi) \tag{4.20}$$

which is always less than that of the Kalman filter. Therefore, at $\Delta = 0$ shrinkage estimator always outperforms Kalman filter. As Δ deviates from 0 the MSE of the SE grows larger and becomes the same as that of the Kalman filter at $\Delta = \ln 4$. Finally, as $\Delta \to \infty$ the MSE of the SE approaches the constant value $\tau\left[1 + \frac{2}{\pi}\right]$ which is approximately 64% larger than the MSE of the Kalman filter. Therefore, for $\Delta < 1.39$ shrinkage estimator is a better choice than the Kalman filter. Otherwise, Kalman filter is the better choice than the shrinkage estimator.

4.1.2. Shrinkage Estimator vs. Preliminary Test Estimator

As the performance of SE depends on Δ alone and that of PTE depends on both α and Δ, it is very hard to pursue any analytical comparison between the MSE functions of the two estimators. In the neighborhood of $\Delta = 0$ both PTE and SE outperform Kalman filter. On the other hand SE outperforms PTE but this domination is not uniform. At $\Delta = 0$ the efficiency of the SE relative to the PTE is given by

$$\xi\left(\hat{\beta}_t^{\mathrm{SE}}; \hat{\beta}_t^{\mathrm{PTE}}\right) = (1 - G_3(\chi_\alpha^2; 0))(1 - 2/\pi)^{-1}. \tag{4.21}$$

The value of $G_3(\chi_\alpha^2; 0)$ is larger for smaller α. However, from the numerical computation it is found that $G_3(\chi_\alpha^2; 0)$ is not less than $2/\pi$ unless α is less than 0.075. As the value of Δ is expected to be very close to 0 but not exactly 0 our main concern is the performance of SE in the neighborhood of $\Delta = 0$. As Δ departs from 0 the MSE of PTE rises more sharply than that of the SE, and hence the performance of SE starts

getting better than that of the PTE up to some large value of Δ given by

$$\Delta^* = \frac{G_3(\chi_\alpha^2; \Delta) - 2/\pi(2\exp(-\Delta/2) - 1)}{2G_3(\chi_\alpha^2; \Delta) - G_3(\chi_\alpha^2; \Delta)}. \tag{4.22}$$

Therefore, for small values of Δ SE outperforms PTE. For very large value of Δ the MSE of SE is larger than that of the PTE but none of the estimators outperforms Kalman filter. From the above discussion it is clear that as an improved estimator of the parameter of linear dynamic regression model shrinkage estimator is a better choice than the preliminary test estimator under certain conditions.

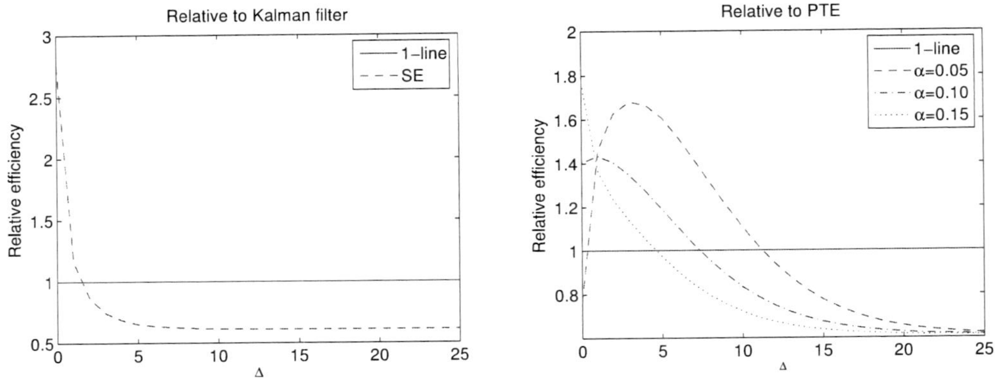

Figure 3. Plot of the relative efficiency of SE of β_{20} against Δ.

5. Concluding Remarks

It is well known that the Kalman filter is an unbiased and minimum MSE estimator in a dynamic state space regression model. The preliminary test and shrinkage estimators for this model, basd on the Kalman filter estimator, were proposed in this paper. Both proposals are biased, but if the non-sample prior information is not too far from the factual value of the parameter, then with respect to the mean square error property, both PTE and SE dominate the Kalman filter. Further the SE dominates both the PTE and Kalman filter estimators in this case. As the prior information is obtained from previous studies or experts knowledge it is expected and assumed in this result that such information would not be too far from the factual value. The results are vastly different in the alternate case. Therefore, the shrinkage estimator ranks as the best estimator followed by the preliminary test estimator, among these three estimators, for the dynamic regression model considered in this paper. A wider class of models can also be considered for the same investigations which is retained for future research.

Acknowledgement

The first author thankfully acknowledges some valuable comments and suggestions of Professor M. Y. El-Bassiouni, UAE University that helped improved an earlier version of the paper.

References

Bancroft, T.A. (1944). On biases in estimation due to the use of the preliminary test of significance, *Annals of Mathematical Statistics*. **15**, 190-204.

Carter C. and Kohn R. (1996). Markov chain Monte Carlo in conditionally Gaussian state space models, *Biometrika*. **83**(3), 589-601.

Duncan, D.B. and Horn, S.D. (1972). Linear Dynamic Recursive estimation from the Viewpoint of Regression Analysis, *Journal of American Statistical Association*. **67**, 815-821.

Gerlach R., Carter C. and Kohn R. (2000). Efficient Bayesian inference in dynamic mixture models, *Journal of American Statistical Association*. **95**(451), 819-828.

Kalman, R.E. (1960). A New Approach to Linear Filtering and Prediction Problems, *Journal of Basic Engineering*. **82**, 34-45.

Kalman, R.E. and Bucy, R.S. (1961). New Results in Filtering and Prediction Theory, *Journal of Basic Engineering*. **83**, 95-108.

Khan, S., Hoque, Z. and Saleh, A.K.Md.E. (2002). Estimation of the slope parameter for linear regression model with uncertain prior information, *Journal of Statistical Research*. **36**(1), 55-73.

Khan, S., Hoque, Z. and Saleh, A.K.Md.E. (2005). Estimation of the intercept parameter for linear regression model with uncertain non-sample prior information, *Statistical Papers*. **46**(3), 379–395.

Meinhold, R.J. and Singpurwalla, N.D. (1983). Understanding the Kalman Filter, *The American Statistician*. **37**(2), 123-127.

In: New Developments in Applied Statistics
Editor: Mohammad Ahsanullah
ISBN 978-1-61324-648-1

Chapter 25

BAYES ESTIMATION OF LORENZ CURVE AND GINI-INDEX FOR CLASSICAL PARETO DISTRIBUTION IN SOME REAL DATA SITUATION

***E.I. Abdul-Sathar*[1] *and E.S. Jeevanand*[2]**
[1]Department of Statistics, Department of Statistics
University of Kerala- 695 581, Kerala, India.
[2]Department of Mathematics and Statistics
Union Christian College
Aluva - 2, Kerala, India

Abstract

This paper presents Bayesian estimation of Lorenz curve and Gini-index in the presence of outliers and using record values for the classical Pareto distribution. Here we considered two important models for accommodating outliers viz the identified outliers model introduced by Veale (1975) and the exchangeable model by Kale (1976). Bayes estimators have been developed under squared error loss function as well as under LINEX loss functions. We demonstrate the use of the proposed outlier estimation procedure with the tournament earnings data given in Arnold (1983). Based on a Monte Carlo simulation study, comparison are made between the proposed estimators.

Keywords: Bayes estimators; outliers; Squared error loss function; LINEX loss function; Lorenz curve; Gini-index; Record values.

1. Introduction

The Pareto distribution was introduced by Pareto (1897) as model for the distribution of income. It has played a very important role in the distribution of wealth, business size, occurrence of natural resources, insurance risk and business failures. Arnold and Press (1983) gave an extensive historical survey of its use in the context of income distribution. In recent years, its model in several different forms have been studied by many authors including Davis and Feldstein (1979), Cohen and Whitten (1988), Grimshaw (1993), among others. The Pareto distribution of the first kind also known as classical Pareto distribution has been found to provide a good model for the income data (see, Arnold (1983)). The probability density function of the classical Pareto distribution is given by

$$f\left(x,\alpha,\sigma\right)=\alpha\sigma^{\alpha}x^{-(\alpha+1)},x\geq\sigma>0,\alpha>1. \tag{1}$$

The Lorenz curve and Gini-index play central role in the analysis of income data and the evaluation of welfare judgements. Also these have been extensively used in the study of inequality of distributions. Let X be a non-negative random variable with distribution function F. The Lorenz curve corresponding to the random variable X, denoted by $L(p)$, is defined (Gastwirth (1971)) by the formula

$$L(p)=\frac{1}{\mu}\int_{0}^{p}F^{-1}(x)dx, \qquad 0\leq p\leq 1$$

where

$$F^{-1}(x)=\sup\{y:F(y)\leq x\}.$$

We assume that the mean μ is finite and positive. If the distribution which is being studied is the income of a certain population, then $L(p)$ denotes the fraction of the total income received by the $100p\%$ of the population which has the lowest income. The Gini index is twice the area between the Lorenz curve and the 'line of equal angle '(also known as 'line of equal distribution'). The Gini-index g is defined as

$$g=1-2\int_{0}^{1}L(p)dp.$$

For the applications of Lorenz curve and Gini-index we refer to Moothathu (1990 b) and Bhattacharjee (1993) and the references therein. These measures have also been found applications in reliability theory. For example, see Chandra and Singpurwalla (1981).

The data related to income or wealth are voluminous and have to be collected directly using sampling methods. In such data the occurrence of outlier or discordant observation is quite natural. An outlier or outlying observation is one that appears to deviate markedly from the other members of the sample in which it occurs and can have a significant effect on any type of statistical inference. The problem of analyzing statistical data in the presence of outliers using various approaches to statistical inference has attracted considerable research in recent years. Kale and Sinha (1971), Joshi (1972), Chikkagoudar and Kunchur (1980) consider the problem of estimating the parameters when an outlier is present in a sample from the exponential model, while Kimber (1983) and Dixit (1989) have investigated the same problem with reference to the gamma population. Jeevanand and Nair (1992) obtained the Bayes estimates of the parameters of classical Pareto distribution on the assumption that a sample of n contains k outliers with k known, but without identifying which of the observations are outliers. Jeevanand and Nair (1993,1996) obtained prediction intervals for the order statistics based on samples that contains a discordant observation. Also, Jeevanand and Sathar (2006) discussed the problem of estimating the survival function of the Pareto distribution, when the samples contains discordant observations.

In the literature concerning estimation of Lorenz curve and Gini-index there are many classical methods for estimating Lorenz curve and inequality measures from grouped data. For example, Kakwani and Podder(1973,1976) proposed a prominent method for the estimation of Lorenz curve and inequality measures with grouped data. Moothathu (1985) discussed the problem of estimation of Lorenz curve and Gini-index of the exponential distribution. Also Moothathu (1990 a) obtained the best estimator of Lorenz curve and Gini-index of Pareto distribution. Further, Chotikapanich and Griffiths (2002) estimated the Lorenz curve using Dirichlet distribution for the error terms. Sathar et al (2005) discussed the problem of estimation of Lorenz curve and Gini-index of the Pareto distribution in the Bayesian frame-work. The present paper focuses attention on estimating Lorenz curve and Gini-index in the presence of outliers and using record values using Bayesian approach when the sample comes from the Pareto distribution specified by (1). Estimates are obtained under two important models for accommodating outliers, the identified outliers model introduced by Veale (1975) and the exchangeable model considered by Kale (1976).

This article is organized as follows. In Section 2, we consider the estimation of Lorenz curve and gini-index under exchangeable model. In Section 3, the estimation of Lorenz curve and gini index assuming identifiable model is discussed. In Section 4, we obtained the estimate of Lorenz curve and Gini-index based on record value statistics. In Section 5, we demonstrate the use of the proposed outlier estimation procedure with the tournament earnings data given in Arnold (1983). Based on a Monte Carlo simulation study, comparison are made between the proposed estimators, on the basis of mean square errors, are discussed in the last section.

2. The Exchangeable Model

The exchangeable model is a modified version of the k- outlier model proposed by David (1979). In this model it was assumed that, prior to the experiment, any set of k random variables out of $(X_1, X_2, ..., X_n)$ has equal probability of $1/\binom{n}{k}$ of being distributed according to a different distribution $G = G(F)$ depend on F, that causes outliers, while the other random variables are distributed according to F and denote the densities of F and G by f and g respectively. The likelihood function corresponding to this set-up can be written as

$$l\left(\underline{x}|f, g\right) = \binom{n}{k}^{-1} \sum_{v \in N_{(k)}} \left\{\prod_{j \in v} g(x_j)\right\} \left\{\prod_{i \notin v} f(x_i)\right\}. \tag{2}$$

Here $v = (v_1, v_2, ..., v_k)$, the indexing subset of observations coming from G is an element of N_k, the collection of all $\binom{n}{k}$ subset of selecting k integers out of n. Accordingly, let $(x_1, x_2, ..., x_n)$ be a random sample containing k outliers which are distributed as Pareto form with density

$$g\left(x, \alpha b, \sigma\right) = \alpha b \sigma^{\alpha b} x^{-(\alpha b+1)}, x \geq \sigma > 0, \alpha > 1 \tag{3}$$

where $b > 1$ and the remaining $(n - k)$ belongs to the population given in (1). In the present discussion b is assumed known. Notice that when $b < 1$ the discordant observation is lower outlier, while $b > 1$ indicates an upper outlier. In accordance with (1),(2) and (3), the likelihood can be written as

$$l\left(\underline{x}|\alpha, b, \sigma\right) = \begin{pmatrix} n \\ k \end{pmatrix}^{-1} \alpha^n b^k \sigma^{(n+(b-1)k)\alpha} \prod_{i=1}^{n} x_i^{-(\alpha+1)} \sum\nolimits_* \prod_{j=1}^{k} x_{A_j}^{-\alpha(b-1)} \tag{4}$$

where

$$\sum\nolimits_* = \sum_{A_1=1}^{n-k+1} \cdots \sum_{A_k=A_{k-1}+1}^{n} .$$

The scale parameter σ which is same for all observations is taken to be known. A convenient prior for (α, b) is obtained by choosing a gamma distribution for α and a vague uniform model for b and taking α and b to be independent.
The joint prior for (α, b) is

$$g\left(\alpha, b\right) = C_1 \alpha^{r-1} \exp\left(-\tau\alpha\right), \alpha, r, \tau > 0, b > 1. \tag{5}$$

The symbol C with various suffixes stands for the normalizing constants. Combining (4) and (5) and using Bayes theorem, the joint posterior density turns out to be

$$f\left(\alpha, b|\underline{x}\right) = C_2 \sum\nolimits_* \alpha^{m-1} b^k \exp\left(-\left[T_A + bt_A\right]\alpha\right), \alpha > 0, b > 1 \tag{6}$$

where

$$m = n + r, T_A = t + \tau - t_A, t_A = \sum_{i=1}^{k} \log\left(\frac{x_{A_i}}{\sigma}\right), t = \sum_{i=1}^{n} \log\left(\frac{x_i}{\sigma}\right).$$

For the model (1), the Lorenz curve simplifies to

$$L(p) = 1 - (1-p)^{1-\alpha^{-1}}, 0 < p < 1. \tag{7}$$

Now using (6) and (7), we get the marginal posterior density of the Lorenz curve as

$$\begin{aligned} f\left(L|\underline{x}\right) = \left[C_3(p,0)\right]^{-1} \sum\nolimits_* \Psi_{m-k}(L,p)\Omega(L,p)\Gamma\left[k+1, \Psi_1(L,p)t_A\right] \\ \exp\left(\Psi_1(L,p)T_A\right) t_A^{-k-1}, \quad 0 < L < p \end{aligned} \tag{8}$$

where

$$\begin{aligned} C_3(p,d) = \int_0^p L^d \sum\nolimits_* \Psi_{m-k-1}(L,p)\Omega(L,p)\Gamma\left[k+1, \Psi_1(L,p)t_A\right] \\ \exp\left(\Psi_1(L,p)T_A\right) t_A^{-k-1} dL \end{aligned} \tag{9}$$

with $\Psi_k(L,p) = \left(\frac{log(1-p)}{log(1-p)-log(1-L)}\right)^k$ and $\Omega(L,p) = \frac{1}{(1-L)log(1-p)}$.
$\Gamma\left[a, z\right]$ is the incomplete Gamma function defined by

$$\Gamma\left(a, z\right) = \int_z^{\infty} e^{-t} t^{z-1} dt.$$

From a decision-theoretic viewpoint, in order to select a single value as representing our best estimator of L, a loss function must be specified. Under squared-error loss, the Bayes estimator of Lorenz curve is the mean of the posterior density given by

$$\hat{L}_{EMS} = E(L|\underline{x}) = \frac{C_3(p,1)}{C_3(p,0)} \tag{10}$$

and the posterior risk (minimum posterior expected loss) of $\hat{L}_{EMS}$ is the posterior variance given by

$$R\left(L, \hat{L}_{EMS}\right) = \frac{C_3(p,2)}{C_3(p,0)} - \left(\hat{L}_{EMS}\right)^2. \tag{11}$$

Another loss function in popular use is the LINEX loss function(LLF) introduced by Variant (1975). LLF is defined as

$$L(\Delta) = b\left(e^{a\Delta} - a\Delta - 1\right), a \neq 0, b > 0, \tag{12}$$

where a and b are the shape parameters of the loss function (12). Obviously, the nature of LLF changes according to the choice of a. Under LLF the Bayes estimator for Lorenz curve is defined as

$$\hat{L}_{EMX} = \frac{\ln\left(\int\limits_0^p e^{aL} f(L|\underline{x})\, dL\right)}{a} \tag{13}$$

where $f(L|\underline{x})$ is the posterior density of the Lorenz curve given by (8). It may be noted here that the Bayes estimator of Lorenz curve under both loss functions are not reducible in nice closed form, however, we seek suitable numerical integration for their evaluation.

Now we derive Bayesian estimator for the Gini-index under this model. For the density given in (1), the Gini- index simplifies to

$$G = (2\alpha - 1)^{-1}. \tag{14}$$

Proceeding analogous to the estimation of Lorenz curve, we get the posterior distribution of Gini-index as

$$f(G|\underline{x}) = [C_4(0)]^{-1} \sum\nolimits_* Z_{m-k}(G)\Gamma[k+1, Z_1(G)t_A] \exp(Z_1(G)T_A)\, t_A^{-k-1}(1+G)^{-2}, \quad 0 < G < 1 \tag{15}$$

where

$$C_4(d) = \int\limits_0^1 G^d \sum\nolimits_* Z_{m-k}(G)\Gamma[k+1, Z_1(G)t_A] \exp(Z_1(G)T_A)\, t_A^{-k-1}(1+G)^{-2}\, dG \tag{16}$$

and

$$Z_k(G) = \left(\frac{G+1}{2G}\right)^k.$$

The Bayes estimate of Gini-index under squared error loss function is

$$\hat{G}_{EMS} = E(G|\underline{x}) = \frac{C_4(1)}{C_4(0)}. \tag{17}$$

and the posterior risk of $\hat{G}_{EMS}$ is given by

$$R\left(G, \hat{G}_{EMS}\right) = \frac{C_4(2)}{C_4(0)} - \left(\hat{G}_{EMS}\right)^2. \tag{18}$$

Under LLF the Bayes estimator for Gini-index is defined as

$$\hat{G}_{EMX} = \frac{\ln\left(\int_0^1 e^{aG} f\left(G|\underline{x}\right) dG\right)}{a}. \tag{19}$$

As mentioned earlier, the integrals involved in (17) and (19) are not solvable analytically and , therefore, we seek numerical integration for their evaluation.

3. Identified Outliers Model

In this section, we employ the identified outlier model, which stipulates that the subset which forms the outliers is known. When uniform prior is assumed over the subsets of k integers out of n, Kale(1976) has shown that for $b > 1$, the last k components of the order statistics of the sample have the largest probability of being outliers. In this case the likelihood can be written as

$$l\left(\underline{x}|f, g\right) = \prod_{i \notin v} f\left(x_i\right) \prod_{j \in v} g\left(x_j\right)$$

for some given $v \in N_k$. In fact this is a two sample set-up, a sample of size $(n-k)$ from F and a sample of size k from G. Accordingly, the likelihood takes the form

$$l\left(\underline{x}|\alpha, b, \sigma\right) = C_5 \alpha^n b^k \exp\left(-(Q + bS)\alpha\right) \tag{20}$$

where

$$S = \sum_{i=1}^{k} \log\left(\frac{x_{n-i+1}}{\sigma}\right), Q = t - S.$$

using (5) and (20), we get the joint posterior density of (α, b) as

$$f\left(\alpha, b|\underline{x}\right) = C_6 \alpha^{m-1} b^k \exp\left(-(Q + \tau + bs)\alpha\right), \alpha > 0, b > 1. \tag{21}$$

Now, we get the posterior density of L as

$$\begin{aligned} f\left(L|\underline{x}\right) = \left[C_7(p,0)\right]^{-1} \Psi_{m-k}(L,p)\Omega(L,p)\Gamma\left[k+1, \Psi_1(L,p)S\right] \\ \exp\left(\Psi_1(L,p)(Q+\tau)\right) S^{-k-1}, \quad 0 < L < p \end{aligned} \tag{22}$$

where

$$C_7(p,d) = \int_0^p L^d \Psi_{m-k}(L,p)\Omega(L,p)\Gamma\left[k+1, \Psi_1(L,p)S\right] \exp\left(\Psi_1(L,p)(Q+\tau)\right) S^{-k-1} dL. \tag{23}$$

Under squared-error loss, the Bayes estimator of Lorenz curve is the mean of the posterior density given by

$$\hat{L}_{IMS} = E(L|\underline{x}) = \frac{C_7(p,1)}{C_7(p,0)} \tag{24}$$

and the posterior risk of $\hat{L}_{IMS}$ is given by

$$R\left(L, \hat{L}_{IMS}\right) = \frac{C_7(p,2)}{C_7(p,0)} - \left(\hat{L}_{IMS}\right)^2. \tag{25}$$

Under LLF the Bayes estimator for Lorenz curve is defined as

$$\hat{L}_{EMS} = \frac{\ln\left(\int_0^p e^{aL} f\left(L|\underline{x}\right) dL\right)}{a}. \tag{26}$$

To evaluate (24) and (26) we seek numerical integration.

Now we derive an estimator for the Gini-index assuming identifiable model. Proceeding analogous to the estimation of Lorenz curve, we get the posterior distribution of G as

$$f\left(G|\underline{x}\right) = \left[C_8(0)\right]^{-1} \sum\nolimits_* Z_{m-k}(G)\Gamma\left[k+1, Z_1(G)S\right] \exp\left(Z_1(G)(Q+\tau)\right) S^{-k-1}\left(1+G\right)^{-2}, \quad 0 < G < 1 \tag{27}$$

where

$$C_8(d) = \int_0^1 G^d \sum\nolimits_* Z_{m-k}(G)\Gamma\left[k+1, Z_1(G)S\right] \exp\left(Z_1(G)(Q+\tau)\right) S^{-k-1}\left(1+G\right)^{-2} dG. \tag{28}$$

The Bayes estimate of Gini-index under squared error loss function is

$$\hat{G}_{IMS} = E(G|\underline{x}) = \frac{C_8(1)}{C_8(0)} \tag{29}$$

and the posterior risk of $\hat{G}_{IMS}$ is given by

$$R\left(G, \hat{G}_{IM}\right) = \frac{C_8(2)}{C_8(0)} - \left(\hat{G}_{IM}\right)^2. \tag{30}$$

Under LLF the Bayes estimator for Gini-index is defined as

$$\hat{G}_{IMX} = \frac{\ln\left(\int_0^1 e^{aG} f\left(G|\underline{x}\right) dG\right)}{a}. \tag{31}$$

To evaluate (29) and (31) we seek numerical integration.

4. Estimation Using Records

The statistical study of record values began with a paper by Chandler in 1952. Let $X_1, X_2, ..., X_n$ be a sequence of independent and identically distributed random variables with common distribution function F. Then X_j is called a (upper) record value in the sequence if it exceeds all the observations that precede it in the sequence. That is if

$$X_j > Max(X_1, X_2, ..., X_{j-1}), j \geq 1.$$

Formally, the nth record in the sequence is denoted by R_n with $R_0 = X_1$ as the initial record value. Elaborate review articles by Glick (1978), Galambos (1978), Nevzarov (1987), Nagaraja (1988), books by Ahsanullah (1995) and Arnold et al.(1988) have highlighted many of the advances in the theory of records.

The most of the literature available in the record values are for characterizations of probability distribution or the inference problem in the classical frame work. Bayesian analysis of record values is of recent origin. It seems that Dunsmore (1983) was the first author to use Bayesian method to derive tolerance region for prediction of the future records from exponential distribution. Gulati (2003) provides a comprehensive description of statistical inference from record braking data, both parametric and non parametric setting, including Batesian inference. The Bayes estimate of the exponential distribution based on record statistics are discussed in Jaheen (2004), Madi (2005), Ahmadi et al (2005), Madi and Raqab (2004) use the records from Pareto distribution to predict the temperature using Bayesian inference.

Now based on the sequence of the first $(n+1)$ records, $\mathbf{R} = (R_0, R_1, ..., R_n)^T$, the joint probability density function of the record sequence $\mathbf{R}$ can be written as (Arnold et al (1988))

$$l\left(\mathbf{R}|\sigma\right) = \alpha^{n+1}\left(\frac{R_n}{\sigma}\right)^{-\alpha}\prod_{i=0}^{n}\frac{1}{R_i}, 1 \leq R_0 < R_1 < ... < R_n < \infty. \tag{32}$$

Using the prior

$$g\left(\alpha, \sigma\right) = C_7\alpha^{p-1}e^{-\alpha\rho}, \alpha \geq 0 \tag{33}$$

The joint posterior density of (α, σ) can be obtained as

$$f\left(\alpha|R\right) = C_8\alpha^{N-1}e^{-\alpha T}, \alpha \geq 0 \tag{34}$$

where $N = n + p$ and $T = \ln\left(\frac{R_n}{\sigma}\right) + \rho$.
Now, we get the posterior density of Lorenz curve as

$$f\left(L|\underline{x}\right) = \left[C_9(p,0)\right]^{-1}\Psi_N(L,p)\Omega(L,p)\exp\left(-T\Psi_1(L,p)\right), 0 < L < p \tag{35}$$

where

$$C_9(p,d) = \int_0^p L^d\Psi_N(L,p)\Omega(L,p)\exp\left(-T\Psi_1(L,p)\right)dL. \tag{36}$$

Under squared-error loss, the Bayes estimator of Lorenz curve is the mean of the posterior density given by

$$\hat{L}_{RS} = E(L|\underline{x}) = \frac{C_9(p,1)}{C_9(p,0)} \tag{37}$$

and the posterior risk of $\hat{L}_{RS}$ is given by

$$R\left(L, \hat{L}_{RS}\right) = \frac{C_9(p,2)}{C_9(p,0)} - \left(\hat{L}_{RS}\right)^2. \tag{38}$$

Under LLF the Bayes estimator for Lorenz is defined as

$$\hat{L}_{RX} = \frac{\ln\left(\int_0^p e^{aL} f\left(L|\underline{x}\right) dL\right)}{a}. \tag{39}$$

where $f\left(L|\underline{x}\right)$ is the posterior density of the Lorenz given by (35). To evaluate (37) and (39) we seek numerical integration.

Now, proceeding analogous to the estimation of Lorenz curve, we get the posterior distribution of G as

$$f\left(G|\underline{x}\right) = \left[C_{10}\left(d\right)\right]^{-1} \frac{\mathrm{Z}_{\mathrm{N}-1}(G)}{g^2} \exp\left(-\mathrm{Z}_1(G)T\right) \tag{40}$$

where

$$C_{10}\left(d\right) = \int_0^1 g^{d-2}\mathrm{Z}_{\mathrm{N}-1}(G) \exp\left(-\mathrm{Z}_{\mathrm{N}-1}(G)T\right) dg. \tag{41}$$

Under squared-error loss, the Bayes estimator of Lorenz curve is the mean of the posterior density given by

$$\hat{G}_{RS} = E(G|\underline{x}) = \frac{C_{10}(1)}{C_{10}(0)} \tag{42}$$

and the posterior risk of $\hat{G}_{RS}$ is given by

$$R\left(G, \hat{G}_{RS}\right) = \frac{C_{10}(2)}{C_{10}(0)} - \left(\hat{G}_{RS}\right)^2. \tag{43}$$

Under LLF the Bayes estimator for Lorenz is defined as

$$\hat{G}_{RX} = \frac{\ln\left(\int_0^1 e^{ag} f\left(G|\underline{x}\right) dg\right)}{a}. \tag{44}$$

where $f\left(G|\underline{x}\right)$ is the posterior density of the Lorenz given by (40). To evaluate (42) and (44) we seek numerical integration.

5. A Numerical Example

To illustrate the usefulness of the proposed outlier estimators with real situations, we considered here the real data-set initially reported by Arnold (1983) representing the life time tournament earnings through 1980 of all those professional golfers whose income exceeded \$ 700,000 during that period. Pareto distribution fits well for this data (see Arnold (1983)). From the data we have $n = 50, \alpha = 2.275$ and $\sigma = 700$. We chose the values

$r = 1, \tau = 2, a = 0.1$ and $p = 0.1$ and using the procedure in Kale and Kale (1992), the observation 3581 in the sample was detected to be an outlier. Based on this data, we evaluate and present the proposed estimates of Lorenz curve and Gini-index in Table 1 and using the above estimated parameters, we obtain the true estmates of Lorenz curve and Gini-index as 0.05734 and 0.2817 respectively.

Table 1. Estimates of Lorenz curve and Gini-index for tournament earnings data

Lor. Estimators	Estimated value	Gini Estimators	Estimated value
$\hat{L}_{EMS}$	0.05241	$\hat{G}_{EMS}$	0.3357
$\hat{L}_{EMX}$	0.05241	$\hat{G}_{EMX}$	0.3359
$\hat{L}_{IDS}$	0.05311	$\hat{G}_{IDS}$	0.3290
$\hat{L}_{IDX}$	0.05312	$\hat{G}_{IDX}$	0.3292

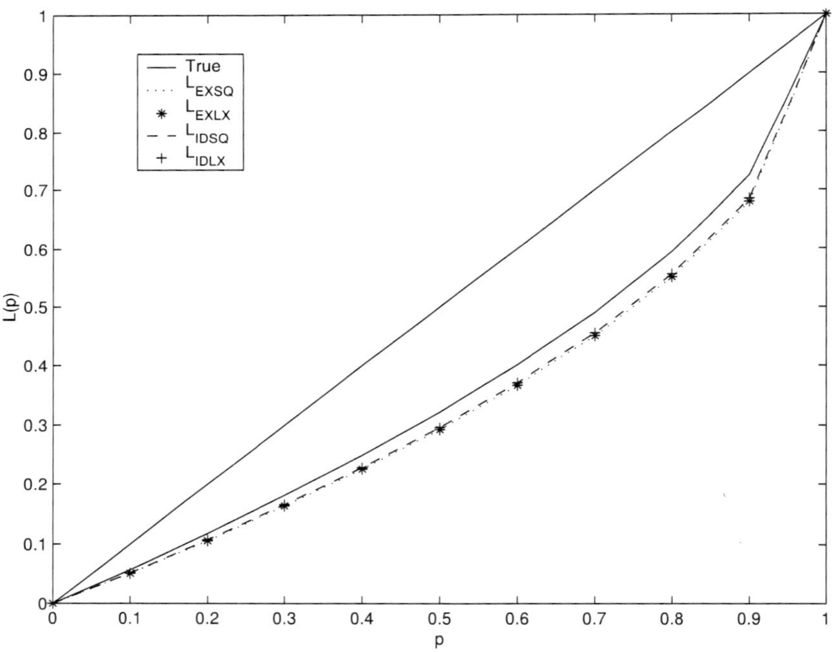

Figure 1. Estimates of Lorenz Curves (Golfers data).

It is clear from Table 1 that the Lorenz curve estimators tends to give lower estimates and the Gini estimators tends to give higher estimates. It is to be noted that the performance of the estimators under both loss functions are more or less similar. In Figure 1, we plot estimates of Lorenz curves, for various values of p. From the estimates of the Gini-index and of the Lorenz Curve, it can be observed that there is a fair amount of disparity in the earnings of different golfers.

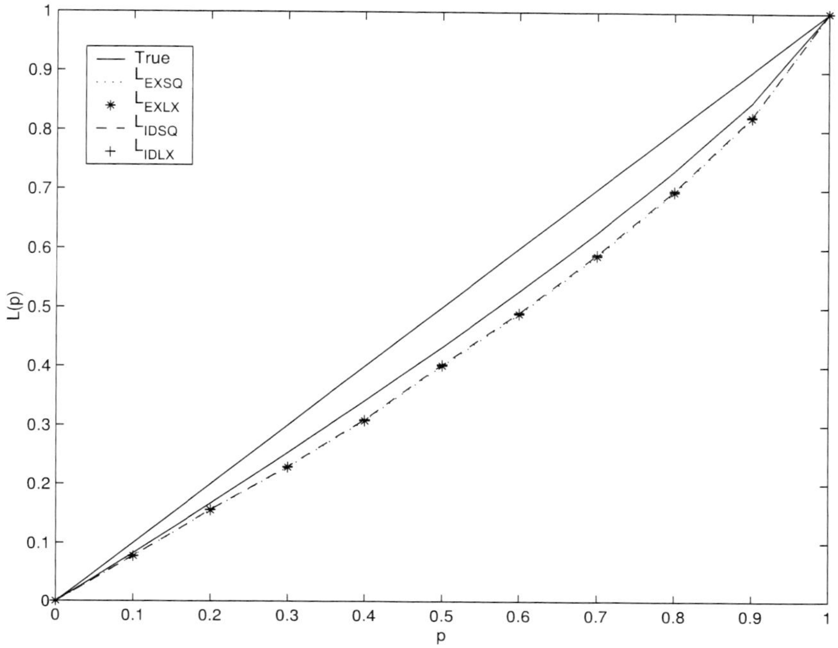

Figure 2. Estimates of Lorenz Curves (Simulated data).

6. Monte Carlo Simulation Study

In order to assess the performance of the estimators, we perform a Monte Carlo simulation study of 2000 samples of sizes $n = 20, 50$ and 100. The Pareto samples were generated from (1) for all combinations of $\sigma = 150$ and $\alpha = 2.5, 3.5, 4.5$ and 5.5. The means and mean squared errors (in parentheses) of the estimators are presented in Tables 2 and 3. It is evident from the Tables 2 and 3 that for smaller values of α, the estimates under exchangeable model have lesser bias and mean square error than the identifiable model and a reverse trend is seen for large values of α. It has been perceived that bias and mean squared error of the proposed estimates increase as the α values increase. The Tables show that for smaller sample values the estimates under squared error loss function performs better than the LINEX loss function. Further, the bias and expected loss decreases as the prior knowledge become very strong. In Figure 2, we plot estimates of Lorenz curves, for simulated data, for various values of p.

Next, for the illustration of estimation under record values, we generated a sample of size 1000 from the Pareto distribution (1) with parameters $\alpha = 6$ and $\sigma = 150$. The observed record values are 173.96, 174.838, 191.018, 213.499, 251.183, 297.701, 353.467, 383.166, 447.576 and 558.051. For this data of $n = 10$ we evaluated the estimators using (27) and (29). The values of estimators of Lorenz curve obtained are $\hat{L}_{RS} = 0.863946$ and $\hat{L}_{RX} = 0.863958$ respectively. Actual value of Lorenz curve obtained is 0.975537. The values of estimators of Gini-index obtained are $\hat{G}_{RS} = 0.0709206$ and $\hat{G}_{RX} = 0.0709548$ respectively. Actual value of Gini index obtained is 0.0909091.

Table 2. Means and MSEs (in parentheses) of the estimates of Lorenz curve

		$\alpha = 2.5$	$\alpha = 3.5$	$\alpha = 4.5$	$\alpha = 5.5$
True Lorenz		0.061259	0.072495	0.078679	0.082592
$n = 20$	$\hat{L}_{EMS}$	0.060318	0.043243	0.051594	0.057525
		(0.000016)	(0.000944)	(0.000767)	(0.000657)
	$\hat{L}_{EMX}$	0.060323	0.043251	0.051600	0.057531
		(0.000016)	(0.000943)	(0.000767)	(0.000656)
	$\hat{L}_{IDS}$	0.062046	0.045214	0.053544	0.059302
		(0.000014)	(0.000833)	(0.000663)	(0.000570)
	$\hat{L}_{IDX}$	0.062050	0.045222	0.053550	0.059307
		(0.000014)	(0.000833)	(0.000663)	(0.000569)
$n = 30$	$\hat{L}_{EMS}$	0.068741	0.050665	0.059226	0.065252
		(0.000061)	(0.000504)	(0.000418)	(0.000317)
	$\hat{L}_{EMX}$	0.068743	0.050670	0.059230	0.065254
		(0.000060)	(0.000503)	(0.000418)	(0.000317)
	$\hat{L}_{IDS}$	0.069664	0.052021	0.060366	0.066263
		(0.000075)	(0.000445)	(0.000373)	(0.000282)
	$\hat{L}_{IDX}$	0.069666	0.052025	0.060369	0.066266
		(0.000075)	(0.000445)	(0.000373)	(0.000282)
$n = 50$	$\hat{L}_{EMS}$	0.075529	0.051460	0.065219	0.071857
		(0.000209)	(0.000472)	(0.000199)	(0.000125)
	$\hat{L}_{EMX}$	0.075529	0.051463	0.065221	0.071857
		(0.000209)	(0.000472)	(0.000199)	(0.000125)
	$\hat{L}_{IDS}$	0.075972	0.052278	0.065813	0.072357
		(0.000222)	(0.000438)	(0.000183)	(0.000114)
	$\hat{L}_{IDX}$	0.075973	0.052281	0.065815	0.072358
		(0.000222)	(0.000437)	(0.000183)	(0.000114)
$n = 100$	$\hat{L}_{EMS}$	0.078886	0.056867	0.068669	0.076014
		(0.000313)	(0.000259)	(0.000104)	(0.000048)
	$\hat{L}_{EMX}$	0.078886	0.056868	0.068669	0.076015
		(0.000313)	(0.000104)	(0.000104)	(0.000048)
	$\hat{L}_{IDS}$	0.079075	0.057237	0.068945	0.076230
		(0.000320)	(0.000247)	(0.000099)	(0.000045)
	$\hat{L}_{IDX}$	0.079075	0.057238	0.068946	0.076230
		(0.000320)	(0.000247)	(0.000099)	(0.000045)

Table 3. Means and MSEs (in parentheses) of the estimates of Gini-index

		$\alpha = 2.5$	$\alpha = 3.5$	$\alpha = 4.5$	$\alpha = 5.5$
True Gini		0.250000	0.166667	0.125000	0.100000
$n = 20$	$\hat{G}_{EMS}$	0.250007	0.415864	0.345515	0.280050
		(0.000817)	(0.065380)	(0.052313)	(0.033912)
	$\hat{G}_{EMX}$	0.250336	0.416829	0.346202	0.280484
		(0.000823)	(0.065887)	(0.052646)	(0.034080)
	$\hat{G}_{IDS}$	0.236655	0.394478	0.327240	0.264769
		(0.000919)	(0.055098)	(0.044229)	(0.028488)
	$\hat{G}_{IDX}$	0.236931	0.395337	0.327835	0.265135
		(0.000916)	(0.055515)	(0.044497)	(0.028618)
$n = 30$	$\hat{G}_{EMS}$	0.196218	0.358446	0.287541	0.209092
		(0.003266)	(0.040811)	(0.028089)	(0.012697)
	$\hat{G}_{EMX}$	0.196328	0.358939	0.287828	0.209222
		(0.003255)	(0.041026)	(0.028190)	(0.012727)
	$\hat{G}_{IDS}$	0.189473	0.345426	0.277228	0.201728
		(0.004011)	(0.035705)	(0.024714)	(0.011092)
	$\hat{G}_{IDX}$	0.189571	0.345869	0.277483	0.201844
		(0.004000)	(0.035887)	(0.024799)	(0.011117)
$n = 50$	$\hat{G}_{EMS}$	0.153241	0.328175	0.224546	0.175755
		(0.009769)	(0.027566)	(0.011234)	(0.006064)
	$\hat{G}_{EMX}$	0.153276	0.328396	0.224634	0.175803
		(0.009764)	(0.027643)	(0.011254)	(0.006072)
	$\hat{G}_{IDS}$	0.150100	0.321048	0.219883	0.172141
		(0.010369)	(0.025240)	(0.010277)	(0.005516)
	$\hat{G}_{IDX}$	0.150133	0.321253	0.219965	0.172186
		(0.010363)	(0.025308)	(0.010294)	(0.005523)
$n = 100$	$\hat{G}_{EMS}$	0.127559	0.291718	0.196472	0.146123
		(0.015106)	(0.016399)	(0.005285)	(0.002244)
	$\hat{G}_{EMX}$	0.127569	0.291794	0.196501	0.146138
		(0.015104)	(0.016419)	(0.005289)	(0.002245)
	$\hat{G}_{IDS}$	0.126256	0.288495	0.194416	0.144641
		(0.015425)	(0.015585)	(0.004991)	(0.002106)
	$\hat{G}_{IDX}$	0.126266	0.288568	0.194444	0.144656
		(0.015422)	(0.015603)	(0.004995)	(0.002108)

References

[1] Arnold, B. C. (1983) *Pareto distributions, International co-operative publishing house.* Maryland,United States.

[2] Arnold, B. C., Balakrishnan, N. and Nagaraja, H.N. (1998) *Records*, John Wiely and Sons, New York.

[3] Arnold, B. C. and PRESS, S. J. (1983) Bayesian inference for Pareto population. *Journal of Econometrics*, **21**, 287-306.

[4] Ahsanullah, H.M. (1995) *Record Statistics*, Nova Science Publishers, New York.

[5] Ahmadi, J., Doostarat, M. and Parsian, A. (2005) Estimation and prediction in a two parameter exponential distribution based on k record values under LINEX loss function, *Communications in Statistics - Theory and Methods,* **34**, 795-805.

[6] Bhattacharjee, M.C. (1993) How rich are rich? Modeling affluence and inequality via reliability theory, *Sankhya B.*, 55, 1-26.

[7] Chandler, K.N. (1952) The distribution and frequency of record values , *Journal of the Royal Statistical Society,B,* **14**, 220-228.

[8] Chandra, M. and Singpurwalla, N.D. (1981). Relationship between some notions which are common to reliability theory and Economics, *Mathematics of operations research*, **6**, 113-121.

[9] Chikkagoudar, M.S. and Kunchur, S.H. (1980) Estimation of the mean of an exponential distribution in the presence of an outlier, *The Canadian Journal of Statistics,* **8**, 59-63.

[10] Chotikapanich, D.and Griffiths, W.E. (2002) Estimating Lorenz curve using a Dirchlet distribution, *Journal of Business and Economic Statistics,* **20**, 290-295.

[11] Cohen, A.C. and Whitten, B.J. (1988) *Parameter Estimation in Reliability and life Span modles.* Marcel Dekker, Inc, New York.

[12] DAVID, H.A. (1979) Robust estimation in the presence of outliers, *Robustness in Statistics,* Edited by R.L.Launer and G.N.Wilkinson, 61-74.

[13] Davis, T.H. and Feldstein, L.M. (1979) The generalised Pareto law as a model for progressively censored survival data, *Biometrica*, **66**(2), 299-306.

[14] Dixit, U.J. (1989) Estimation of parameters of the gamma distribution in the presence of outliers, *Communications in Statistics, A,* **18**, 3071-3085.

[15] Dunsmore, I.R. (1983) The future occurrence of records, *Annals of the Institute of Statistical Mathematics*, 35267-35277.

[16] Galambos, J. (1978) *The asymptotic theory of extreme order statistics*, Wiley, New York.

[17] Gastwirth, J.L. (1971) A general definition of the Lorenz curve, *Econometrica*, **39**, 1037-1039.

[18] Gather, U and Kale, K. (1992) *Outlier generating models a review,* in Venugopal N (Ed) Contribution to Stochastic, Wiley Eastern, New Delhi, 57-86.

[19] Glick, N. (1978) Breaking records and breaking boards , *American Mathematical Monthly* , **85**, 2-25.

[20] Grimshaw, D.S. (1993) Computing maximum likelihood estimates for the generalized Pareto distribution , *Technometrics*, **35**(2), 185-191.

[21] Gulati, S. (2003) *Parametric and Non-Parametric Inference from Record Breaking data*, Springer Verlag, New York.

[22] Jaheen, Z.F. (2004) Empirical Bayes inference for generalized exponential distribution based on records, *Communications in Statistics - Theory and Methods,* **33**(8), 1851-1861.

[23] Jeevanand, E. S. and Nair, N. U. (1992) Bayes estimates of Pareto parameters in the presence of outliers, *Proceedings of Symposium on Statistical Inference* , 53-70.

[24] Jeevanand, E. S. and Nair, N. U. (1993) Prediction of future observations from Pareto populations in the presence of outliers.*Statistica*,**53**, 171-176.

[25] Jeevanand, E. S. and Nair, N. U.(1996) Bayes prediction interval for the exponential and Pareto populations in the presence of an outlier, *Far East Journal of Mathematical Sciences* , Professor Bhattacharya Memorial Volume, 81-95.

[26] Jeevanand, E. S. and Sathar, A.E.I. (2006) Estimation of Pareto Survival Function in the presence of Outlying Observations. *Economic Quality Control,* **21(2)**, 165-174.

[27] Joshi, P.C.,(1972) Efficient Estimation of the mean of an exponential distribution when an outlier is present, *Technometrics*, **14**, 137-143.

[28] Kakawani, N.C. and Podder, N. (1973) On the estimation of Lorenz curves from grouped observations, *International Economic Review,* **14**, 278-292.

[29] Kakawani, N.C. and Podder, N. (1976) Efficient estimation of the Lorenz curve and associated inequality measures from grouped data, *Econometrica*, **44**, 137-148.

[30] Kale, B. K. (1976) Detection of outliers, *Sankhya B,* **38**, 356-363.

[31] Kale, V.B. and Kale, B.K. (1992) Outliers in exponential sample-a Bayesian approach, *Gujarat Statistical Review,* Professor Khatri Memorial.

[32] Kale, B. K. and Sinha, S. K. (1971) Estimation of expected life in the presence of outlier observation, *Technometrics*, **13**, 755-759.

[33] Kimber, A.C. (1983) Trimming in gamma sample, *Applied Statistics,* **32**, 7-14.

[34] Madi, M.T. (2005) Decision theoretic estimation using record statistics, *Metrika*, **63**, 91-97.

[35] Madi, M.T. and Raqab, M.Z. (2004) Bayesian prediction of temparature records using Pareto model, *Environmetrics*, **14**(7),701-710.

[36] Moothathu, T.S.K. (1985) Distributions of maximum likelihood estimators of Lorenz curve and Gini-index of exponential distribution, *Ann.Inst. Statist. Math. A,* **37**, 473-479.

[37] Moothathu, T.S.K. (1990 a). The best estimator and a strongly consistent asymptotically normal unbiased estimator of Lorenz curve, Gini-index and Theil entropy index of Pareto distribution, *Sankhya. B,* **52**, 115-127.

[38] Moothathu, T.S.K. (1990 b). Lorenz curve and Gini-index, *Culcutta Statistical Association Bulletin A,* **40**, 157-160.

[39] Nagaraja, H.N. (1988) Record values and related statistics - A Review, *Communications in Statistics - Theory and Methods,* **17**, 2223-2231.

[40] Nevzarov, V.B. (1987) *Records, Theory of probability and Applications*, **32** (2), 201-228.

[41] Pareto, V. (1897) *Cours d'Economie Politique*, Rouge et Cie, Paris.

[42] Sathar, A.E.I., Jeevanand, E.S. and NAIR, K.R.M. (2005) Bayesian estimation of Lorenz curve, Gini-index and variance of logarithms in a Pareto distribution, *Statistica*, **65** (2), 193-205.

[43] Varian, H.R. (1975) A Bayesian approach to Real Estate Assessment. In *Studies in Bayesian Econometrics and Statistics in honour of L.J.Savage,* Eds. S.E.Feinberg and A. Zellner, Amsterdam, North Holland, 195-208.

[44] Veale, J.R. (1975) Improved estimation of expected life when one identified spurious observation may be present, *Jour.Amer.Stat.Assn.,* **70**, 398-401.

INDEX

A

B

C

D

E

F

G

H

I

T

U

V

W

Y